Introductory Special Relativity

Introductory Special Relativity

W.G.V. ROSSER

University of Exeter, UK

Taylor & Francis
London • New York • Philadelphia
1991

UK Taylor & Francis Ltd, 4 John St., London WC1N 2ET
USA Taylor & Francis Inc., 1900 Frost Road, Suite 101, Bristol PA 19007

British Library Cataloguing in Publication Data
Rosser, W.G.V. (William Geraint Vaughan)
Introductory special relativity.
1. Physics. Special theory of relativity
I. Title
530.11

ISBN 0-85066-838-7
ISBN 0-85066-839-5 Pbk.

Library of Congress Cataloging-in-Publication Data
Rosser, W.G.V. (William Geraint Vaughan)
Introductory special relativity / W.G.V. Rosser.
p. cm.
Includes bibliographical references and index.
ISBN 0-85066-838-7 — ISBN 0-85066-839-5 (Pbk.)
1. Special relativity (Physics) 2. Mathematical physics.
I. Title.
QC173.65.R65 1991 90-22665
530. 1'1—dc20 CIP

Printed in Great Britain by Burgess Science Press, Basingstoke on paper which has a specified pH value on final paper manufacture of not less than 7.5 and is therefore 'acid free'.

Preface

This book is based on a course the author has given to first-year physics undergraduates over a period of twenty years at the University of Exeter. The course was based originally on the author's earlier book *Relativity and High Energy Physics*, which was one of the Wykeham Science Series and was written as introductory reading at the level of the school/university interface. When *Relativity and High Energy Physics* went out of print, it was decided to expand it into a full undergraduate textbook on special relativity by adding extra material, including a far wider range of applications, and adding extra problems for solution. Most of the new book is written at the level an average first-year undergraduate student of experimental physics should understand. It is written by an experimental physicist, and the predictions of the theory are always related to experiments. Some sections, marked with a star ★, go beyond the needs of a first course on special relativity and include introductions to some of the material the students are likely to meet in later courses. These sections can be omitted in a first reading. This book should also be suitable as an introductory course for students of theoretical physics, who can, after the experimental approach of this book, go on to concentrate on the mathematical aspects of the theory. It is also hoped that the book will be useful for mathematicians and engineers interested in special relativity.

After a first chapter on Newtonian mechanics and the Galilean transformations, this book follows the approach of *Relativity and High Energy Physics*. The theory of the motion of charges, moving in electric and magnetic fields at speeds comparable to the speed of light, is developed from experiments in Chapter 2. This approach shows that Newtonian mechanics and the Galilean transformations are completely inadequate at speeds comparable to the speed of light. It is then shown that, if it is assumed that these new experimentally confirmed laws, developed for high-speed particles, obey the principle of relativity then the Galilean transformations must be replaced by the Lorentz transformations. This experimental approach prepares the reader psychologically for the changes in the concepts of space and time necessitated by the Lorentz transformations. Following a short historical introduction in Chapter 3, the Lorentz transformations are derived in Chapter 4 from the principle of the constancy of the speed of light using radar methods and Bondi's method of the K calculus. Following a discussion of the velocity transformations in Chapter 5, a comprehensive account of relativistic mechanics is given in Chapter 6. The author has abandoned the concept of the variation of relativistic mass with velocity, which was used in *Relativity and High Energy Physics*, and, following contemporary trends, has only used the invariant mass, previously called rest mass, which can be measured in the zero-velocity limit using Newtonian mechanics, or at any speed using the equations of relativistic mechanics. Relativistic mechanics is applied to a wide range of phenomena. Following an introduction to 4-vectors in Chapter 7, there is a brief introduction in Chapter 8 to relativistic electromagnetism. Finally, Chapter 9 gives an account and interpretation of the clock paradox. A brief account of the principle of equivalence is given in an Appendix to illustrate some of the limitations of special relativity and inertial reference frames in the presence of gravitational fields.

Contents

1

Newtonian Mechanics and the Principle of Relativity

1.1 INTRODUCTION

Measurements of the positions of events must be made with respect to an origin and axes (i.e. a coordinate system), which we shall call a reference frame. A typical Cartesian coordinate system is shown in Fig. 1.1. The position of an event, taking place at the time t, can be specified by the coordinates x, y and z. Suitable units must be chosen for measuring length, mass and time. Until we reach Section 4.3 in Chapter 4, we shall find it convenient to define the metre independently of the speed of light, and we shall adopt the 1960 definitions of length, mass and time, which are fully consistent with Newtonian mechanics. They are as follows:

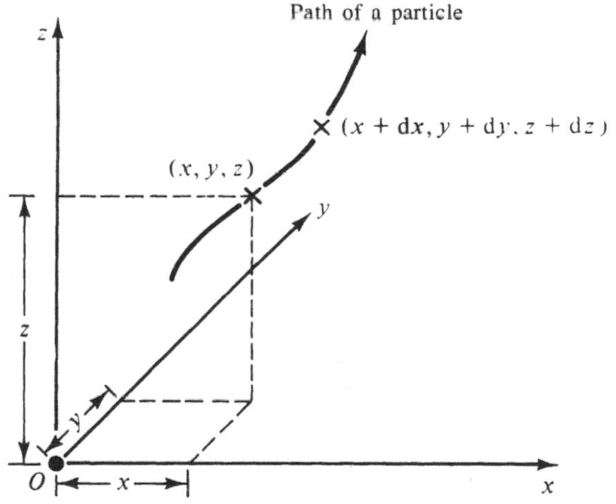

Fig. 1.1 A Cartesian coordinate system is used to plot the path of an accelerating particle.

the metre (m) is equal to 1 650 763.73 wavelengths of the orange-red line of the krypton-86 atom;

the kilogram (kg) is defined as the mass of a cylinder of platinum-iridium kept at Sèvres, near Paris;

the second (s) is the duration of 9 192 631 770 periods of the radiation corresponding to the transition between the two hyperfine levels of the ground state of the caesium-133 atom.

Consider the measurement of the velocity of the moving particle shown in Fig.1.1. By a particle, we mean a body of finite mass, but of such exceedingly small dimensions that it can be considered to be at one point of space at one instant of time. Such an idealization is generally a satisfactory approximation for atomic particles, such as electrons and protons. Let the position of the particle at time t be (x, y, z), and let its position a very short time dt later be $(x + dx, y + dy, z + dz)$ as shown in Fig. 1.1. The velocity $\boldsymbol{u}$ of the particle is defined as a vector having components u_x, u_y and u_z given by

$$u_x = \frac{dx}{dt}, \qquad u_y = \frac{dy}{dt}, \qquad u_z = \frac{dz}{dt}. \tag{1.1}$$

Bold italic type is used to denote vectors. The suffixes x, y and z are used in equations (1.1) to denote the directions of the components of a vector. For example, u_x is the component of the velocity $\boldsymbol{u}$ of the particle in the $+x$ direction.

The acceleration $\boldsymbol{a}$ of the particle is defined as the rate of change of velocity:

$$\boldsymbol{a} = \frac{d\boldsymbol{u}}{dt}. \tag{1.2}$$

1.2 A CRITIQUE OF NEWTON'S LAWS OF MOTION

1.2.1 Newton's laws of motion

It is all too easy in a first course on Newtonian mechanics to go on to the applications of Newton's laws of motion, without fully appreciating the definitions and interpretations of the quantities, such as mass and force, that appear in the theory. Before discussing the changes in the laws of mechanics that are necessitated by the theory of special relativity, we shall give a brief survey of the interpretation of Newtonian mechanics. It will be assumed that the reader has already had a course on mechanics leading up to Newton's laws of motion, which will be taken as our starting point. Newton's laws of motion are as follows.

(i) Every body continues in its state of rest or of uniform motion in a straight line unless it is compelled to change that state by an external impressed force.

(ii) The rate of change of momentum is proportional to the impressed force, and takes place in the direction of the force.

(iii) Action and reaction are equal and opposite. That is, between two bodies the force exerted by one on the other is equal in magnitude to the force it experiences from the other and in the opposite direction.

These laws were developed and have been applied to describe the mechanical behaviour of macroscopic bodies, such as the orbits of the planets and the collisions of billiard balls, all moving at speeds very much less than the speed of light in empty space, which is $c = 2.99792 \times 10^8$ m s^{-1} (we shall generally use the approximate value of $c \approx 3 \times 10^8$ m s^{-1} in our calculations). The speed of an aeroplane going at 3000 km h^{-1} is less than $3 \times 10^{-6}\, c$. It will be shown in Chapters 2 and 6 that when the speeds of particles are comparable to that of light, Newtonian mechanics must be replaced by relativistic mechanics. The only particles that can be accelerated to speeds close to the speed of light in the laboratory are atomic particles, such as protons and electrons—so that small size generally goes with high speed. This is well outside the range of our everyday experiences, and special relativity should be approached in Chapter 4 with an open mind, with the realization that it will require the development of new concepts on the part of the reader.

1.2.2 Newton's first law and inertial reference frames

The question arises as to how one can select a coordinate system in which Newton's laws of motion are valid. If in a reference frame a particle under the influence of no forces (e.g. a particle far away from any other particles capable of exerting forces) travels in a straight line with constant speed then Newton's first law (the principle of inertia) is valid in that frame, which is then considered suitable for the application of Newton's laws of motion. Such a reference frame is called an *inertial reference frame*, or *inertial frame* or sometimes a *Galilean reference frame*. An experimenter at rest in an inertial reference frame is called an *inertial observer*. The same definition of an inertial reference frame will be used in the theory of special relativity. Owing to the rotation of the Earth, the laboratory frame, that is a reference frame fixed to the Earth, is strictly not an inertial reference frame, and effects associated with the Earth's rotation are sometimes important. For example, consider a spaceship coasting with uniform velocity relative to the fixed stars. There are no applied forces acting on the spaceship. If its position is plotted on a coordinate system fixed to the Earth then, owing to the Earth's rotation, the spaceship will appear to travel in a spiral path, going around through 360° every day. According to Newton's first law, since there are no forces acting on the spaceship, it should travel in a straight line in all inertial reference frames. This shows that the *laboratory frame*, i.e. a reference frame fixed to the Earth, is *not* an inertial reference frame, and effects associated with the Earth's rotation are sometimes important even in terrestrial experiments, such as long-range naval gunnery and Foucault's pendulum experiment. In these cases we get a better approximation to an inertial reference frame by taking a frame at rest relative to the solar system or the fixed stars. However, the angular velocity of rotation of the Earth is only 7.3×10^{-5} rad s^{-1} (the Earth turns through 360° in 24 h), so that the effects associated with the Earth's rotation about its axis are generally very small. Effects due to the Earth's rotation about the Sun are much smaller still (the Earth goes through 360° around the Sun in a year). Effects associated with phenomena such as the rotation of the Galaxy are very much smaller again. For most low-speed mechanical experiments a reference frame fixed to the Earth (which will be called the laboratory frame) is generally a satisfactory approximation to an inertial reference frame. The theory of special relativity is generally applied to optics, electricity and magnetism, and atomic and nuclear physics. The effect of the Earth's rotation is usually negligible in all these cases, and the laboratory frame is almost invariably a satisfactory approximation to an inertial reference frame in which to apply the laws of optics, electricity and magnetism and atomic and nuclear physics.

1.2.3 The measurement of force and mass

In Newtonian mechanics the linear momentum $\boldsymbol{p}$ of a particle is defined as a vector, of magnitude equal to the product of the inertial mass m and the velocity $\boldsymbol{u}$ of the particle:

$$\boldsymbol{p} = m\boldsymbol{u} \,. \tag{1.3}$$

Using Newton's second law, the force $\boldsymbol{f}$ acting on a particle is defined as the rate of change of the momentum of the particle:

$$\boldsymbol{f} = \frac{\mathrm{d}}{\mathrm{d}t}(m\boldsymbol{u}). \tag{1.4}$$

Assuming that the inertial mass of the particle is independent of its velocity, equation (1.4) can be written as

$$\boldsymbol{f} = m\frac{\mathrm{d}\boldsymbol{u}}{\mathrm{d}t} = m\boldsymbol{a}, \tag{1.5}$$

where $\boldsymbol{a}$ is the acceleration of the particle. Hence force can be measured experimentally in terms of the acceleration that it produces on a body of known inertial mass; but before it can be measured in this way, one must know the inertial mass of the body. If equal forces act on two bodies of inertial masses m_1 and m_2, producing accelerations a_1 and a_2 respectively, then, according to equation (1.5), $m_1 a_1 = m_2 a_2$, so that

$$\frac{m_1}{m_2} = \frac{a_2}{a_1} \quad \text{(provided that } f_1 = f_2\text{)}. \tag{1.6}$$

Newton's third law states that action and reaction are equal and opposite. Hence, if the two bodies collide, the instantaneous values of the forces acting on the two bodies should be equal and opposite. Let a body of inertial mass m_1 moving with velocity u_1 collide head on with a body of inertial mass m_2 that is at rest before the collision. Let both the bodies m_1 and m_2 move in the same direction after the collision with velocities u_3 and u_4 respectively. If f_1 and f_2 are the instantaneous values of the forces acting on m_1 and m_2 at any instant during the collision then, according to Newton's third law,

$$f_1 = -f_2 .$$

Integrating over the time of the collision, we obtain

$$\int f_1 \, dt = -\int f_2 \, dt ;$$

that is,

$$\int m_1 \frac{du_1}{dt} \, dt = -\int m_2 \frac{du_2}{dt} \, dt .$$

Hence,

$$\int_{u_1}^{u_3} m_1 \, du_1 = -\int_0^{u_4} m_2 \, du_2,$$

$$m_1 (u_3 - u_1) = - m_2 u_4 ;$$

that is,

$$m_1 u_1 = m_1 u_3 + m_2 u_4 . \tag{1.7}$$

This is the *law of conservation of momentum*. It is important to realize that it follows automatically from Newton's laws of motion. From equation (1.7),

$$\frac{m_1}{m_2} = \frac{u_4}{u_1 - u_3} . \tag{1.8}$$

Hence, if the instantaneous values of the velocities just before and just after the collision can be measured, for example using high-speed photography, the ratio m_1/m_2 can be determined. If one of the masses, say m_1, is the primary standard of mass then the inertial mass of the other particle can be determined. This mass can then be used as a secondary standard of inertial mass. The theory of Newtonian mechanics has to be used before one can say how the inertial masses of two bodies can be compared. Having defined and prescribed how the inertial mass of a body can be measured, the force acting on a body can be measured by observing the acceleration that the force produces on a body of known inertial mass. Equation (1.5) can then be used to calculate the force. Thus the precise meanings of some quantities such as force and mass depend on the theory being used, and must be measured in practice as prescribed by that theory. If the experimental evidence forces one to replace a theory by a new theory then these quantities must be redefined and measured as prescribed by the new theory, and these definitions and measurement procedures may differ from those of the old theory. It will be found that this will have to be the case when one comes to replace Newtonian mechanics by relativistic mechanics. As new theories are developed, the meaning of words such as mass and force keeps changing, and the meaning to be attached to any word must be interpreted within the context of the theory being used.

When presented in the above way, Newton's second law in the form $\boldsymbol{f} = m\boldsymbol{a}$ might appear, at first sight, to be no more than a definition of force. In practice, it is assumed that the force $\boldsymbol{f}$ is known from a *force law*, such as Newton's law of universal gravitation. As an example, consider two isolated particles 1 and 2 of known masses m_1 and m_2 respectively, situated a distance r apart. The gravitational force of attraction between the particles accelerates the particles. These accelerations can be determined from the changes in the motions of the particles. If the masses of the particles are known, the force acting on each particle at separation r can be calculated using $\boldsymbol{f} = m\boldsymbol{a}$. The experiment can be repeated for different values of r, the separation of the particles. A general pattern will emerge, which can be summarized in Newton's law of universal gravitation, according to which the gravitational force of attraction between the two particles is given by

$$f = \frac{Gm_1m_2}{r^2}, \tag{1.9}$$

where G is the gravitational constant. According to equation (1.9), every particle in the Universe attracts every other particle with a force given by this relationship. Strictly m_1 and m_2 in equation (1.9) are the gravitational masses of the particles (see Section C.2 in Appendix C). If we are later presented with a gravitational problem, such as the motion of a particle in the Earth's gravitational field, then, on the basis of our previous experimental investigations, summarized in equation (1.9), we can use that equation to determine the total gravitational force $\boldsymbol{f}$ acting on a particle of mass m from the positions of the other particles. According to Newton's second law, the acceleration of the particle is equal to $\boldsymbol{f}/m$, and the subsequent motion of the particle can be calculated. This illustrates how, when Newton's second law of motion is generally applied in practice, it is assumed that the force acting on a particle is known from a force law, such as Newton's law of universal gravitation. In the case of a charge moving in an electrostatic field the appropriate force law is Coulomb's law.

1.2.4 Conservation of linear momentum

It was shown in Section 1.2.3, for the special case of a head-on collision, that it follows from Newton's laws that linear momentum is conserved in the collision.

It is straightforward to extend the method to show that, in the general case of the collision of two particles 1 and 2 of masses m_1 and m_2, with velocities $\boldsymbol{u}_1$ and $\boldsymbol{u}_2$ before the collision and velocities $\boldsymbol{u}_3$ and $\boldsymbol{u}_4$ respectively after the collision, provided that no external forces act on the particles, we have

$$m_1\boldsymbol{u}_1 + m_2\boldsymbol{u}_2 = m_1\boldsymbol{u}_3 + m_2\boldsymbol{u}_4 . \tag{1.10}$$

Therefore

$$\boldsymbol{p}_1 + \boldsymbol{p}_2 = \boldsymbol{p}_3 + \boldsymbol{p}_4 , \tag{1.11}$$

where $\boldsymbol{p}_1 = m_1\boldsymbol{u}_1$ and $\boldsymbol{p}_2 = m_2\boldsymbol{u}_2$ are the momenta of particles 1 and 2 before the collision and $\boldsymbol{p}_3 = m_1\boldsymbol{u}_3$ and $\boldsymbol{p}_4 = m_2\boldsymbol{u}_4$ are the respective momenta after the collision. Equation (1.10) expresses the general case of the law of conservation of momentum for the collision of two particles. It is a vector equation. The momentum of a particle is a vector quantity *defined* as the product of the mass and velocity of the particle. The law of conservation of momentum is valid for both elastic and inelastic collisions. It is often more convenient in practice to use this law in collision problems rather than Newton's second law, since it is not always necessary to know the forces between the particles when the law of conservation of momentum is applied.

1.2.5 Kinetic energy

The work $\mathrm{d}W$ done by an applied force $\boldsymbol{f}$ acting on a particle of mass m, which is displaced a distance $\mathrm{d}\boldsymbol{l}$ is given by

$$\mathrm{d}W = \boldsymbol{f}\bullet\mathrm{d}\boldsymbol{l} , \tag{1.12}$$

where $\boldsymbol{f}\bullet\mathrm{d}\boldsymbol{l}$ is the scalar product of the vectors $\boldsymbol{f}$ and $\mathrm{d}\boldsymbol{l}$. ($\boldsymbol{f}\bullet\mathrm{d}\boldsymbol{l} = f\,\mathrm{d}l\cos\theta$, where θ is the angle between $\boldsymbol{f}$ and $\mathrm{d}\boldsymbol{l}$.) In the special case when $\mathrm{d}\boldsymbol{l}$ is parallel to $\boldsymbol{f}$ and $\mathrm{d}\boldsymbol{l}$ is in the $+x$ direction,

$$\mathrm{d}W = f\,\mathrm{d}x .$$

Using equation (1.5) to substitute for f, in the absence of friction,

$$\mathrm{d}W = m\frac{\mathrm{d}u}{\mathrm{d}t}\mathrm{d}x = m\frac{\mathrm{d}x}{\mathrm{d}t}\mathrm{d}u = mu\,\mathrm{d}u .$$

The total work done by the force f in a finite displacement of the particle, during which the particle is accelerated from rest to a velocity u in the $+x$ direction, is

$$W = \int f\,\mathrm{d}x = \int_0^u mu\,\mathrm{d}u = \tfrac{1}{2}mu^2 .$$

If all of this work goes into increasing the kinetic energy of the particle from zero to a final value T then

$$T = \int f \, dx = \tfrac{1}{2} mu^2 . \qquad (1.13)$$

The unit of kinetic energy is the joule (J).

If the particle started with kinetic energy T_1 and velocity u_1 and was accelerated by the force f to kinetic energy T_2 and velocity u_2 then, by changing the limits of integration in equation (1.13),

$$T_2 - T_1 = \int_1^2 \boldsymbol{f} \bullet d\boldsymbol{l} = \tfrac{1}{2} mu_2^2 - \tfrac{1}{2} mu_1^2 . \qquad (1.14)$$

1.2.6 Potential energy

In many cases the force f acting on a particle depends only on the position of the particle. This is true, for example, for a charged particle in an electrostatic field and for a mass in a gravitational field. It is convenient in such cases to introduce the potential energy function, which will be denoted by U, and which is a function of the coordinates x, y and z. The difference in the potential energy that a particle would have at two points can be defined operationally as the work that would have to be done to move the particle from one point to the other at zero speed. If the force acting on the particle is $\boldsymbol{f}$ then a force $\boldsymbol{f}_0$ that is equal and opposite to $\boldsymbol{f}$ must be applied to move the particle at zero speed. Equating the change of potential energy to the work done by the force $\boldsymbol{f}_0$ in moving the particle from one position to the other, we have

$$U_2 - U_1 = \int_1^2 \boldsymbol{f}_0 \bullet d\boldsymbol{l} = - \int_1^2 \boldsymbol{f} \bullet d\boldsymbol{l} . \qquad (1.15)$$

Equation (1.15) is sometimes used as the definition of the difference of potential energy. For the potential energy U at a point to have a unique value, the line integral $\int \boldsymbol{f} \bullet d\boldsymbol{l}$ must be independent of the path taken from position 1 to position 2. This is the condition that must be satisfied if the force is to be *conservative*. The gravitational force and the electrostatic force satisfy this condition, but frictional forces do not. The choice of the zero of potential energy is arbitrary. For example, for the motion of a particle in the Earth's gravitational field, it is sometimes convenient to assume that the potential energy is zero at the Earth's surface. In this case the potential energy of a particle of mass m at a small height h above the Earth's surface is mgh, where g is the acceleration due to gravity. In other cases it is convenient to say that the potential energy is zero at an infinite distance from the Earth, in which case the potential energy of a particle of mass m that is outside the Earth at a distance r from the Earth's centre is $-GMm/r$, where M is the Earth's mass.

1.2.7 Conservation of energy

From equation (1.15), the change in the potential energy of a particle, acted upon by a conservative force f, when the particle moves from position 1 to position 2, is

$$U_2 - U_1 = -\int_1^2 f \bullet dl . \qquad (1.15)$$

If all the work done by the force f acting on the particle goes into changing the kinetic energy of the particle from T_1 to T_2 then it follows from equation (1.14) that

$$T_2 - T_1 = +\int_1^2 f \bullet dl. \qquad (1.14)$$

Adding equations (1.15) and (1.14), we obtain

$$U_2 - U_1 + T_2 - T_1 = 0 .$$

Therefore

$$U_2 + T_2 = U_1 + T_1 . \qquad (1.16)$$

This shows that the sum of the kinetic energy and the potential energy of a particle acted upon by conservative forces is constant. Gravitational forces and electrostatic forces are examples of conservative forces.

1.2.8 Discussion

Newton's laws for particles can be used to develop the theory of the dynamics of rigid bodies, leading up to the concepts of the kinetic energy and the angular momentum of rotating bodies and the law of conservation of angular momentum. Newton's laws can also be used to develop Lagrange's equations and Hamilton's equations. None of these topics will be needed in our introductory development of special relativity. The interested reader is referred to a textbook on mechanics, such as that by Kibble (1985).

1.3 THE PRINCIPLE OF RELATIVITY

Consider a large ocean liner moving out to sea with *uniform* velocity on a calm day. It is quite possible for the passengers to play table tennis in the games room in such conditions. There is no need for the players to stop and wonder between each stroke which way the ship is going. They can play their normal game, just as if they were playing on land, and they can ignore the motion of the ship. Of course, in rough seas the sudden accelerations of the ship would affect their normal game of table tennis, and they would have to try to allow for such accelerations.

On the basis of mechanical experiments on large bodies carried out inside a ship moving with *uniform* velocity, one would conclude that Newton's laws held to a very good approximation relative to the ship. Without looking at anything external to the ship, one could not determine, on the basis of these experiments, whether it was even moving. If one were told that it was moving, one could not determine the speed of the ship without looking at something external to it. This is an example of the *principle of relativity*, according to which the laws of physics are the same in all inertial reference frames. Though the laws are the same, it does not mean that the measurements of particular quantities are the same in all inertial reference frames. For example, if a ball is rolling at a speed of, say, $10\,\mathrm{m\,s^{-1}}$ relative to the ship in the direction of motion of the ship, then, according to the Galilean velocity transformations, the speed of the ball relative to the Earth should be $10+v$, where v is the speed of the ship relative to the Earth. The reader will probably accept it as quite natural that the speed of the ball will be different when the standard of rest is changed.

We shall now proceed to see how the values of various other quantities, measured relative to one inertial reference frame, can be changed (or transformed) into the values measured relative to an inertial reference frame moving with uniform velocity relative to the first.

1.4 THE GALILEAN TRANSFORMATIONS

Consider again an ocean liner going out to sea with *uniform* velocity on a calm day. The inertial reference frame Σ is a Cartesian coordinate system fixed on the Earth, whereas Σ' is a coordinate system fixed on the ship, as shown in Fig. 1.2. (The symbol Σ is the Greek capital letter sigma). The x' axis of Σ' coincides with the x axis of Σ, and both are in the direction in which the ship is moving with uniform velocity v relative to the Earth; the y' axis of Σ' is parallel to the y axis of Σ (Fig. 1.2). Primes will be used on variables such as x', y' and z' to show that they are measured in Σ'. The origins O and O' of Σ and Σ' coincide at time $t=0$.

Consider the measurement of the position of a mass m that is moving in the (x, y) plane of Σ at the time of observation t, as shown in Fig. 1.2. It is assumed in Newtonian mechanics that *time is absolute*. For example, in his *Principia* Newton wrote

> Absolute, true and mathematical time of itself and from its own nature flows equably without relation to anything external.

The same symbol t is generally used to denote the time measured relative to both the Earth (Σ) and the ship (Σ'). If time were absolute, an observer at rest relative

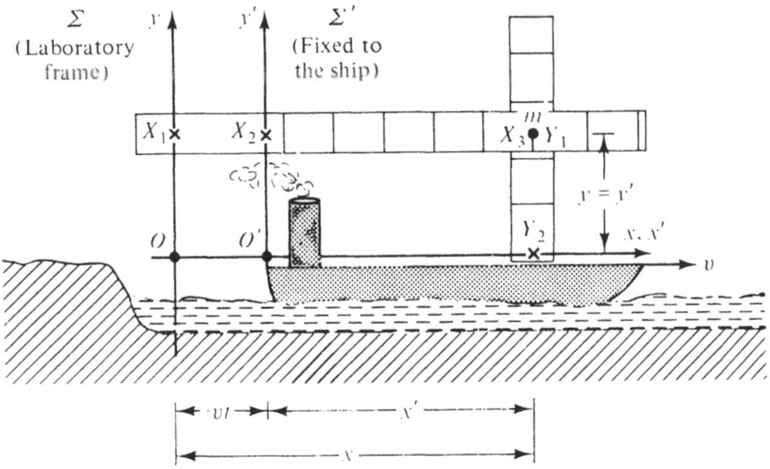

Fig. 1.2 The coordinate system Σ is fixed to the Earth. The coordinate system Σ′ is fixed to the ship. The position of the mass m is determined at time t. The origins of Σ and Σ′ coincide at $t = 0$.

to the Earth (Σ) and an observer at rest relative to the ship (Σ′) should agree that at the time t the y axis of Σ, the y' axis of Σ′ and the mass m coincided with the marks X_1, X_2 and X_3 on the ruler, which is parallel to the x and x' axes as shown in Fig. 1.2 (the ruler may be moving relative to both Σ and Σ′). Hence, according to Newtonian mechanics, observers at rest relative to the Earth and the ship should agree that in Fig. 1.2

$$X_1X_3 = X_1X_2 + X_2X_3 \, .$$

Since the speed of the ship is v, in a time t the origin O' of Σ′ moves a distance $X_1X_2 = vt$ relative to the laboratory frame Σ. Since $X_1X_3 = x$ and $X_2X_3 = x'$ (Fig. 1.2),

$$x = x' + vt \, .$$

Similarly, according to Newtonian mechanics, observers at rest relative to the Earth (Σ) and the ship (Σ′) should agree that the x and x' axes of Σ and Σ′ coincide with the point Y_2 on the vertical ruler at the same time t that the mass m coincides with the position Y_1. Hence the observers should agree on the vertical position of the mass m at time t, so that $y = y'$. Hence, if it is *assumed* that time is absolute — that is, that the observers on the Earth (Σ) and the ship (Σ′) agree on the times of all events — then the coordinates (x, y, z) of an event at time t in Σ are related to the coordinates of the *same* event at (x', y', z') at time t' in Σ′ by the following relations:

$$x = x' + vt', \qquad x' = x - vt, \tag{1.17}$$

$$y = y', \qquad y' = y, \tag{1.18}$$

$$z = z', \qquad z' = z, \tag{1.19}$$

$$t = t', \qquad t' = t. \tag{1.20}$$

These are the Galilean coordinate transformations. The same units of length, time and mass must be used in both Σ and Σ'. For example, the inertial observers in Σ and Σ' could use the definitions of the metre and the second given in Section 1.1. It is straightforward to show that, according to the Galilean transformations, the length of a moving object is the same in all inertial reference frames (see Problem 1.4).

Consider again the moving mass m in Fig. 1.2. Differentiating equations (1.17), (1.18) and (1.19) with respect to time, we have, since the ship's speed v is assumed to be constant,

$$\frac{dx}{dt} = \frac{dx'}{dt} + v, \qquad \frac{dy}{dt} = \frac{dy'}{dt}, \qquad \frac{dz}{dt} = \frac{dz'}{dt};$$

that is,

$$u_x = u_x' + v, \tag{1.21}$$

$$u_y = u_y', \tag{1.22}$$

$$u_z = u_z'. \tag{1.23}$$

These are the Galilean velocity transformations.

Differentiating equations (1.21), (1.22) and (1.23) with respect to time, since v is constant, we have

$$\frac{du_x'}{dt} = \frac{du_x}{dt}, \qquad \frac{du_y'}{dt} = \frac{du_y}{dt}, \qquad \frac{du_z'}{dt} = \frac{du_z}{dt};$$

that is,

$$\boldsymbol{a}' = \boldsymbol{a}, \tag{1.24}$$

where $\boldsymbol{a}'$ is the acceleration of the particle relative to Σ' and $\boldsymbol{a}$ is its acceleration relative to Σ. Provided that v, the relative velocity of Σ and Σ', is constant, the acceleration is the same in Σ and Σ'. It is assumed in Newtonian mechanics that the mass of a particle is an invariant. Multiplying both sides of equation (1.24) by the invariant mass m of the particle, we have

$$m\boldsymbol{a}' = m\boldsymbol{a}. \tag{1.25}$$

Since the force acting on a particle can be defined as the product of the mass of the particle and the acceleration produced by the force $m\boldsymbol{a}'$ is equal to the force producing the acceleration of the particle measured in Σ', while $m\boldsymbol{a}$ is the value of the force acting on the particle measured in Σ. Hence

$$\boldsymbol{f} = \boldsymbol{f}'. \tag{1.26}$$

Thus the force on the particle, measured in terms of the acceleration it produces, should have the same value in both Σ and Σ′.

1.4.1 Example of the Galilean velocity transformations

Consider the inelastic collision of two particles 1 and 2, each of mass 2 kg. In the inertial reference frame Σ shown in Fig. 1.3(a) particles 1 and 2 are moving parallel to the x axis with velocities $u_1 = 20\ \mathrm{m\ s^{-1}}$ and $u_2 = 10\ \mathrm{m\ s^{-1}}$ respectively. The particles collide and stick together to form a composite particle of mass 4 kg, which moves with velocity $\bar{u}$ parallel to the x axis of Σ after the collision, as shown in Fig. 1.3(a). Applying the law of conservation of momentum in Σ gives

$$m_1u_1 + m_2u_2 = (m_1 + m_2)\bar{u},$$

$$(2 \times 20 + 2 \times 10)\ \mathrm{kg\ m\ s^{-1}} = 4\bar{u}.$$

Therefore

$$\bar{u} = 15\ \mathrm{m\ s^{-1}}.$$

Now, consider the collision from the reference frame Σ′, which is moving with uniform velocity $v = 5\ \mathrm{m\ s^{-1}}$ along the x axis of Σ, as shown in Fig. 1.3(b). Application of the Galilean velocity transformation, equation (1.21), gives for the velocities u_1' and u_2' of particles 1 and 2 in Σ′ before the collision and for $\bar{u}'$, the velocity of the composite particle in Σ′ after the collision, the following values:

$$u_1' = u_1 - v = (20 - 5)\ \mathrm{m\ s^{-1}} = 15\ \mathrm{m\ s^{-1}},$$

$$u_2' = u_2 - v = (10 - 5)\ \mathrm{m\ s^{-1}} = 5\ \mathrm{m\ s^{-1}},$$

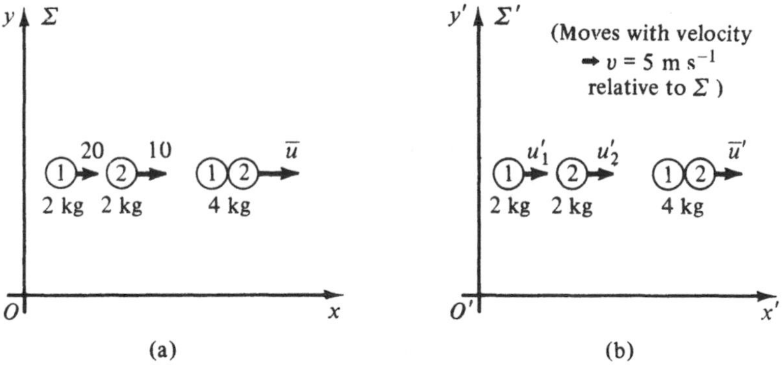

Fig. 1.3 The inelastic collision of two particles considered (a) relative to Σ and (b) relative to Σ′, which moves with velocity $v = 5\ \mathrm{m\ s^{-1}}$ relative to Σ.

$$\bar{u}' = \bar{u} - v = (15 - 5)\ \mathrm{m\ s^{-1}} = 10\ \mathrm{m\ s^{-1}}.$$

These results illustrate how a problem can be solved using Newtonian mechanics in Σ, and how the results appropriate to Σ' can then be derived using the Galilean transformations. Note that the total momentum in Σ' before the collision, which is $(2 \times 15 + 2 \times 5)\ \mathrm{kg\ m\ s^{-1}} = 40\ \mathrm{kg\ m\ s^{-1}}$, is equal to the total momentum in Σ' after the collision, which is $4\ \mathrm{kg} \times 10\ \mathrm{m\ s^{-1}} = 40\ \mathrm{kg\ m\ s^{-1}}$. This illustrates how the law of conservation of momentum is valid in Σ' if it is valid in Σ, provided that the Galilean transformations are used to transform the velocities of the particles from Σ to Σ'.

Consider now a reference frame Σ^0 that is moving with uniform velocity $v = 15\ \mathrm{m\ s^{-1}}$ relative to Σ. In this case application of the Galilean velocity transformations gives $u_1^0 = (20 - 15)\ \mathrm{m\ s^{-1}} = 5\ \mathrm{m\ s^{-1}}$, $u_2^0 = (10 - 15)\ \mathrm{m\ s^{-1}} = -5\ \mathrm{m\ s^{-1}}$ and $\bar{u}^0 = (15 - 15)\ \mathrm{m\ s^{-1}} = 0$ in Σ^0. In this example particles 1 and 2 approach each other with equal and opposite speeds of $5\ \mathrm{m\ s^{-1}}$ and equal and opposite momenta of $2 \times 5 = 10\ \mathrm{kg\ m\ s^{-1}}$ before the collision. The composite particle is at rest in Σ^0 after the collision. Since the total momentum is zero in Σ^0 before and after the collision, Σ^0 is called the *zero-momentum frame* (in older textbooks it is called the *centre-of-mass system*). It is sometimes more convenient to solve problems first in the zero-momentum frame Σ^0 and then use the Galilean transformations to determine the appropriate results in the laboratory frame Σ.

★ 1.5 THE TRANSFORMATION OF NEWTON'S LAWS OF MOTION

Consider a particle that is *not* acted upon by any applied force and that is moving in a straight line in the (x, y) plane of the inertial reference frame Σ with *uniform* velocity $\boldsymbol{u}$ having components u_x, u_y, and $u_z = 0$. Consider a reference frame Σ' moving with uniform velocity v relative to Σ along their common x axis. According to the Galilean velocity transformations, equations (1.21), (1.22) and (1.23), the components of the velocity $\boldsymbol{u}'$ of the particle measured in Σ' are

$$u_x' = u_x - v, \qquad u_y' = u_y, \qquad u_z' = u_z = 0.$$

Therefore

$$u'^2 = (u_x - v)^2 + u_y^2.$$

Since u_x, u_y and v are all constant, the speed u' of the particle in Σ' is constant. The particle moves at an angle ϕ' to the x' axis of Σ' given by

$$\tan\phi' = \frac{u_y'}{u_x'} = \frac{u_y}{u_x - v}.$$

Since u_x, u_y and v are all constant, so is ϕ'. Hence in Σ' the particle moves with a uniform velocity $\boldsymbol{u}'$ in a straight line, inclined at a constant angle ϕ' to the x' axis of Σ'. Thus Newton's first law is valid in Σ' as well as in Σ, if the Galilean transformations are used to transform from Σ to Σ'. According to the definition of an inertial reference frame given in Section 1.2.2, Σ' is also an inertial reference frame.

Consider now a different particle of mass m, acted upon by a resultant force $\boldsymbol{f}$, measured in Σ, and that has an acceleration $\boldsymbol{a}$ in Σ. According to Newton's second law, equation (1.5), in Σ we have

$$\boldsymbol{f} = m\boldsymbol{a}. \tag{1.27}$$

From equation (1.26), $\boldsymbol{f}=\boldsymbol{f}'$, where $\boldsymbol{f}'$ is the value of the force acting on the particle, measured in Σ', while from equation (1.24), $\boldsymbol{a}=\boldsymbol{a}'$, where $\boldsymbol{a}'$ is the acceleration of the particle in Σ'. The mass of a particle is assumed to be an invariant in Newtonian mechanics, so that $m=m'$. Hence equation (1.27) can be rewritten in the form

$$\boldsymbol{f}' = m'\boldsymbol{a}'. \tag{1.28}$$

This shows that, using the Galilean transformations, which follow from the assumption of an absolute time, Newton's second law is valid in Σ' if it is valid in Σ. The velocity v of Σ' relative to Σ does not appear in equation (1.28).

If two particles labelled 1 and 2 collide, if Newton's third law is valid in Σ, the forces $\boldsymbol{f}_1$ and $\boldsymbol{f}_2$ acting on particles 1 and 2 during the collision are equal and opposite; that is, $\boldsymbol{f}_1=-\boldsymbol{f}_2$ in Σ. Since from equation (1.26), $\boldsymbol{f}_1=\boldsymbol{f}_1'$ and $\boldsymbol{f}_2=\boldsymbol{f}_2'$, in Σ' we have $\boldsymbol{f}_1'=-\boldsymbol{f}_2'$. Hence if Newton's third law is valid in Σ and equation (1.26), which was derived using the Galilean transformations, is used to transform the forces then Newton's third law must also be valid in Σ'.

In summary, it has been shown that if Newton's three laws of motion are valid in an inertial reference frame Σ then, if the Galilean transformations are used, all three of Newton's laws should be valid in any other reference frame Σ' moving with any uniform velocity v relative to Σ. Conversely, the Galilean transformations are the appropriate coordinate and time transformations if Newton's laws of motion obey the principle of relativity—that is, if Newton's laws hold in all inertial reference frames.

REFERENCE

Kibble, T.W.B. (1985) *Classical Mechanics*, 3rd edn. Longman, London.

PROBLEMS

1.1 If you were locked in the hold of a ship, how would you prove that the ship was (a) moving with uniform velocity, (b) accelerating relative to the Earth?

1.2 A ship is travelling due east at a speed of 25 m s^{-1}.

(a) A ball is rolled due north on the deck of the ship at a speed of 5 m s^{-1} relative to the ship. What is its velocity relative to the Earth?

(b) If the ball is rolled 30° east of north at a speed of 5 m s^{-1} relative to the ship, what is its speed relative to the Earth?

1.3 A stone is dropped from rest from the mast of a ship moving with a velocity of 15 m s^{-1} relative to the Earth. Choose the origins of Σ (the laboratory frame) and Σ' (the coordinate system in which the ship is at rest) such that these origins both coincide with the stone at the instant $t = 0$ when the stone is dropped. If the acceleration due to gravity is 9.8 m s^{-2}, find the position of the stone (a) relative to the ship, (*b*) relative to the laboratory frame after 2 s. Show how the results can be related using the Galilean transformations. The ship is moving parallel to the x axis of Σ.

1.4 Assume that at time t the two ends of a moving rod are at (x_1, y_1, z_1) and (x_2, y_2, z_2) in the inertial reference frame Σ, so that the length of the rod in Σ is $[(x_2 - x_1)^2 + (y_2 - y_1)^2 + (z_2 - z_1)^2]^{1/2}$. Use the Galilean transformations to determine the coordinates of the ends of the rod at time $t' = t$ in an inertial reference frame Σ' that is moving with uniform velocity v relative to Σ along their common x axis. The origins of Σ and Σ' coincide at $t = t' = 0$. Hence show that the length of the moving rod is the same in Σ and Σ'.

2

Motion of Charges in Electric and Magnetic Fields

2.1 INTRODUCTION

In this chapter we shall consider experiments that show that at very high speeds, close to the speed of light, Newtonian mechanics is completely inadequate. For example, in Section 2.5 an experiment will be described which shows clearly that there is a limiting speed for accelerated electrons, which is equal to the speed of light in empty space, whereas, according to Newtonian mechanics, the speed of an accelerated electron should increase continuously towards infinity. It will be shown in this chapter that if momentum is redefined using equation (2.1) then it is possible to develop a new theory for the motions of high-speed electric charges in electric and magnetic fields that is in excellent agreement with the experimental results. This new theory is generally called *relativistic mechanics*. The adjective "relativistic" is often used to describe quantities appropriate to particles moving at speeds close to the speed of light. For example, a particle moving at a velocity close to the velocity of light is often said to be moving at a relativistic velocity. Some of the implications of assuming that the new experimentally confirmed laws of relativistic mechanics obey the principle of relativity will be discussed in Section 2.13, where it will be shown that the Galilean transformations must be replaced by the Lorentz transformations. A full discussion of relativistic mechanics is given in Chapter 6.

2.2 RELATIVISTIC MOMENTUM, FORCE AND KINETIC ENERGY

In Newtonian mechanics the momentum of a particle of inertial mass m moving with velocity $\boldsymbol{u}$ was defined as the product $m\boldsymbol{u}$. In relativistic mechanics the momentum $\boldsymbol{p}$ of the particle is redefined as

$$\boldsymbol{p} = \alpha m \boldsymbol{u}, \tag{2.1}$$

where

$$\alpha = \frac{1}{(1 - u^2/c^2)^{1/2}}, \tag{2.2}$$

u is the speed of the particle and c the speed of light in empty space. Writing equation (2.1) in component form, we have

$$p_x = \alpha m u_x, \qquad p_y = \alpha m u_y, \qquad p_z = \alpha m u_z. \tag{2.3}$$

Compared with the definition of momentum in Newtonian mechanics, the only change in equation (2.1) is the addition of an extra relativistic factor α, which depends only on the *speed of the particle* and the *speed of light* in empty space. When $u \ll c$, α tends to unity, and equation (2.1) reduces to the expression for momentum used in Newtonian mechanics. It is assumed in relativistic mechanics that the mass m appearing in equation (2.1) is an invariant.† We shall treat m as a quantity defined and measured using the laws of relativistic mechanics (see Sections 2.12 and 6.14). It is the same as the (inertial) mass m defined and measured in the zero-velocity limit using Newtonian mechanics, for example using equation (1.8).

The fastest aeroplanes have speeds of about $3000\,\text{km h}^{-1}$, which is only $3 \times 10^{-6}\,c$, where $c = 3 \times 10^8\,\text{m s}^{-1}$ is the speed of light in empty space. In this case, with $u/c = 3 \times 10^{-6}$, the value of α given by equation (2.2) is

$$[1 - (3 \times 10^{-6})^2]^{-1/2} = (1 - 9 \times 10^{-12})^{-1/2}.$$

According to the binomial theorem, for $x \ll 1$,

$$(1 + x)^n \approx 1 + nx,$$

and therefore

$$\alpha = 1.000\,000\,000\,004\,5.$$

It can thus be seen that, for everyday mechanical phenomena, we can put $\alpha = 1$ in equation (2.1) and use Newtonian mechanics.

The differences between relativistic mechanics and Newtonian mechanics are extremely important when the speeds of the particles are comparable to c. The only particles that can be accelerated to speeds comparable to the speed of light are atomic particles, such as protons and electrons. Thus small size is generally associated with high speed. Both the high speeds and the small sizes involved in high-energy physics go beyond the realm of direct perception and beyond our "common sense", which is almost invariably based on the concepts of Newtonian mechanics. The reader must approach the development of high-energy physics and special relativity with an open mind, and be guided by how well the theory agrees with experiments.

---

† In older textbooks the mass m in equation (2.1) is often called the *rest mass* or *proper mass* of the particle and generally denoted by m_0. Such books often call the product αm_0 the relativistic mass and denote it by m. There is no need in our approach to call the mass m in equation (2.1) the rest mass. Apart from the historical note in Section 6.14, we shall not introduce any type of mass other than the invariant mass m.

As in Newtonian mechanics, the *force* acting on a particle will be put equal to the rate of change of the momentum of the particle; that is,

$$\boldsymbol{f} = \frac{\mathrm{d}\boldsymbol{p}}{\mathrm{d}t} = \frac{\mathrm{d}}{\mathrm{d}t}(\alpha m\boldsymbol{u})$$

$$= \frac{\mathrm{d}}{\mathrm{d}t}\left[\frac{m\boldsymbol{u}}{(1 - u^2/c^2)^{1/2}}\right]. \qquad (2.4)$$

Owing to the presence of the factor α, which depends on the speed u, we have, since the mass m is an invariant and $\mathrm{d}\boldsymbol{u}/\mathrm{d}t = \boldsymbol{a}$,

$$\boldsymbol{f} = \alpha m\boldsymbol{a} + m\boldsymbol{u}\frac{\mathrm{d}\alpha}{\mathrm{d}t}. \qquad (2.5)$$

If the acceleration $\boldsymbol{a}$ leads to a change in the speed of the particle, $\mathrm{d}\alpha/\mathrm{d}t$ is finite. Hence, because of the presence of α in the first term on the right-hand side of equation (2.5) and the presence of $\mathrm{d}\alpha/\mathrm{d}t$ in the second term, the equation "force equals mass times acceleration" is not valid in relativistic mechanics.

The expression for the kinetic energy of a particle moving at relativistic speeds will now be derived. If the total force f acting on a particle gives rise to a displacement $\mathrm{d}l$ in the direction of the force then the work done by the force is $f\,\mathrm{d}l$. If it is assumed that all the work done goes into increasing the kinetic energy of the particle then that increase is given by

$$\mathrm{d}T = f\,\mathrm{d}l = f\frac{\mathrm{d}l}{\mathrm{d}t}\,\mathrm{d}t$$

$$= fu\,\mathrm{d}t, \qquad (2.6)$$

where T is the kinetic energy of the particle and $u = \mathrm{d}l/\mathrm{d}t$ is the velocity of the particle. If we put

$$u = c\sin\theta \qquad (2.7)$$

then

$$\alpha = \frac{1}{(1 - u^2/c^2)^{1/2}}$$

$$= \frac{1}{(1 - \sin^2\theta)^{1/2}} = \frac{1}{\cos\theta} = \sec\theta, \qquad (2.8)$$

and so

$$p = \alpha m u = \sec\theta\, mc\sin\theta = mc\tan\theta. \qquad (2.9)$$

Hence

$$f = \frac{\mathrm{d}p}{\mathrm{d}t} = \frac{\mathrm{d}}{\mathrm{d}t}(mc\tan\theta)$$

$$= mc \sec^2 \theta \frac{d\theta}{dt}. \tag{2.10}$$

Substituting for f and u in equation (2.6), we have

$$dT = mc \sec^2 \theta \frac{d\theta}{dt} c \sin \theta \, dt = mc^2 \frac{\sin \theta}{\cos^2 \theta} d\theta.$$

Assuming the particle starts from rest, when $T = 0$ and $\theta = 0$, we have

$$\int_0^T dT = mc^2 \int_0^\theta \frac{\sin \theta}{\cos^2 \theta} d\theta. \tag{2.11}$$

Putting $\cos \theta = x$, so that $\sin \theta \, d\theta = -dx$ and

$$\int \frac{\sin \theta \, d\theta}{\cos^2 \theta} = -\int \frac{dx}{x^2} = \frac{1}{x},$$

and substituting in equation (2.11), we have

$$[T]_0^T = mc^2 \left[\frac{1}{\cos \theta} \right]_0^\theta = mc^2 (\sec \theta - 1).$$

Now, from equation (2.8), $\sec \theta = \alpha$, and hence the kinetic energy T of the particle is

$$T = mc^2 (\sec \theta - 1) = mc^2 (\alpha - 1); \tag{2.12}$$

that is,

$$T = mc^2 \left[\frac{1}{(1 - u^2/c^2)^{1/2}} - 1 \right]. \tag{2.13}$$

Expanding the right-hand side of equation (2.13) using the binomial theorem, we have

$$T = mc^2 \left(1 + \tfrac{1}{2} \frac{u^2}{c^2} + \tfrac{3}{8} \frac{u^4}{c^4} + \ldots \right) - mc^2$$

$$= \tfrac{1}{2} mu^2 + \tfrac{3}{8} mu^2 \left(\frac{u^2}{c^2} \right) + \ldots .$$

If $u \ll c$ then $u^2 \ll c^2$ and $u^2/c^2 \ll 1$, so that

$$T \approx \tfrac{1}{2} mu^2. \tag{2.14}$$

Hence, when the speed of the particle is very much less than the speed of light, equation (2.13) for the kinetic energy of a particle approximates very closely to the expression for kinetic energy derived using Newtonian mechanics; but at speeds comparable to c, Newtonian mechanics is completely inadequate. Note that as u tends to c, the kinetic energy tends to infinity, and thus an infinite amount of work must be done to accelerate a particle of finite mass up to the speed of light. This illustrates how, according to equation (2.13), the speed of light c should be the limiting speed for an accelerating particle.

2.3 UNITS

The equations developed in Section 2.2 can be applied in any coherent set of units. In SI units the speed of light $c = 2.99792 \times 10^8\ \mathrm{m\ s^{-1}}$. In calculations we shall generally use the approximate value of $3.00 \times 10^8\ \mathrm{m\ s^{-1}}$.

Equations valid for high-speed particles (i.e. relativistic mechanics) are generally used in atomic, nuclear and elementary-particle physics, where it is often convenient to use a system of units based on the electron volt eV, which is defined as the work done in moving a charge equal to the electronic charge of 1.602×10^{-19} C through a potential difference of 1 V. Since the work done in moving one coulomb through one volt is one joule (J), the work done in taking 1.602×10^{-19} C through 1 V is 1.602×10^{-19} J. Hence

$$1\ \mathrm{eV} = 1.602 \times 10^{-19}\ \mathrm{J} = 1.602 \times 10^{-12}\ \mathrm{erg}. \tag{2.15}$$

The following multiples of the electron volt are often used:

$$1\ \mathrm{keV} = 10^3\ \mathrm{eV} = 1.602 \times 10^{-16}\ \mathrm{J}, \tag{2.16}$$

$$1\ \mathrm{MeV} = 10^6\ \mathrm{eV} = 1.602 \times 10^{-13}\ \mathrm{J}, \tag{2.17}$$

$$1\ \mathrm{GeV} = 10^9\ \mathrm{eV} = 1.602 \times 10^{-10}\ \mathrm{J}, \tag{2.18}$$

$$1\ \mathrm{TeV} = 10^{12}\ \mathrm{eV} = 1.602 \times 10^{-7}\ \mathrm{J}. \tag{2.19}$$

2.4 THE LORENTZ FORCE

Consider a positive charge $+q$ (in coulombs) moving between the plates of a parallel plate capacitor (Fig. 2.1a). If a potential difference V (in volts) is applied across the plates of the capacitor, which are a distance d (in metres) apart, there is an electric field of strength V/d (in volts per metre) across the capacitor as shown in Fig. 2.1(a). Experiments have shown that the electric force $\boldsymbol{f}_{\mathrm{elec}}$ on the charge $+q$ is

$$\boldsymbol{f}_{\mathrm{elec}} = +q\boldsymbol{E}. \tag{2.20}$$

For a positive charge this force acts in the direction of the electric field, and leads to the deflection of the moving electric charge, as shown in Fig. 2.1(a). It is assumed that the electric force on a moving charge is independent of the velocity of the charge.

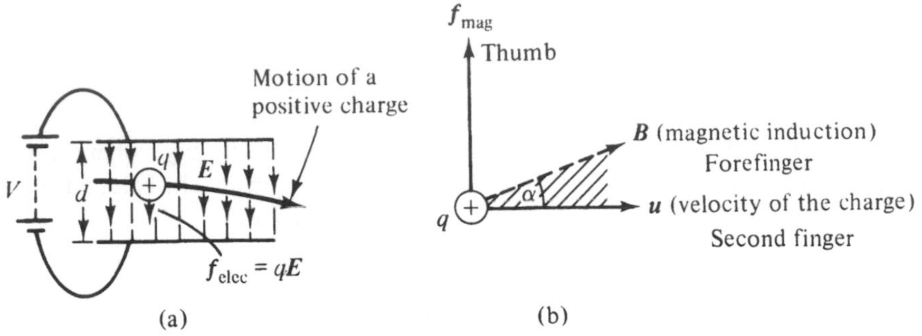

Fig. 2.1 (a) Deflection of a moving charge in an electric field. (b) The left-hand motor rule.

The motion of a moving charge in a magnetic field will now be considered. Let a charge $+q$ move with velocity $\boldsymbol{u}$ at an angle α to the direction of the magnetic field $\boldsymbol{B}$ as shown in Fig. 2.1(b). The magnetic field $\boldsymbol{B}$ is generally called the *magnetic induction* or the *magnetic flux density*. The magnetic force on the moving charge, $\boldsymbol{f}_{mag}$, is given by the vector product

$$\boldsymbol{f}_{mag} = q\boldsymbol{u} \times \boldsymbol{B}. \tag{2.21}$$

The magnitude of the magnetic force f_{mag} is given by

$$f_{mag} = quB \sin \alpha. \tag{2.22}$$

If q is in coulombs, u in metres per second and $\boldsymbol{B}$ in teslas (webers per square metre) then f_{mag} is in newtons. The direction of this magnetic force is given by the left-hand motor rule. If the forefinger of the left hand points in the direction of the magnetic field $\boldsymbol{B}$ and the second finger in the direction of the velocity of the charge $\boldsymbol{u}$ (for a positive charge) then the thumb of the left hand points in the direction of the magnetic force acting on the moving charge, as shown in Fig. 2.1(b). This magnetic force is perpendicular to the plane containing $\boldsymbol{u}$ and $\boldsymbol{B}$, as shown in Fig. 2.1(b).

When a charge moves in both an electric and a magnetic field, there is both an electric force given by equation (2.20) and a magnetic force given by equation (2.21) acting on the moving charge. The total force acting on the moving charge is the vector sum of these forces:

$$\boldsymbol{f}_{total} = \boldsymbol{f}_{elec} + \boldsymbol{f}_{mag} = q\boldsymbol{E} + q\boldsymbol{u} \times \boldsymbol{B}. \tag{2.23}$$

The total force given by equation (2.23) is generally known as the Lorentz expression for the total force acting on a moving charge, or sometimes just the *Lorentz force law*. Using equation (2.4), we have

$$f_{\text{total}} = \frac{\mathrm{d}}{\mathrm{d}t}(\alpha m\boldsymbol{u})$$

$$= \frac{\mathrm{d}}{\mathrm{d}t}\left[\frac{m\boldsymbol{u}}{(1 - u^2/c^2)^{1/2}}\right] = q\boldsymbol{E} + q\boldsymbol{u} \times \boldsymbol{B}\,. \tag{2.24}$$

When equation (2.24) is used to determine the changes in the motion of the charge q, it is assumed that the values of $\boldsymbol{E}$ and $\boldsymbol{B}$ are known. For example, $\boldsymbol{E}$ and $\boldsymbol{B}$ can be determined from the charge and current distributions using the laws of electromagnetism, which are based on previous experiments.

It is also assumed when equation (2.24) is applied that the value of the charge q is an invariant, i.e. independent of the velocity of the charge. There is now independent experimental evidence in favour of the principle of constant charge. For example, if the charge q on a particle varied according to the equation

$$q = q_0\,(1 - u^2/c^2)^{1/2}, \tag{2.25}$$

where u is the velocity of the charge and q_0 the value of the charge when it is at rest, then hydrogen molecules would not be electrically neutral, since the negative electrons move in orbits around the atomic nuclei in the molecules, and on average move faster than the positive nuclei (protons in this case) relative to the laboratory. If the charge did vary with velocity then hydrogen molecules would have a resultant electric charge and would be deflected in electric fields, for example of the type shown in Fig. 2.1(a). In 1960 King showed that the charges on the electrons and the protons in hydrogen molecules were numerically equal to within one part in 10^{20}. We therefore conclude that the charge on a particle is independent of its velocity.

It will now be shown that predictions based on equation (2.24) are in excellent agreement with experiment, whereas predictions based on Newtonian mechanics are completely at variance with experimental results obtained when the speed of the moving charge is comparable to the speed of light.

2.5 MOTION IN AN ELECTRIC FIELD

It will be assumed that there is a potential difference V between the plates of a large parallel plate capacitor, as shown in Fig. 2.2. Let the separation of the plates be d. Away from the ends of the capacitor, there is a uniform electric field of magnitude $E = V/d$ between the plates. Let the x axis of a coordinate system be parallel to the direction of the electric field, and let the origin O of the coordinate system coincide with the positive plate, as shown in Fig. 2.2. Let a charge $+q$ of mass m be released from rest from O at time $t = 0$ such that it accelerates from the positive to the negative plate of the capacitor. For motion in an electric field the force on the moving charge is qE. When the velocity of the charge is u, we have from equation (2.24)

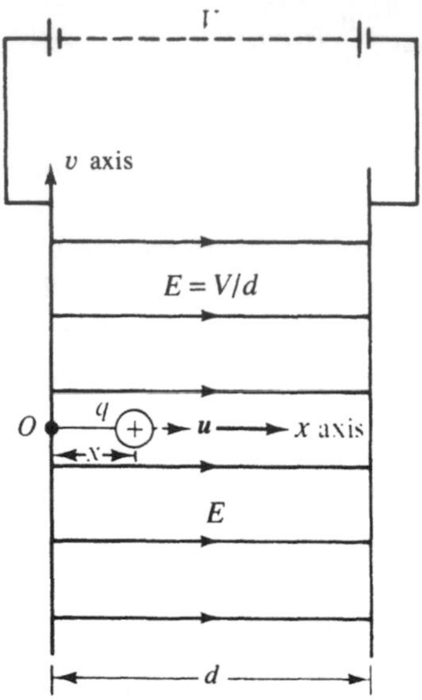

Fig. 2.2 Motion of an electric charge in a uniform electric field, for example in the electric field inside a parallel plate capacitor.

$$\frac{d}{dt}\left[\frac{mu}{(1-u^2/c^2)^{1/2}}\right] = qE.$$

Since the mass m is constant,

$$\frac{d}{dt}\left[\frac{u}{(1-u^2/c^2)^{1/2}}\right] = \frac{qE}{m}. \tag{2.26}$$

Integrating equation (2.26) with respect to time, if the velocity of the charge after a time t is u then, since it starts from rest at $t = 0$, we have

$$\left[\frac{u}{(1-u^2/c^2)^{1/2}}\right]_0^u = \left[\frac{qE}{m}t\right]_0^t,$$

$$\frac{u}{(1-u^2/c^2)^{1/2}} = \frac{qEt}{m}. \tag{2.27}$$

Squaring, we obtain

$$u^2 = \left(1 - \frac{u^2}{c^2}\right)\frac{q^2E^2t^2}{m^2},$$

so that

$$u^2 = \frac{q^2E^2t^2}{m^2(1+q^2E^2t^2/m^2c^2)},$$

and

$$u = \frac{qEt}{m\,(1 + q^2E^2t^2/m^2c^2)^{1/2}} \,. \tag{2.28}$$

Multiplying the numerator and denominator of the right-hand side by mc/qEt, we have

$$u = \frac{c}{(1 + m^2c^2/q^2E^2t^2)^{1/2}} \,. \tag{2.29}$$

If the time t is very long (that is, if the charge continues to accelerate in the electric field for a long time) then the quantity $m^2c^2/q^2E^2t^2$ becomes small compared with unity, so that as t tends to infinity, u tends to c, though u always remains less than c. Hence the accelerating charge has a limiting speed, which is equal to the speed of light in empty space.

According to Newtonian mechanics, since the electric force acting on the charge is qE, its acceleration a is given by

$$a = \frac{f}{m} = \frac{qE}{m} \,. \tag{2.30}$$

Hence, according to Newtonian mechanics,

$$u = at = \frac{qE}{m}\,t, \tag{2.31}$$

and thus the velocity u tends to infinity as t tends to infinity, and there should be no limiting speed. If $t << mc/qE$ (that is, if the time is not long enough for the charge to gain very much speed) then, since $qEt/mc << 1$ and $q^2E^2t^2/m^2c^2 << 1$, equation (2.28) becomes

$$u \approx \frac{qEt}{m} \,,$$

in agreement with Newtonian mechanics, which is a satisfactory approximation at low speeds. However, equations (2.28) and (2.31) differ markedly at high speeds.

An experimental determination of the speeds of electrons accelerated by electric fields was carried out by Bertozzi (1964). The principle of the experiment is outlined in Fig. 2.3. Electrons were accelerated by a van de Graaff generator up to kinetic energies of 1.5 MeV. Higher energies were obtained using an electron linear accelerator. The electrons were accelerated in bursts approximately 3 ns long. The times of flight of electrons of various kinetic energies were measured over a distance of 8.4 m. (If the electrons did travel at the speed of light, the time of flight would be $8.4\ \text{m}/3 \times 10^8\ \text{m s}^{-1} = 28$ ns which is long compared with the length of each burst of electrons.) The time of flight was measured by displaying signals from the first and second detectors on an oscilloscope screen, using connecting cables that took the same time to transmit

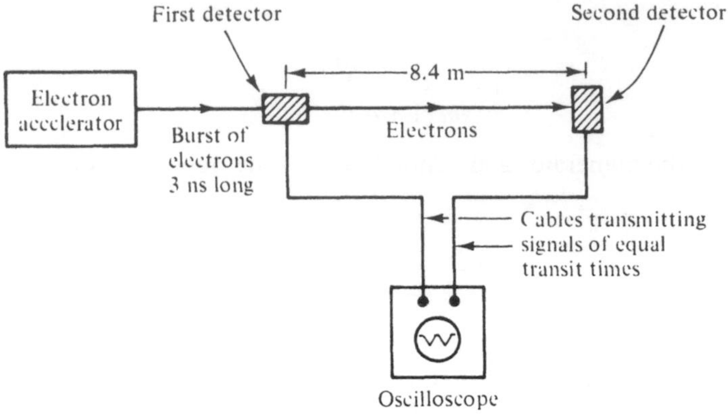

Fig. 2.3 Basic principles of the experimental arrangement used by Bertozzi (1964) to measure the time of flight of electrons over a distance of 8.4 m.

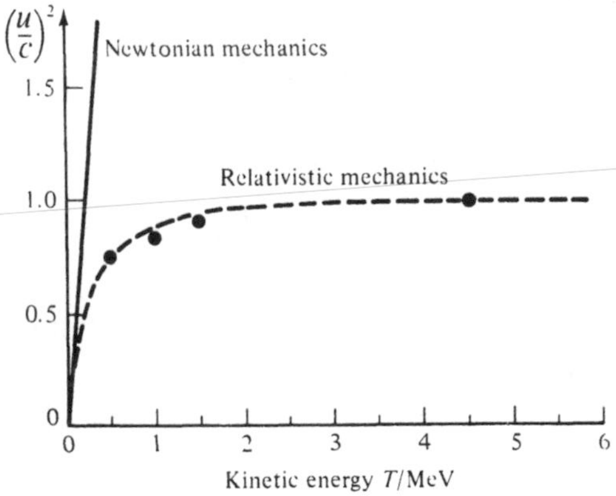

T / MeV	T/mc^2	$(u/c)_{obs}$	$(u/c)^2_{obs}$
0.5	1	0.867	0.752
1.0	2	0.910	0.828
1.5	3	0.960	0.922
4.5	9	0.987	0.974
15.0	30	1.0	1.0

Fig. 2.4 Experimental results obtained by Bertozzi (1964). The solid line represents the prediction of $(u/c)^2$ according to Newtonian mechanics: $(u/c)^2 = 2T/mc^2$, where $T = \frac{1}{2}mu^2$. The dashed curve is the prediction of relativistic mechanics: $(u/c)^2 = 1 - [mc^2/(T + mc^2)]^2$, where m is the mass of the electron and $c = 3 \times 10^8$ m s^{-1} is the speed of light in vacuum. The solid circles are the experimental points. They agree with the relativistic theory of high-speed particles and show that c is the maximum speed for electrons. (By courtesy of *Am. J. Phys.*)

signals from the detectors to the oscilloscope. The separation of the electrical pulses from the two detectors on the oscilloscope screen gave the time of flight of the electrons t (in s) and their velocity (in m s^{-1}) was $8.4/t$. Fuller details of the experiment are given by Bertozzi (1964). The results obtained by Bertozzi are shown in Fig. 2.4. From equation (2.13),

$$\frac{mc^2}{(1-u^2/c^2)^{1/2}} = T + mc^2,$$

where T is the kinetic energy. Rearranging,

$$\frac{u^2}{c^2} = 1 - \left(\frac{mc^2}{T+mc^2}\right)^2. \tag{2.32}$$

Equation (2.32) is plotted in Fig. 2.4. The values of kinetic energy shown in Fig. 2.4 were calculated by assuming that the kinetic energy was equal to qV, where $q = 1.602 \times 10^{-19}$ C and V is the potential difference through which the electrons were accelerated. The experimental results of Bertozzi are in excellent agreement with equation (2.32), and show clearly that there is a limiting speed for accelerated electrons and that this limiting speed is equal to the speed of light in empty space. This is in agreement with the theory developed from equation (2.24), which follows from the redefinition of momentum as αmu in equation (2.1). The predictions of Newtonian mechanics are completely at variance with the experimental results.

Returning to the problem of the charge accelerating between the plates of the capacitor illustrated in Fig. 2.2, since the charge is moving along the x axis, its velocity is equal to $\mathrm{d}x/\mathrm{d}t$. Substituting in equation (2.28), we obtain

$$u = \frac{\mathrm{d}x}{\mathrm{d}t} = \frac{qEt}{m\left(1+\dfrac{q^2E^2t^2}{m^2c^2}\right)^{1/2}}. \tag{2.33}$$

Integrating, assuming that the charge leaves the point $x = 0$ at time $t = 0$ and is at the point x at a time t, we have

$$\int_0^x \mathrm{d}x = \int_0^t \frac{qEt\,\mathrm{d}t}{m\left(1+\dfrac{q^2E^2t^2}{m^2c^2}\right)^{1/2}}. \tag{2.34}$$

Put

$$1 + \frac{q^2E^2t^2}{m^2c^2} = w,$$

so that after differentiating

$$\frac{2q^2E^2t\,\mathrm{d}t}{m^2c^2} = \mathrm{d}w,$$

or

$$\frac{qEt\,\mathrm{d}t}{m} = \frac{mc^2\,\mathrm{d}w}{2qE}.$$

Therefore

$$\int \frac{qEt\,\mathrm{d}t}{m\,(1 + q^2E^2t^2/m^2c^2)^{1/2}} = \frac{mc^2}{2qE}\int \frac{\mathrm{d}w}{w^{1/2}}$$

$$= \frac{mc^2}{2qE}\,\frac{w^{1/2}}{\frac{1}{2}} = \frac{mc^2}{qE}\left(1 + \frac{q^2E^2t^2}{m^2c^2}\right)^{1/2}$$

Substituting in equation (2.34), we obtain

$$[x]_0^x = \left[\frac{mc^2}{qE}\left(1 + \frac{q^2E^2t^2}{m^2c^2}\right)^{1/2}\right]_0^t;$$

that is,

$$x = \frac{mc^2}{qE}\left[\left(1 + \frac{q^2E^2t^2}{m^2c^2}\right)^{1/2} - 1\right]. \tag{2.35}$$

Equation (2.35) gives the distance x travelled in time t. After a little algebra, it can be rewritten as

$$\left(x + \frac{mc^2}{qE}\right)^2 - c^2t^2 = \frac{m^2c^4}{q^2E^2}. \tag{2.36}$$

This is the equation of a hyperbola, and thus the graph of x against ct is a hyperbola.

According to Newtonian mechanics, the acceleration of the charge in Fig. 2.2 is given by equation (2.30), so that

$$x = \frac{1}{2}at^2 = \frac{1}{2}\frac{qE}{m}t^2.$$

If $t \ll mc/qE$, so that the electron does not have enough time to gain very much speed, expanding equation (2.35) using the binomial theorem we have:

$$x = \frac{mc^2}{qE}\left(1 + \frac{1}{2}\frac{q^2E^2t^2}{m^2c^2} + \ldots - 1\right) \approx \frac{1}{2}\frac{qE}{m}t^2.$$

Even though Newtonian mechanics is completely inadequate at speeds comparable to the speed of light, it is always a very satisfactory approximation at speeds very much less than the speed of light. Even if the speed of an aeroplane is as high as 3000 km h^{-1}, u/c is only about 3×10^{-6} and the deviations from Newtonian mechanics are completely negligible. In such cases it is an unnecessary over-elaboration to use the equations for high-speed particles (i.e. relativistic mechanics), even though we know that they give a better account of the behaviour of nature than Newtonian mechanics.

2.6 MOTION IN A MAGNETIC FIELD (NO ELECTRIC FIELD)

According to equation (2.21), the magnetic force on a charge of magnitude q moving with velocity $\boldsymbol{u}$ in a magnetic field $\boldsymbol{B}$ is given by the vector product

$$\boldsymbol{f} = q\boldsymbol{u} \times \boldsymbol{B}. \tag{2.37}$$

The magnetic force $\boldsymbol{f}$ is always perpendicular to the velocity $\boldsymbol{u}$ of the charge and to the magnetic field $\boldsymbol{B}$. According to equation (1.12), the work dW done by the magnetic force in a displacement d$\boldsymbol{l}$ of the charge is given by the scalar product $\boldsymbol{f} \bullet \mathrm{d}\boldsymbol{l}$. Hence the rate at which the magnetic force $\boldsymbol{f}$ does work on the moving charge is

$$\frac{\mathrm{d}W}{\mathrm{d}t} = \boldsymbol{f} \bullet \frac{\mathrm{d}\boldsymbol{l}}{\mathrm{d}t} = \boldsymbol{f} \bullet \boldsymbol{u} \cdot \tag{2.38}$$

Since, from (2.37), $\boldsymbol{f}$ is always perpendicular to $\boldsymbol{u}$, the scalar product $\boldsymbol{f} \bullet \boldsymbol{u}$ is always zero, showing that a magnetic field does no work on a charge moving in a magnetic field. Hence in the absence of an electric field, the kinetic energy and therefore also the speed of a charge moving in an arbitrary time-independent magnetic field are constant. The magnetic force changes the direction of the velocity $\boldsymbol{u}$, but not its magnitude. Therefore, for motion in a magnetic field, $\alpha = (1 - u^2/c^2)^{-1/2}$ remains constant, and equation (2.24) reduces to

$$\boldsymbol{f} = \frac{\mathrm{d}}{\mathrm{d}t}(\alpha m\boldsymbol{u}) = \alpha m \frac{\mathrm{d}\boldsymbol{u}}{\mathrm{d}t}$$
$$= \alpha m\boldsymbol{a} = q\boldsymbol{u} \times \boldsymbol{B}, \tag{2.39}$$

where $\boldsymbol{a}$ is the acceleration of the charge.

Consider a particle of charge q moving in a uniform magnetic field $\boldsymbol{B}$. Let the initial velocity of the charge be perpendicular to the direction of the magnetic field, as shown in Fig. 2.5(a), and let the magnetic field be in the negative direction—that is, downwards away from the reader in Fig. 2.5(a). According to equation (2.22), since in this case the angle between $\boldsymbol{u}$ and $\boldsymbol{B}$ is 90°, the magnitude of the magnetic force is quB. Substituting in equation (2.39), we have

$$a = quB/\alpha m.$$

Since q, u, B, α and m are all constants, the magnitude of the acceleration $\boldsymbol{a}$ is a constant. Since $\boldsymbol{a}$ is always perpendicular to the velocity $\boldsymbol{u}$, the charge moves in a circular orbit of radius ρ, as shown in Fig. 2.5(a). It can be seen from Fig. 2.5(a, b) that the centripetal acceleration a is equal to u^2/ρ. Hence

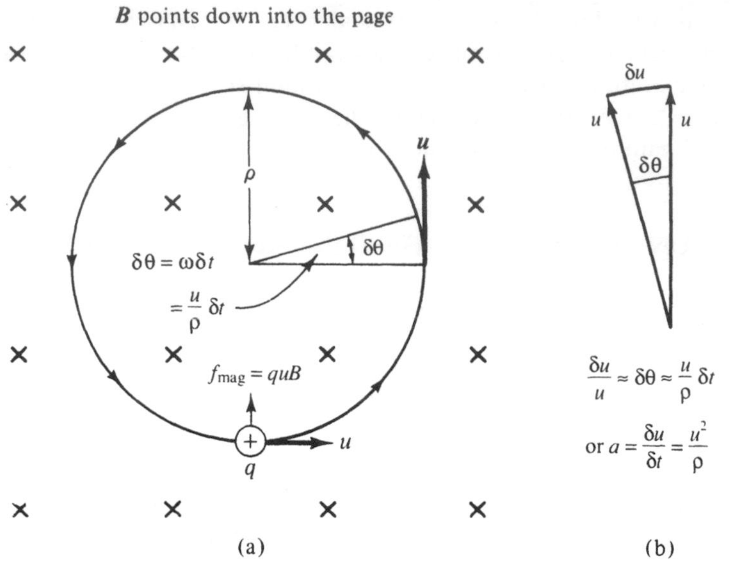

Fig. 2.5 The motion of a positively charged particle in a uniform magnetic field $\boldsymbol{B}$ is circular if the initial velocity of the particle is perpendicular to $\boldsymbol{B}$.

$$a = \frac{quB}{\alpha m} = \frac{u^2}{\rho},$$

and therefore

$$\rho = \frac{\alpha m u}{qB}. \tag{2.40}$$

From equation (2.1), $\alpha m u$ is equal to the momentum p of the moving charge, so that equation (2.40) can be rewritten in the form

$$p = \alpha m u = qB\rho. \tag{2.41}$$

According to equation (2.41), if the radius of curvature ρ of the orbit of a charged particle, of known charge q, is measured when the charge is moving in a direction perpendicular to a known magnetic field $\boldsymbol{B}$ then the momentum of the charged particle is $qB\rho$.

For a charged particle moving in a circle of radius ρ with velocity $\boldsymbol{u}$ in a uniform magnetic field $\boldsymbol{B}$, the time T to complete one revolution is equal to $2\pi\rho/u$. Since, from equation (2.40), $\rho/u = \alpha m/qB$, we have

$$T = \frac{2\pi\rho}{u} = \frac{2\pi\alpha m}{qB}. \tag{2.42}$$

The number of revolutions per second, sometimes called the *Larmor frequency*, is

$$n = \frac{qB}{2\pi\alpha m}. \tag{2.43}$$

If the velocity of the charged particle is not perpendicular to the uniform magnetic field then the velocity of the particle can be resolved into two components, one parallel to the magnetic field and the other perpendicular to it. If there is no electric field present, the component of the velocity parallel to the magnetic field is unchanged in magnitude and direction. This velocity must be compounded with the circular motion perpendicular to the magnetic field. The resulting motion is a helix.

Equation (2.39) can also be used to calculate the motions of charged particles in non-uniform magnetic fields.

2.7 THE ACCELERATION OF PROTONS TO HIGH ENERGIES

A brief outline of the physical principles of some proton accelerators will now be given to illustrate the applicability of the laws developed for high-speed particles. One simple method of accelerating charged particles is to use a high-voltage rectifier circuit. The charged particles are then accelerated by the electric field between the positive and negative terminals. Another method is to use a van de Graaff generator. In this case also the charges are accelerated by an electric field, and the analysis of Section 2.5 is applicable.

Most high-energy proton accelerators are developments of the cyclotron, which is shown in schematic form in Fig. 2.6. The cyclotron consists of two flat hollow conducting semicircular boxes, generally called *dees*, placed in a vacuum. The two dees are placed between the poles of an electromagnet, which provides a magnetic field perpendicular to the planes in which they lie. Protons are injected from the source S, which is at the centre between the dees. A potential difference is applied between the dees so that there is an electric field in the gap between them. This electric field accelerates the protons towards one of the dees. As the latter are hollow conductors, the electric field inside them is negligible (cf. Faraday's ice pail and Faraday's cage experiments). Hence, once the protons are inside a dee, their velocities remain constant, and they move in circular orbits in the magnetic field, which penetrates the dees. From equation (2.40), the radius of the orbit is

$$\rho = \frac{\alpha m u}{qB} = \frac{mu}{qB\,(1 - u^2/c^2)^{1/2}}, \tag{2.44}$$

where m is the (invariant) mass, q the charge and u the velocity of the proton in the magnetic field B. After going through a semicircle the protons reach the gap between the dees again, as shown in Fig. 2.6. According to equation (2.42), the time to go through a semicircle is $\pi\rho/u = \pi\alpha m/qB$. If by this time the potential difference between the dees is reversed, the electric field is in the opposite direction to its previous one, and the protons are accelerated by it and gain energy again as they cross the gap. As a result of this acceleration the velocity u of the protons is increased, so that, according to equation (2.44), they move in a

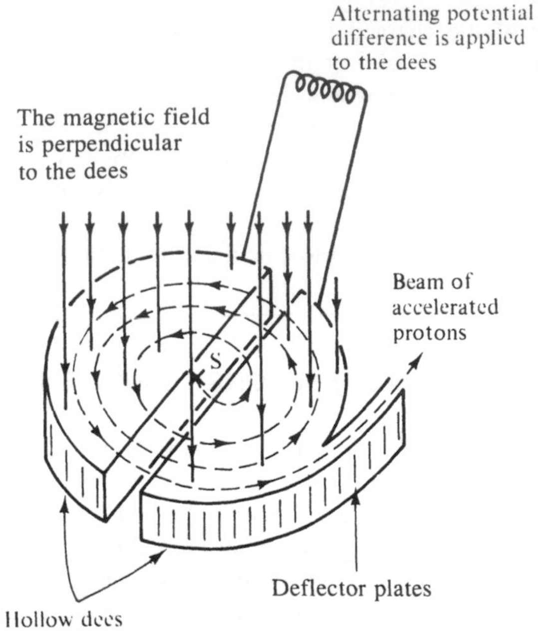

Fig. 2.6 Simplified diagram of a cyclotron; S is the proton source. The path of a typical proton is shown dashed.

semicircle of larger radius before reaching the gap again, as shown in Fig. 2.6. The time $\pi\alpha m/qB$ taken to go around the second semicircle is the same as the time for the previous semicircle, provided that α can be treated as constant. Hence if the potential difference applied to the dees reverses direction at time intervals $\pi\,\alpha m/qB$ the protons are accelerated on each crossing of the gap between the dees. This reversal can be achieved by applying an alternating potential difference to the dees. If it is to reverse in direction every time interval $\pi\,\alpha m/qB$ the period of the alternating potential difference must be $2\pi\,\alpha m/qB$, so that its frequency n must be

$$n = \frac{qB}{2\pi\,\alpha m} = \frac{qB\,(1 - u^2/c^2)^{1/2}}{2\pi m}\,. \tag{2.45}$$

Provided that the variation of α with velocity u can be neglected, the protons can be accelerated by applying a potential difference of fixed frequency to the dees. As the energy of the proton, and hence its velocity, increases, the radius of its orbit is increased. Since, from equation (2.41),

$$p = qB\rho, \tag{2.46}$$

for a fixed value of magnetic field, the maximum momentum that can be obtained from the cyclotron is proportional to the radius ρ of the dees. For a discussion of typical numerical values see Problem 2.9. The cyclotron has been used to accelerate protons to about 40 MeV (electrons can only be accelerated up to about 50 keV in a cyclotron). At higher energies the variation of α with velocity becomes important, and as u, and hence α, increases, according to equation (2.42), the time taken to go in a semicircle also increases, so that if n and B were fixed then after every half-cycle the protons would tend to get further and further behind the maximum potential difference across the dees. This loss of resonance can be overcome in two ways. From equation (2.45), as u increases, the frequency necessary for resonance goes down. In the 184 inch frequency-modulated cyclotron at the University of California the frequency of the potential difference applied to the dees was reduced continuously after injection of the protons. By reducing the frequency from 22.9 MHz at injection to 15.8 MHz when the protons reached the outside of the dees, it was possible to accelerate protons up to 350 MeV. The other method of overcoming resonance is to increase the strength of the magnetic field in such a way that as α increases, B/α remains constant. One can then keep the frequency n constant. This is the principle of the synchrotron, which has been used to accelerate electrons up to energies of more than 400 MeV. In modern proton synchrotrons both n and B are varied in such a way that not only is equation (2.45) satisfied, but in equation (2.46) the radius of the orbit remains constant. This means that, instead of dees, one can use doughnut-shaped vacuum chambers, and the magnetic field need only extend over the dimensions of the chamber. The NIMROD proton synchrotron of the Rutherford – Appleton laboratory was able to accelerate protons up to kinetic energies of 7 GeV. The tevatron at the Fermi laboratory in the USA can accelerate protons up to kinetic energies of 1000 GeV, i.e. 1 TeV. The successful design of such high-energy accelerators shows that the theory developed in Section 2.4 is satisfactory up to very high proton energies. Equation (2.42) breaks down if the accelerating charge emits electromagnetic radiation. Whenever an accelerating charge emits a photon, the charge recoils to conserve momentum. This phenomenon is called *radiation reaction*. The effect is generally very small for protons moving in circular orbits in a magnetic field, but it is extremely important for electrons moving in circular orbits.

2.8 THE TOTAL ENERGY OF A PARTICLE

According to equation (2.12), the kinetic energy T of a particle is given by

$$T = \alpha mc^2 - mc^2, \tag{2.47}$$

where m is the mass of the particle, $\alpha = (1 - u^2/c^2)^{-1/2}$ and u is the speed of the particle. Equation (2.47) can be rewritten in the form:

$$T + mc^2 = \alpha mc^2, \tag{2.48}$$

The quantity on the left-hand side of equation (2.48) is called the *total energy* of the particle and is denoted by E; that is,

$$E = T + mc^2. \tag{2.49}$$

Thus the total energy E is equal to the sum of the *kinetic energy* T of the particle and the quantity mc^2, which will be identified in Chapter 6 with the *rest energy* of the particle. From equation (2.48),

$$E = \alpha mc^2 = \frac{mc^2}{(1 - u^2/c^2)^{1/2}} . \tag{2.50}$$

Putting

$$u/c = \sin\theta, \tag{2.51}$$

as in Section 2.2, we have

$$\alpha = (1 - \sin^2\theta)^{-1/2} = \sec\theta, \tag{2.52}$$

$$E = \alpha mc^2 = mc^2 \sec\theta. \tag{2.53}$$

Many writers refer to the total energy E as the energy of the particle. No confusion should arise if this convention is adopted, since all other forms of energy are given an adjective such as "kinetic" or "potential". We shall continue to refer to E as the total energy of the particle, but we shall place the adjective "total" inside parentheses to indicate that its use is optional.

It follows from equation (2.53) that $E/mc^2 = \alpha = \sec\theta$. Hence the variation of (total) energy E, measured in units of mc^2, with $u/c = \sin\theta$ can be obtained by plotting $\sec\theta$ against $\sin\theta$, as shown in Fig. 2.7. From equation (2.12),

$$\frac{T}{mc^2} = \sec\theta - 1. \tag{2.54}$$

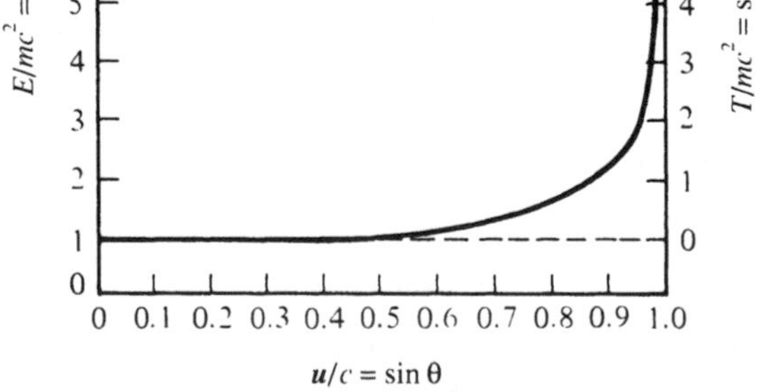

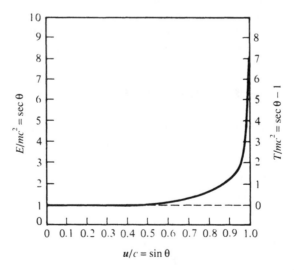

Fig. 2.7 Plot of $\sec\theta$ against $\sin\theta$. This is equivalent to plotting $E/mc^2 = (1 - u^2/c^2)^{-1/2}$ against u/c.

The variation of the kinetic energy T, measured in units of mc^2, with u/c can be obtained by plotting $\sec\theta - 1$ against $\sin\theta$. This is achieved in Fig. 2.7 by moving the scale of E/mc^2 up by one unit, as shown along the right-hand ordinate. It can be seen that the kinetic energy T tends to infinity as u tends to c.

Squaring equation (2.53), we have

$$E^2 = m^2c^4 \sec^2\theta = m^2c^4(\tan^2\theta + 1).$$

Using equation (2.9) for the momentum of the particle, we find

$$E^2 = c^2p^2 + m^2c^4. \tag{2.55}$$

Equation (2.55) is very important. It relates the total energy E and the momentum $\boldsymbol{p}$ of a particle. Using equation (2.49),

$$\begin{aligned} T &= E - mc^2 \\ &= (c^2p^2 + m^2c^4)^{1/2} - mc^2. \end{aligned} \tag{2.56}$$

Equation (2.56) relates the momentum p of a particle to its kinetic energy T.

It follows from equation (2.55) and the converse of Pythagoras' theorem that the quantities E, cp and mc^2 can be represented by the sides of a right-angled triangle, as shown in Fig. 2.8, from which we have

$$\sin\theta = \frac{cp}{E} = \frac{c\alpha mu}{\alpha mc^2} = \frac{u}{c}, \tag{2.57}$$

$$cp = mc^2 \tan\theta. \tag{2.58}$$

$$E = mc^2 \sec\theta. \tag{2.59}$$

Many readers may find it easier to use these trigonometrical relations in calculations than the alternatives

$$\alpha = \frac{1}{(1 - u^2/c^2)^{1/2}}, \tag{2.2}$$

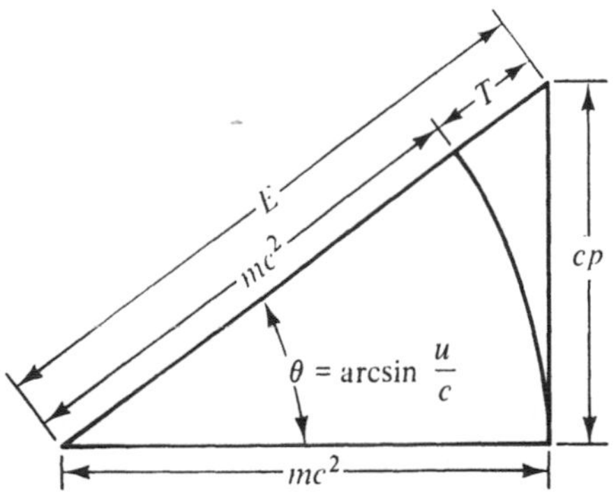

Fig. 2.8 Mnemonic diagram to illustrate the relations between E, cp and mc^2.

$$\mathbf{p} = \alpha m\mathbf{u} = \frac{m\mathbf{u}}{(1 - u^2/c^2)^{1/2}}, \tag{2.1}$$

$$E = \alpha mc^2 = \frac{mc^2}{(1 - u^2/c^2)^{1/2}}, \tag{2.50}$$

$$T = mc^2(\alpha - 1) . \tag{2.60}$$

The following is useful in calculations:

$$E^2 = c^2p^2 + m^2c^4. \tag{2.55}$$

Since

$$\frac{cp}{E} = \frac{c\alpha mu}{\alpha mc^2} = \frac{u}{c},$$

we have

$$u = \frac{c^2p}{E} . \tag{2.61}$$

2.9 PHOTONS

Electromagnetic radiation is emitted in the form of discrete quanta of energy called *photons*. According to quantum theory, if the frequency of the radiation is ν, the energy of each quantum, or photon, is $E = h\nu$, where $h = 6.625 \times 10^{-34}$ J s is Planck's constant. If the energy of the quantum or photon is in the MeV range, it is generally referred to as a γ ray. Einstein used the photon concept to interpret the photoelectric effect. Compton used the idea of individual photons of energy $h\nu$ carrying momentum to interpret the scattering of X rays. The existence of individual photons travelling at the speed of light is now a generally accepted experimental fact. For a particle of mass m, we have, from equations (2.1) and (2.50),

$$p = \frac{mu}{(1 - u^2/c^2)^{1/2}} \qquad E = \frac{mc^2}{(1 - u^2/c^2)^{1/2}} .$$

As u tends to c, the denominator tends to zero in each case. If the mass m also tends to zero as u tends to c then both p and E can remain finite. If $m \to 0$ as $u \to c$, such that $m/(1 - u^2/c^2)^{1/2} = k$ then

$$p = kc; \qquad E = kc^2.$$

From Planck's relation, for a photon $E = h\nu$, so that k must equal $h\nu/c^2$. This suggests that a photon of energy $h\nu$ should have a linear momentum

$$p = h\nu/c. \tag{2.62}$$

This result is confirmed by experiments on the Compton effect (see Section 6.4).

It is often stated that a photon has zero rest mass. Since it always travels at the speed of light, one can never find a reference frame in which it is at rest, so that the use of the phrase "rest mass" is inappropriate in relation to the photon. However, since for a photon $E = h\nu$ and $p = h\nu/c$, in this case we must put $m = 0$ in the equation $E^2 = c^2p^2 + m^2c^4$, which then reduces to $E = cp$.

2.10 USE OF ENERGY UNITS IN CALCULATIONS

The equations of relativistic mechanics can be applied in any consistent set of units, such as SI or c.g.s. units. In practice, it is often convenient to use energy units based on the electron volt (eV). It can be seen from equation (2.55) that the quantities mc^2 and cp have the dimensions of energy, so that both mc^2 and cp can be expressed in eV or MeV. When using energy units, the quantities mc^2 and cp are employed instead of m and p. This approach is used extensively in high-energy physics. For an electron of mass 9.1083×10^{-31} kg

$$\begin{aligned} mc^2 &= 9.108\,3 \times 10^{-31} \times (2.997\,93 \times 10^8)^2 \\ &= 8.186\,16 \times 10^{-14}\ \text{J} \\ &= 0.510\,98\ \text{MeV}. \end{aligned} \tag{2.63}$$

Some textbooks quote the masses of particles in MeV, but what they are actually giving is the rest energy mc^2. The value of the mass is sometimes quoted in MeV/c^2. For example, the mass of an electron is sometimes written in the form

$$m = 0.510\,98\ \text{MeV}/c^2.$$

When solving problems using energy units, the reader should *rewrite this in the form* $mc^2 = 0.510\,98$ MeV *and work in* MeV.

From equation (2.41),

$$p = \alpha mu = qB\rho. \tag{2.41}$$

Hence the momentum of a particle of charge q can be determined from the radius of curvature ρ of its orbit when it is moving perpendicular to a magnetic field B. If ρ is in metres, B in teslas and q is Z times the electronic charge,

$$\begin{aligned} cp &= qB\rho c = Z \times 1.602 \times 10^{-19}\, B\, \rho \times 3.00 \times 10^8\ \text{J} \\ &= 300ZB\rho\ \text{MeV}. \end{aligned} \tag{2.64}$$

This is sometimes rewritten as

$$p = 300ZB\rho\ \text{MeV}/c. \tag{2.65}$$

When solving problems using energy units, the reader *should convert momenta expressed in terms of* MeV/c *into* cp *and work in* MeV.

Example

The radius of curvature of the circular orbit of an electron moving perpendicular to a uniform magnetic field of magnetic induction 1 T is 0.020 m. Determine the momentum and kinetic energy of the electron.

Solution From equation (2.65),

$$cp = 300\,B\rho = 300 \times 1 \times 2 \times 10^{-2}\ \text{MeV} = 6.0\ \text{MeV}\ ;$$

that is,

$$p = 6\ \text{MeV}/c.$$

Since for an electron $mc^2 = 0.510\,98$ MeV, using equation (2.55)

$$E^2 = c^2p^2 + m^2c^4 = (6^2 + 0.511^2)\ \text{MeV}^2$$
$$= 36.26\ \text{MeV}^2.$$

Hence

$$E = 6.02\ \text{MeV},$$
$$T = E - mc^2 = (6.02 - 0.51)\ \text{MeV} = 5.51\ \text{MeV}.$$

Alternatively, since $cp = 6$ MeV and $mc^2 = 0.511$ MeV, from equation (2.58) we have

$$\tan\theta = \frac{cp}{mc^2} = \frac{6}{0.511} = 11.74.$$

Hence $\theta = 85.13°$, giving $u/c = \sin\theta = 0.996$ and $E = mc^2 \sec\theta = 6.02$ MeV.

To convert these results into SI units, we have

$$cp = 6\ \text{MeV} = 6 \times 1.602 \times 10^{-13}\ \text{J} = 9.61 \times 10^{-13}\ \text{J},$$

so that

$$p = \frac{9.61 \times 10^{-13}\ \text{J}}{3 \times 10^8\ \text{m s}^{-1}} = 3.20 \times 10^{-21}\ \text{kg m s}^{-1}\ ,$$

$$E = 6.02\ \text{MeV} = 6.02 \times 1.602 \times 10^{-13}\ \text{J} = 9.64 \times 10^{-13}\ \text{J}.$$

If one uses SI units from the outset then, from equation (2.41),

$$p = qB\rho = 1.602 \times 10^{-19}\ \text{C} \times 1\ \text{T} \times 0.02\ \text{m} = 3.204 \times 10^{-21}\ \text{kg m s}^{-1},$$

so that

$$cp = 3 \times 10^8\ \text{m s}^{-1} \times 3.204 \times 10^{-21}\ \text{kg m s}^{-1} = 9.61 \times 10^{-13}\ \text{J}.$$

From equation (2.63),

$$mc^2 = 8.186 \times 10^{-14}\ \text{J}.$$

Substituting in equation (2.55),

$$E^2 = (cp)^2 + (mc^2)^2 = (9.61 \times 10^{-13}\ \mathrm{J})^2 + (8.186 \times 10^{-14}\ \mathrm{J})^2,$$

so that

$$E = 9.64 \times 10^{-13}\ \mathrm{J}.$$

Further problems of this sort are given at the end of the chapter.

2.11 CONSERVATION OF ENERGY IN AN ELECTROSTATIC FIELD

The arguments given in Section 1.2.7 leading to the conclusion that the sum of the kinetic and potential energies of a particle acted upon by a conservative force (e.g. the electrostatic force) is constant are equally valid in relativistic mechanics. The decrease in the potential energy of a charge of magnitude q when it is accelerated through a potential difference V is qV. If the charge starts from rest and is accelerated in a vacuum then, according to the law of conservation of energy, the final kinetic energy of the charged particle is qV. Using equation (2.12), we have

$$\begin{aligned} T = qV &= mc^2\,(\alpha - 1) \\ &= mc^2\,(\sec\theta - 1). \end{aligned} \tag{2.66}$$

As a numerical example, consider an electron accelerated through a potential difference of -0.511×10^6 V. According to the law of conservation of energy, the kinetic energy of the electron is 0.511 MeV. Applying equation (2.66), with $mc^2 = 0.511$ MeV for an electron, we have

$$0.511 = 0.511\,(\alpha - 1) = 0.511\,(\sec\theta - 1),$$

giving $\alpha = \sec\theta = 2$, $\theta = 60°$ and $u/c = \sin\theta = 0.866$.

In a typical television tube the electrons are accelerated through a potential difference of about 25 kV, so that the kinetic energy of an electron reaching the TV screen is typically about 25 keV = 0.025 MeV. Applying equation (2.66), with $mc^2 = 0.511$ MeV for an electron, we have

$$0.025 = 0.511\,(\alpha - 1) = 0.511\,(\sec\theta - 1),$$

giving $\alpha = \sec\theta = 1.049$, $\theta = 17.6°$ and $u/c = \sin\theta = 0.302$. This example shows that the final velocity of an electron in a TV tube is typically of order $0.3c$. Since $\alpha = 1.049$ for such an electron, the deviations from Newtonian mechanics are significant for the electrons in a typical TV tube.

2.12 MASS SPECTROMETER

A *simplified* form of a typical mass spectrometer is shown in Fig. 2.9. Positive ions are produced at the ion source by bombarding atoms of a gas of an element with electrons. The positive ions are then accelerated through a potential difference V, gaining kinetic energy

$$T = qV. \tag{2.67}$$

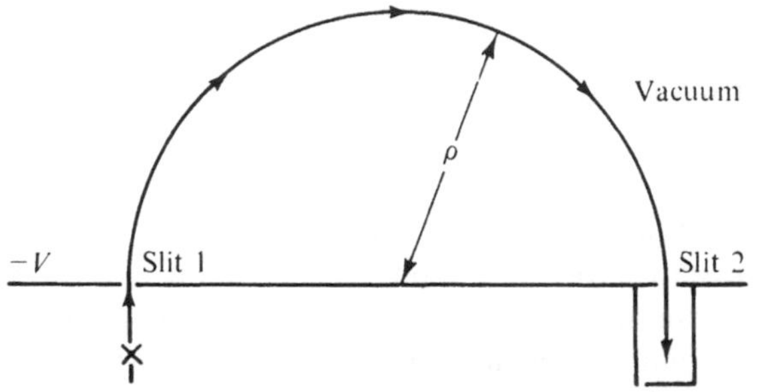

Fig. 2.9 Simplified diagram of a typical mass spectrometer.

After passing through slit 1, the ions move in circular orbits in a uniform magnetic field of magnetic induction $\boldsymbol{B}$. The strength of the magnetic field is varied until the positive ions pass through slit 2, and are recorded by the ion detector. For the simple case shown in Fig. 2.9 the radius ρ of the circular orbit is equal to half the distance between slits 1 and 2. From equation (2.41), for a charge of magnitude q and momentum p moving in a circular orbit of radius ρ perpendicular to a uniform magnetic field of magnetic induction B

$$p = qB\rho. \tag{2.68}$$

Putting $E = T + mc^2$ in equation (2.55), we have

$$E^2 = (T + mc^2)^2 = c^2p^2 + m^2c^4.$$

Therefore

$$m = \frac{c^2p^2 - T^2}{2Tc^2}. \tag{2.69}$$

Substituting for p using equation (2.68) and putting $T = qV$, we find that the mass m of the ion is given by

$$m = \frac{qB^2\rho^2}{2V} - \frac{qV}{2c^2}. \tag{2.70}$$

Equation (2.70) is valid whatever the final speed of the ion.

If Newtonian mechanics is used, equation (2.67) can be rewritten in the form

$$T = qV = \tfrac{1}{2} mu^2 ;$$

therefore

$$m^2u^2 = 2mT = 2mqV. \tag{2.71}$$

For motion in a magnetic field, in the low-velocity limit when α tends to unity, equation (2.40) becomes

$$quB = mu^2/\rho,$$

and

$$mu = qB\rho .$$

Substituting for mu in equation (2.71) and rearranging, using Newtonian mechanics we obtain

$$m = \frac{qB^2\rho^2}{2V} . \tag{2.72}$$

Comparing equations (2.70) and (2.72), it can be seen that the difference between the predictions of relativistic mechanics and Newtonian mechanics is the extra term $-qV/2c^2$ on the right-hand side of equation (2.70). This term will be called the *relativistic correction*. It is negligible when $qV/2c^2 << m$; that is, if $qV << 2mc^2$.

As an example, assume that $q = 1.602 \times 10^{-19}$ C, $B = 1$T, $V = 50$ kV and $\rho = 0.0644$ m. Substitution into equation (2.70) gives

$$m = (6.644 \times 10^{-27} - 4.45 \times 10^{-32})\ \text{kg} \tag{2.73}$$

$$\approx 6.644 \times 10^{-27}\ \text{kg};$$

so that

$$mc^2 = 3733\ \text{MeV}. \tag{2.74}$$

In atomic mass units, with 1 u $= 931.5$ MeV/c^2, we have

$$m = 4.007\ \text{u} .$$

Within experimental error, these results correspond to a singly ionized helium atom, denoted by ${}^4_2\text{He}^+$. It can be seen that the contribution of the relativistic correction to equation (2.70) is -4.45×10^{-32} kg. The error incurred in using Newtonian mechanics instead of relativistic mechanics for an accelerating potential difference of 50 kV is about one part in 1.5×10^5. Modern mass spectrometers have an accuracy of better than one part in 10^6, so that for an accelerating potential difference of 50 kV equation (2.70), based on relativistic mechanics, should be used. More accurate values for q and c should also be used. For an accelerating potential difference of $V = 1$ kV the relativistic correction $-qV/2c^2$

is only -8.9×10^{-34} kg, which is only about one part in 1.5×10^{7}, showing that in this case Newtonian mechanics is a sufficiently good approximation. If $V=10^{9}$ V the relativistic correction is -8.9×10^{-28} kg, which is one part in 7.46. In this last example the use of relativistic mechanics is essential. The above discussion shows how the mass m of an ion can be measured, *whatever the speed of the ion*, provided that the equations of relativistic mechanics are used. It also shows how Newtonian mechanics can be used as an excellent approximation to determine m in the zero-velocity limit.

The masses of the stable isotopes of the elements have been determined accurately using mass spectrometers. The results obtained will be used in Chapter 6 to interpret the energetics of nuclear reactions.

2.13 HIGH-SPEED PARTICLES AND THE PRINCIPLE OF RELATIVITY

2.13.1 Introduction

In this chapter the laws applicable to high-speed particles (relativistic mechanics) have been obtained on the basis of twentieth-century laboratory experiments. This approach shows that such modern *experiments* show that Newtonian mechanics is completely inadequate at high speeds, and one must be prepared to reinterpret all concepts based on Newtonian mechanics, including the Galilean transformations. In this section it will be *indicated* how, if the new experimentally confirmed laws, developed for high-speed particles, are to obey the principle of relativity, then the coordinates and time must be transformed, using the Lorentz transformations and not the Galilean transformations. The full development and interpretation of the Lorentz transformations will be deferred until Chapter 4, when the conventional approach to special relativity, based on the principle of the constancy of the speed of light, will be developed. The arguments in the rest of this section may appear a little concise and subtle at first reading. The reader can return to this section after reading Chapter 4. All we are trying to indicate here is that the Lorentz transformations follow if it is assumed that the new experimentally confirmed laws developed for high-speed particles in this chapter obey the principle of relativity—that is, they hold in all inertial reference frames.

2.13.2 Relativistic particles and the principle of relativity

Consider an astronaut inside a spaceship coasting with uniform velocity relative to the fixed stars. The astronaut is an inertial observer. If the astronaut carried out experiments on high-speed particles, it seems reasonable to assume that on the basis of such experiments the astronaut would obtain the same experimental laws for high-speed particles as we have obtained in this chapter on the basis of terrestrial experiments.

In the theory of special relativity it is *assumed* that all the laws of physics, including those of high-speed particles, obey the principle of relativity. Some of the consequences of extending the principle of relativity to the experimental laws of high-speed particles will now be discussed.

Consider a spaceship moving with uniform velocity v along the x axis relative to the laboratory frame Σ, as shown in Fig. 2.10. If the laws for high-speed particles obey the principle of relativity then, if the astronaut on the spaceship repeated Bertozzi's experiment (described in Section 2.5), the astronaut should find that the accelerated electrons had a limiting speed relative to the spaceship. If space is isotropic—that is, if all directions in space are equivalent—the astronaut should find experimentally that the limiting speed of accelerated electrons was the same in all directions, relative to the spaceship.

A plausibility argument will now be given to show that the limiting speed of accelerated electrons should have the same numerical value relative to the spaceship as relative to the Earth, provided that the same units of length and time are used in both reference frames. For purposes of discussion only, it will be assumed initially that the numerical value of the limiting speed of accelerated electrons, measured relative to the spaceship in Fig. 2.10, depends on the speed of the spaceship relative to the Earth. The astronaut can transmit a radio message to the base station on the Earth giving the measured value of the limiting speed of accelerated electrons relative to the spaceship. An observer on the Earth or the astronaut can measure v, the speed of the spaceship relative to the Earth (e.g. using radar methods). The experiment can be repeated for various values of v and a table made, relating the limiting speeds of electrons measured relative to the spaceship to the speed of the spaceship relative to the Earth. This information can be sent by radio to the astronaut. Thereafter, by measuring the limiting speeds of accelerated electrons relative to the spaceship and using the table, the astronaut could determine the speed of the spaceship relative to the Earth,

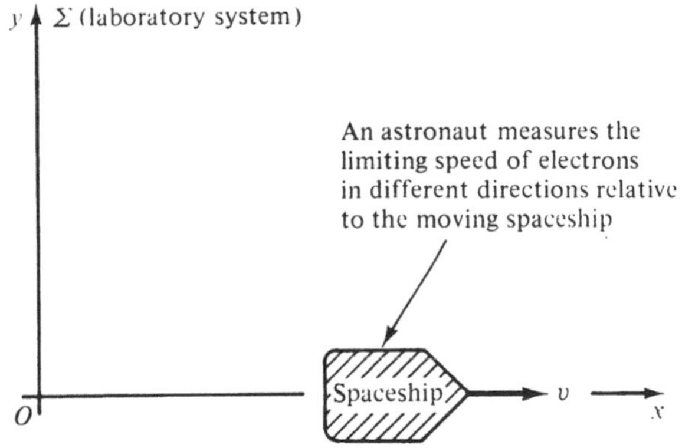

Fig. 2.10 The limiting speed of accelerated electrons is measured inside the moving spaceship. If space is isotropic, the limiting speed should be the same in all directions relative to the spaceship.

without looking at anything external to the spaceship. This is contrary to the assumption that the laws of physics are the same in all inertial reference frames, and that they do not involve the speed of the astronaut relative to any absolute reference frame. The way out of this dilemma is to conclude that the numerical value of the limiting speed of accelerated electrons relative to the spaceship is the same, and equal to the terrestrial value of 3.0×10^8 m s^{-1}, whatever the speed of the spaceship relative to the Earth. This principle will be called the *principle of the constancy of the limiting speed of particles*, according to which the numerical value of the limiting speed of accelerated particles is the same in all directions in all inertial reference frames and is equal to $c = 3.0 \times 10^8$ m s^{-1}. This principle will now be used to develop the Lorentz transformations.

2.13.3 The Lorentz transformations

Consider two inertial reference frames Σ and Σ', such as two spaceships passing each other in outer space, with Σ' moving with a high uniform velocity v along the x axis relative to Σ, as shown in Fig. 2.11. Let the origins coincide at times $t = t' = 0$, as shown in Fig. 2.11(a). Let an electron pass the origins of Σ and Σ' at this instant and move along the $+x'$ axis of Σ' with a speed extremely close to the limiting speed c relative to Σ'. Suppose that the electron is detected at an event at P at x' at a time t' relative to Σ', as shown in Fig. 2.11(b). Since the electron moves with a speed very close to c relative to Σ', it moves a distance very nearly equal to ct' in a time t' relative to Σ', so, that relative to Σ',

$$x' \approx ct'. \tag{2.75}$$

Now, Σ' moves with uniform velocity v along the x axis relative to Σ, so that one would expect the speed of the electron, which is moving along the $+x$ axis, to be greater relative to Σ than Σ', but still less than the limiting value c. Hence the speed of the electron must be extremely close to the limiting value c relative to Σ also. If the electron is detected at the event at P at x at a time t relative to Σ by the detector, which may be moving relative to both Σ and Σ', then, relative to Σ,

$$x \approx ct. \tag{2.76}$$

Since in Fig. 2.11 the origin of Σ' moves relative to the origin of Σ during the time the electron moves from the origins of Σ and Σ' until it is detected at P, we have $x' < x$. Hence t' must be less than t if equations (2.75) and (2.76) are both to be satisfied. Thus time cannot be absolute, as it was assumed to be in Newtonian mechanics. The concept of absolute time, which leads to the Galilean transformations must therefore be abandoned if the experimental laws of high-speed particles are to obey the principle of relativity.

Let the electron passing the origins of Σ and Σ' at $t = t' = 0$, not now move along the x axis of Σ, but rather let it travel in an arbitrary direction at a speed extremely close to c relative to both Σ and Σ' as shown in Fig. 2.12(a). Suppose that the electron is detected at an event with coordinates (x', y', z') at time t'

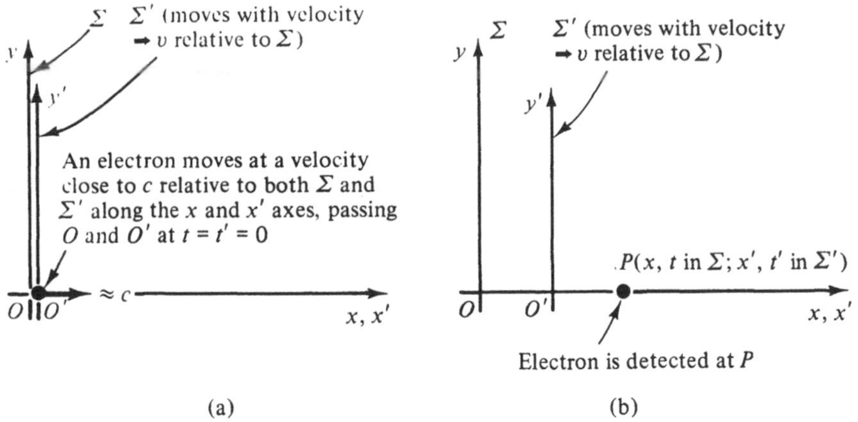

Fig. 2.11 (a) An electron travelling at a speed very close to c along the common x axis relative to both Σ and Σ' passes the origins O and O' at $t=t'=0$. (b) The electron is detected at P, which is measured to be at x at a time t relative to Σ and at x' at a time t' relative to Σ'.

relative to Σ' and coordinates (x, y, z) at time t relative to Σ, as shown in Fig. 2.12(b). In time t' the electron travels a distance $(x'^2+y'^2+z'^2)^{1/2}$ at a speed very close to c relative to Σ', so that, in Σ',

$$(x'^2+y'^2+z'^2)^{1/2} \approx ct',$$

or

$$x'^2+y'^2+z'^2-c^2t'^2 \approx 0. \tag{2.77}$$

Similarly, relative to Σ,

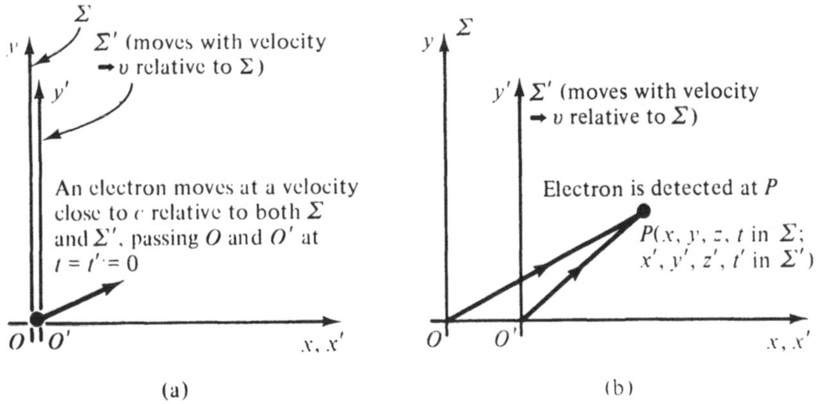

Fig. 2.12 (a) An electron travelling at a speed very close to the limiting speed c relative to both Σ and Σ' passes O and O' at $t=t'=0$. (b) The electron appears to go along the path OP relative to Σ and along the path $O'P$ relative to Σ'.

$$(x^2 + y^2 + z^2)^{1/2} \approx ct\,,$$

or

$$x^2 + y^2 + z^2 - c^2t^2 \approx 0. \tag{2.78}$$

Note that the same value c has been used for the limiting speed of the electron in both equations (2.77) and (2.78). The appropriate transformations which transform equation (2.77) into equation (2.78) are the Lorentz transformations:

$$x' = \frac{x - vt}{(1 - v^2/c^2)^{1/2}} = \gamma\,(x - vt), \tag{2.79}$$

$$y' = y, \tag{2.80}$$

$$z' = z, \tag{2.81}$$

$$t' = \frac{t - vx/c^2}{(1 - v^2/c^2)^{1/2}} = \gamma\,(t - vx/c^2), \tag{2.82}$$

where v is the velocity of Σ' relative to Σ, and

$$\gamma = \frac{1}{(1 - v^2/c^2)^{1/2}}\,. \tag{2.83}$$

For example,

$$\begin{aligned} x'^2 + y'^2 + z'^2 - c^2t'^2 &= \gamma^2\,(x - vt)^2 + y^2 + z^2 - c^2\gamma^2\,(t - vx/c^2)^2 \\ &= \gamma^2\,(x^2 - 2xvt + v^2t^2 - c^2t^2 + 2vxt - v^2x^2/c^2) \\ &\quad + y^2 + z^2 \\ &= \gamma^2x^2\,(1 - v^2/c^2) + y^2 + z^2 - c^2\gamma^2t^2\,(1 - v^2/c^2) \\ &= x^2 + y^2 + z^2 - c^2t^2. \end{aligned}$$

Hence the Lorentz transformations transform equation (2.77) into equation (2.78). The full development of these transformations and their physical interpretation will be deferred until Chapter 4. In this section we have merely tried to indicate how the Lorentz transformations follow if it is assumed that the experimentally confirmed laws, found for high-speed particles, obey the principle of relativity.

Returning to equations (2.75) and (2.76), it will be assumed that equation (2.76) is valid; that is, $x = ct$. Using the Lorentz transformations and $x = ct$, we have

$$\frac{x'}{t'} = \frac{\gamma\,(x - vt)}{\gamma\,(t - vx/c^2)} = \frac{ct - vt}{t - vct/c^2} = \frac{c\,(ct - vt)}{ct - vt} = c.$$

Hence, if the coordinates and time are transformed using the Lorentz transformations, it is possible to satisfy both $x = ct$ and $x' = ct'$ in Fig. 2.11. In addition to different measures of the distances x and x' measured relative to Σ and Σ' respectively, the times t and t' must also differ such that equations (2.75) and (2.76) are both valid. It will be illustrated in Chapter 4 how this new concept of time is in accord with experiment.

The symbol γ, defined by equation (2.83), should not be confused with α, defined by equation (2.2). In the expression $\gamma = (1 - v^2/c^2)^{-1/2}$, v is the velocity of the inertial reference frame Σ' relative to the inertial reference frame Σ. On the other hand, in the definition $\alpha = (1 - u^2/c^2)^{-1/2}$, u is the speed of the particle measured in one particular reference frame, namely Σ. Neither u nor α has anything to do with the Lorentz transformations and transforming from one inertial reference frame to another. The quantities α and γ are completely different except in the *special case* when a particle is at rest in Σ', in which case the particle moves with speed $u = v$ in Σ, and in this *special case* $\alpha = \gamma$. The notation $\beta = u/c$ will sometimes be used for the ratio of the speed of the particle to the speed of light in empty space.

Historically, the theory of special relativity evolved from classical optics and electricity and magnetism. A brief outline of the historical development of special relativity is given in Chapter 3. This has been preceded by the account of high-speed particles in this chapter so as to convince the reader that Newtonian mechanics, and concepts based on it, such as the Galilean transformations, are completely inadequate at high speeds. There is generally a natural bias in people's minds, based on their everyday experiences, in favour of the concepts of Newtonian mechanics, which makes the necessity for the theory of special relativity difficult to appreciate when it is approached in the traditional way via optics and electromagnetism, where the experimental deviations from the Galilean transformations are generally extremely small. With historical hindsight, the reader can now approach Chapter 3 and the theory of special relativity, already familiar with the inadequacies of Newtonian mechanics and the Galilean transformations, when they are applied to interpret the motions of charged particles, moving with speeds comparable to the speed of light in empty space in electric and magnetic fields, and already familiar with the need to reinterpret the concept of time, which was *assumed* to be absolute in Newtonian mechanics.

REFERENCE

Bertozzi, W. (1964) *Am. J. Phys.* **32**, 551.

PROBLEMS

The following values should be used in the calculations:

velocity of light $c = 3.00 \times 10^8\ \mathrm{m\ s^{-1}}$;
electron charge $e = 1.602\,06 \times 10^{-19}\ \mathrm{C}$;
electron mass $m_e \equiv 9.1083 \times 10^{-31}\ \mathrm{kg}$
$\equiv 0.510\,976\ \mathrm{MeV}/c^2$;
proton mass $m_p = 1.672\,39 \times 10^{-27}\ \mathrm{kg}$
$\equiv 938.211\ \mathrm{MeV}/c^2$.

2.1 Find the values of $\alpha = (1 - u^2/c^2)^{-1/2}$ for particles of speeds (a) 300 km h^{-1}; (b) 3000 km h^{-1}; (c) 0.1 c; (d) 0.5 c; (e) 0.9 c; (f) 0.99 c; (g) 0.999 c; (h) 0.9999 c.

2.2 Calculate the velocities of electrons accelerated through potential differences of (a) 10 kV; (b) 100 kV; (c) 1 MV. [*Hint*: Use equation (2.66).]

2.3 Calculate the velocities and momenta of protons of kinetic energies (a) 100 MeV; (b) 10 GeV; (c) 1 TeV. What is the (total) energy $E = T + mc^2$ in each case?

2.4 Calculate the amount of work that must be done to increase the velocity of an electron from zero to (a) 0.5 c; (b) 0.75 c.

2.5 Through what potential difference must an electron fall if, according to Newtonian mechanics, its velocity is to equal the velocity of light? What speed does the electron actually acquire according to the theory of high-speed particles?

2.6 Show that the mass of a particle is given by

$$m = \frac{p^2c^2 - T^2}{2Tc^2},$$

where p is its momentum and T its kinetic energy. [*Hint*: Put $E = T + mc^2$ in $E^2 = c^2p^2 + m^2c^4$.] Calculate the mass of a particle (a) if its momentum is 130 MeV/c when its kinetic energy is 50 MeV; (b) if its momentum is 1090 MeV/c when its kinetic energy is 500 MeV.

2.7 A particle of charge 1.60×10^{-19} C is accelerated through a potential difference of 5×10^8 V. It is then passed through a uniform magnetic field of strength 1 T. The radius of curvature of the orbit of the particle in the magnetic field is 3.62 m when the plane of the orbit is perpendicular to the magnetic field. Calculate the mass and velocity of the particle. [*Hint*: Calculate the kinetic energy and the momentum of the particle; then use the method of Problem 2.6.]

2.8 A large parallel plate capacitor, in vacuum, is charged to a potential difference of 10^6 V. The plates are 0.05 m apart. Calculate how long it will take an electron, starting from rest, to travel from the negative to the positive plate. [*Hint*: Derive and use equation (2.35).]

2.9 In a cyclotron the strength of the magnetic field is 1.5 T. Show, using equation (2.45), that the frequency for resonance is 23 MHz. Using equation (2.65), show that if the radius of the dees is 0.4 m, the protons have momenta of 180 MeV/c when they reach the outside of the dees. Show that their velocity is 5.6×10^7 m s^{-1}. Use equation (2.12) to show that this corresponds to a kinetic energy of 17 MeV.

3

Historical Development of the Theory of Special Relativity

3.1 INTRODUCTION

Newtonian mechanics was developed mainly in the seventeenth century, whereas the wave theory of physical optics and the theory of classical electromagnetism were developed mainly in the nineteenth century. For example, Young reintroduced wave theory to interpret interference experiments in 1801, and Oersted discovered the magnetic effect due to an electric current in 1820. The enormous successes of Newtonian mechanics naturally led people to assume that the Galilean transformations were correct. After reading Chapter 2 the reader will realize that twentieth-century experiments have shown that Newtonian mechanics is inadequate at high speeds, and by now is presumably prepared to accept that the Galilean transformations must therefore be replaced. However, if we try to cast our minds back to the nineteenth century, at that time no experiments on very high-speed particles had been performed, and for centuries Newton's theories had appeared to be the perfect examples for all physical theories. Thus it was natural in the nineteenth century to assume that the Galilean transformations, which were based on Newtonian mechanics, were correct, and could be applied to the laws of optics and of electricity and magnetism. It will now be illustrated how this assumption led to difficulties that could only be resolved by abandoning the Galilean transformations.

Experimenters in different inertial reference frames must agree to use the same fundamental units of length and time. In 1983 the metre was redefined in terms of the velocity of light in empty space. Until then, it had been defined independently of the velocity of light. For example, when the historic experiments leading to special relativity were first performed, and when Einstein (1905) first developed special relativity, the metre was defined as the distance between two lines on a platinum – iridium bar kept at Sèvres, near Paris. In those days the velocity of light was considered to be a quantity whose value had to be determined by experiment. Hence, to put the historical background to special relativity into its proper perspective, it will be assumed, until we reach Section 4.3, that inertial observers in different inertial reference frames agree to define the metre independently of the speed of light. Since it would not be convenient for them all to borrow the platinum–iridium bar from Sèvres, the different inertial observers could agree to use the 1960 definition, and define the metre as 1 650 763.73 wavelengths of the orange–red line of krypton-86. They could also agree to define the second as the duration of 9 192 631 770 periods of the radiation corresponding to the transition between the two hyperfine levels of the ground state of the caesium-133 atom. Using these units, the accepted experimental value of the velocity of light, measured in the laboratory frame in 1964, was

$$c = (2.997\,925 \pm 0.000\,003) \times 10^8 \text{ m s}^{-1}.$$

3.2 OPTICAL EXPERIMENTS AND THE PRINCIPLE OF RELATIVITY

Consider an *absolute* inertial reference frame Σ in which the speed of light in empty space is the same, and equal to c, in all directions of empty space, as shown in Fig. 3.1(a). In the nineteenth century the absolute frame Σ was identified with the reference frame in which, it was assumed, the luminiferous ether was at rest. Consider a reference frame Σ', such as the laboratory frame, that is moving with uniform velocity v relative to the absolute frame Σ along their common x axis, as shown in Fig. 3.1(b). The Galilean velocity transformations will be used to calculate the speed of light in various directions of empty space

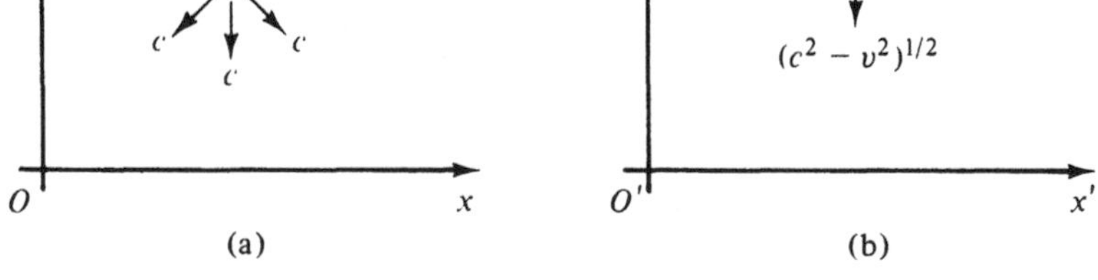

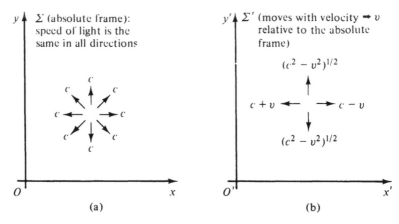

Fig. 3.1 It is assumed that there is an absolute frame Σ in which the speed of light in empty space is the same in all directions, as shown in (a). If it is assumed that the Galilean velocity transformations are correct then the speed of light in empty space should be different in different directions in the moving system Σ', as shown in (b).

relative to the reference frame Σ', which is moving relative to the absolute frame Σ. For light moving in the $+x$ direction of Σ, $u_x = c$ and $u_y = 0$. Applying the Galilean velocity transformations, equations (1.21) and (1.22), we have, relative to Σ',

$$u_x' = u_x - v = c - v,$$

$$u_y' = 0,$$

as illustrated in Fig. 3.1(b). For light moving in the negative x direction in the absolute frame Σ, $u_x = -c$ and $u_y = 0$. Using the Galilean velocity transformations, equations (1.21) and (1.22), relative to Σ', we have

$$u_x' = u_x - v = -c - v = -(c + v),$$

$$u_y' = 0,$$

as illustrated in Fig. 3.1(b). Light moving parallel to the y' axis of Σ' must have a velocity c having components $u_x = +v$ and $u_y = (c^2 - v^2)^{1/2}$ in Σ, so that using equations (1.21) and (1.22), we should then have in Σ', $u_x' = 0$ and $u_y' = (c^2 - v^2)^{1/2}$, as illustrated in Fig. 3.1(b). Thus, if the Galilean velocity transformations were correct, the speed of light in empty space would be different in different directions in a reference frame Σ' moving relative to the absolute frame Σ, in which the speed of light in empty space was the same in all directions. In the nineteenth century it was argued that if there were such an absolute reference frame, it would be extremely unlikely that the Earth was always at rest relative to it. Even if the Earth happened to be at rest relative to the absolute frame at one instant of time then, owing to the rotation of the Earth about its axis and its orbital motion around the Sun, it would not remain at rest relative to the absolute frame at all periods of the day and year. Hence, if the Galilean transformations were correct, the speed of light in empty space would not be the same in all directions relative to the Earth at all times, so that, by measuring the speed of light in different directions of empty space relative to the Earth, it should have been possible to estimate v, the speed of the Earth relative to the absolute frame in which the speed of light in empty space was the same in all directions. For example, the maximum and minimum speeds relative to the Earth should have been $c + v$ and $c - v$ respectively, which would give c and v. An experiment of this type was attempted by Arago in 1810. The most famous of the optical experiments was performed by Michelson and Morley in 1887 (an account of this experiment is given in Appendix A). All such optical experiments failed to find any differences in the speed of light in different directions and failed to establish the existence of any absolute frame for the laws of optics. Michelson and Morley looked upon their experiment as a failure. According to the Galilean transformations, they should have been able to determine the speed of the Earth relative to the absolute frame, but experimentally they found nothing, even though their apparatus was sensitive enough to detect the effect, if it were present. The way in which Einstein (1905) resolved this dilemma was to conclude that the laws of optics obeyed the principle of relativity, even though this assumption meant abandoning the Galilean transformations and the concept of absolute time. Thus, according to Einstein, if one performed optical experiments inside an ocean liner (or spaceship) moving with uniform velocity relative to the Earth, it would be impossible to determine its speed relative to the Earth or any other inertial reference frame without looking at something external to the ocean liner (or spaceship). If the laws of optics obey the principle of relativity, it is impossible to detect any hypothetical absolute frame by means of optical experiments, and the whole idea of an absolute frame for the laws of optics becomes superfluous.

3.3 THE LAW OF ELECTRICITY AND THE PRINCIPLE OF RELATIVITY

The laws of electricity and magnetism, such as Coulomb's law for the force between two stationary point charges, the Biot–Savart law for the magnetic field of an electric current and Faraday's law of electromagnetic induction, can be summarized in a concise way using Maxwell's equations, which were developed by Maxwell in the second half of the nineteenth century. From time to time, we shall refer to the laws of electricity and magnetism collectively as Maxwell's equations. These equations have solutions corresponding to electromagnetic waves, which are nowadays identified with light waves and radio waves. Physical optics can now be interpreted in terms of electromagnetic waves using Maxwell's equations, so that, after Maxwell's work, classical optics and electricity and magnetism were unified into the same theory.

If the Galilean transformations were correct then the generally accepted laws of electricity and magnetism would not agree with the principle of relativity. If these laws did not obey this principle then they could only hold in one absolute reference frame. The system in which the Earth is at rest is not likely to be this absolute frame. Thus, if the Galilean transformations were correct, it should have been possible to determine the speed of the Earth relative to this absolute frame by means of electrical experiments, such as the Trouton–Noble experiment, an account of which is given by Rosser (1964). All such electrical experiments failed to determine the existence of any absolute frame for the laws of electricity and magnetism. Einstein's way out of this dilemma was to conclude, that the laws of electricity and magnetism, which by 1905 incorporated the laws of optics, obeyed the principle of relativity in all inertial reference frames, even though this meant abandoning the Galilean transformations. If the laws of electricity and magnetism obey the principle of relativity, it is impossible to determine the existence of any absolute reference frame for these laws. One could not determine the speed of an ocean liner (or spaceship) moving with uniform velocity relative to the Earth by means of electrical or optical experiments confined to the inside of the ocean liner (or spaceship). According to the principle of relativity, experimenters inside an ocean liner (or spaceship) should confirm the validity of Maxwell's equations in inertial reference frames moving with uniform velocity relative to the Earth.

In addition to the principle of relativity, as his second postulate Einstein (1905) could have postulated that the correct laws of classical optics and electromagnetism were Maxwell's equations. These postulates lead to the Lorentz transformations. However, Einstein chose the principle of the constancy of the speed of light as his second postulate, according to which the speed of light in empty space is the same in all inertial reference frames. It will be shown in Section 3.4 that the principle of constancy of the speed of light follows if it is assumed that the laws of electricity and magnetism (Maxwell's equations) obey the principle of relativity. The choice by Einstein of the principle of the constancy of the speed of light in empty space as his second postulate enabled him to analyse the concept of measurement of the times of distant events before he developed the Lorentz transformations. The reinterpretation of the measurement of the time of distant events is the key new idea in Einstein's 1905 paper. For example, in an account of some conversations he had with Einstein, Shankland (1963) wrote:

> I asked Professor Einstein how long he had worked on the Special Theory of Relativity before 1905. He told me that he had started at age 16 and worked for 10 years; first as a student when, of course, he could only spend part-time on it, but the problem was always with him. He abandoned many fruitless attempts, "until at last it came to me that time was suspect!" Only then, after all his earlier efforts to obtain a theory consistent with the experimental facts had failed, was the development of the Special Theory of Relativity possible.

After deriving the Lorentz transformations using the principle of the constancy of the speed of light, Einstein (1905) went on to show, for the special case of a field point in empty space, that the Lorentz transformations did transform Maxwell's equations in one inertial reference frame into Maxwell's equations in a reference frame moving with uniform velocity relative to the first. He also went on to commence the reinterpretation of Newtonian mechanics necessitated by the replacement of the Galilean transformations by the Lorentz transformations. For example, Einstein derived equation (2.13) of Chapter 2 for the kinetic energy of a moving particle.

3.4 THE PRINCIPLE OF THE CONSTANCY OF THE SPEED OF LIGHT

Some of the evidence in favour of the principle of the constancy of the speed of light will now be reviewed. It will first be shown that if it is assumed that the accepted laws of electricity and magnetism (Maxwell's equations) obey the principle of relativity then the principle of the constancy of the speed of light follows.

It is shown in textbooks on electricity and magnetism that Maxwell's equations have wave solutions. In empty space the velocity of these electromagnetic waves, measured in the laboratory frame Σ, is

$$c = (\mu_0 \epsilon_0)^{-1/2}, \qquad (3.1)$$

where μ_0 is the magnetic constant (permeability of free space) and ϵ_0 is the electric constant (permittivity of free space). This velocity is identified with the speed of light in empty space. It is independent of the velocities of the accelerating electric charges giving rise to the electromagnetic waves, illustrating how, according to Maxwell's equations, the velocity of light in empty space is independent of the velocity of the source of light.

If it is assumed that the laws of electricity and magnetism (Maxwell's equations) obey the principle of relativity then in an inertial reference frame Σ' that is moving with uniform velocity v relative to Σ, according to the laws of electricity and magnetism, the velocity of electromagnetic waves measured in Σ' should be

$$c' = (\mu_0' \epsilon_0')^{-1/2}, \tag{3.2}$$

where μ_0' and ϵ_0' are the values of the magnetic and electric constants in Σ'.

It can be shown, using the Biot–Savart law, that the force of attraction per unit length on each of two infinitely long, thin, parallel wires a distance r apart in empty space and carrying currents I_1 and I_2 is $\mu_0 I_1 I_2/2\pi r$. The ampere is defined as

> that unvarying current which, if present in each of two infinitely thin parallel conductors of infinite length and 1 m apart in empty space, causes each conductor to experience a force of exactly 2×10^{-7} newton per metre of length.

Putting $I_1 = I_2 = 1$ and $r = 1$ in the expression $\mu_0 I_1 I_2/2\pi r$, we find that it follows from the definition of the ampere that, in SI units, $\mu_0 = 4\pi \times 10^{-7}$ H m^{-1}.

If Maxwell's equations are valid in both Σ and Σ' then the force between two parallel currents should be given by the Biot–Savart law in both Σ and Σ'. If the ampere is defined in the same way in Σ and Σ', this implies that $\mu_0 = \mu_0' = 4\pi \times 10^{-7}$ by definition. For the velocity of light to have the same numerical value in Σ and Σ' one must also have $\epsilon_0 = \epsilon_0'$. In principle, according to classical physics, one could place two protons a known distance, say 1 m, apart in empty space and measure the force between them when they are at rest in the inertial frame Σ. Similarly, one could place two protons 1 m apart and measure the force between them when they are at rest relative to the inertial frame Σ'. If Maxwell's equations are valid in both Σ and Σ' then Coulomb's law should be applicable in both cases. For ϵ_0 to be equal to ϵ_0' it would require that the force between the protons at rest in Σ (measured in Σ) would have the same numerical value as the force between the two protons at rest in Σ' (measured in Σ'). Though it is not always stated specifically, it is generally assumed in the theory of special relativity, that, if two experiments are carried out under *identical* conditions in two *inertial* frames Σ and Σ' that are equivalent in every way then they give the same numerical results, within experimental error. If this assumption is made then ϵ_0 should equal ϵ_0'. Since μ_0 is equal to μ_0' by definition, then $(\mu_0 \epsilon_0)^{-1/2}$ should be equal to $(\mu_0' \epsilon_0')^{-1/2}$. According to Maxwell's theory, these latter expressions are equal to the velocities of electromagnetic waves in free space in Σ and Σ′ respectively so that $c = c'$. Thus if it is assumed that Maxwell's equations are correct and obey the principle of relativity then the principle of the constancy of the velocity of light follows. The simultaneous measurement of the velocity of light in Σ and Σ′ is not carried out under *identical* conditions in Σ and Σ′, since the velocity of the source of light is not the same relative to inertial observers at rest in Σ and Σ′ respectively, so that *a priori* there is no reason why the same numerical value for the velocity of light should be obtained; but, according to Maxwell's equations, it should.

The principle of the constancy of the speed of light can be looked upon as a special case of the principle of the constancy of the limiting speed of particles, developed in Section 2.13. It was illustrated in Section 2.9 that light quanta (photons) can be treated as particles travelling at the speed of light. Photons can therefore be treated as particles travelling *at* the limiting speed. If the limiting speed of particles has the same numerical value in all inertial reference frames then particles travelling at the limiting speed of 3.0×10^8 m s^{-1} must have that speed in all inertial reference frames, which for photons is just the principle of the constancy of the speed of light.

There is now direct experimental evidence in favour of this principle. Among the particles produced in the high-energy collisions of atomic particles are charged and neutral pions. The neutral pion, denoted π^0, has a mass of 135 MeV/c^2, which is 264 times the mass of the electron. They are unstable particles, with an average lifetime of about 10^{-16} s when at rest, before decaying into two γ rays, i.e. high-energy light quanta (photons). In the inertial reference frame Σ′ in which the decaying π^0 is at rest, the two γ rays are emitted in opposite directions, so as to conserve linear momentum, as shown in Fig. 3.2(b). In this

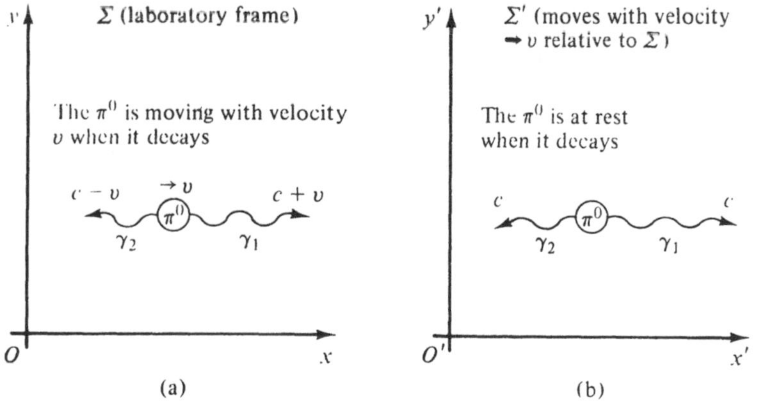

Fig. 3.2 The π^0 is moving with uniform velocity v when it decays in the laboratory frame Σ, as shown in (a). In Σ′ the π^0 is at rest when it decays, so that the speeds of the photons γ_1 and γ_2 are both equal to c in Σ′, as shown in (b). If the Galilean velocity transformations were correct, the speeds of γ_1 and γ_2 in the laboratory frame Σ would be $c + v$ and $c - v$ respectively.

reference frame the speeds of the γ rays should be equal to the accepted value of $c = 2.9979 \times 10^8$ m s^{-1} for the speed of light emitted by a stationary light source.

Consider a π^0 moving with a velocity v relative to the laboratory frame Σ when it decays into two γ rays, labelled γ_1 and γ_2 (Fig. 3.2a). In the inertial frame Σ', moving with velocity v relative to Σ, the π^0 is at rest when it decays. Consider the special case shown in Fig. 3.2(b), in which the two γ rays γ_1 and γ_2 are emitted parallel and antiparallel to the x' axis of Σ', with speeds equal to c, the speed of light emitted by a stationary source. If the Galilean velocity transformations were correct then, using equation (1.21), in the laboratory frame Σ we should have for the velocity of γ_1

$$u_x = u_x' + v = c + v,$$

and for the velocity of γ_2

$$u_x = u_x' + v = -c + v = -(c - v),$$

as shown in Fig. 3.2(a). Thus, if the Galilean velocity transformations were correct, the speeds of the γ rays arising from the decay of a *moving* π^0 would differ from c, the speed of light emitted by a stationary source.

In 1964 Alväger, Farley, Kjellman and Wallin measured the speeds of γ rays from the decay of neutral pions (π^0) of energy greater than 6 GeV relative to the laboratory. The π^0 were produced by protons of momenta 19.2 GeV/c from the CERN proton synchrotron. In this case the speed of the light source, that is the π^0, measured relative to the laboratory was 0.999 75 c, if calculated using the equations derived in Chapter 2. The measured speed of the γ rays relative to the laboratory was $(2.9977 \pm 0.0004) \times 10^8$ m s^{-1}. This was the same within experimental error, as the accepted value of 2.9979×10^8 m s^{-1} for the speed of light emitted by a stationary light source. These results show clearly that the speed of light quanta (photons) emitted by a moving source is always equal to c, and the results illustrate vividly the inadequacy of the Galilean velocity transformations.

Some of the evidence in favour of the constancy of the speed of light has been reviewed in this section. The proof of the pudding is, however, in the eating. Some of the best evidence in favour of the principle of the constancy of the speed of light is the indirect evidence that the predictions made in Chapters 4, on the assumption that the principle is correct, are invariably in agreement with experimental results. The principle of the constancy of the speed of light will be taken as *axiomatic* in Chapter 4.

REFERENCES

Alväger, T., Farley, F.J.M., Kjellman, J. and Wallin, I. (1964) *Phys. Lett.* **12**, 260.

Einstein, A. (1905) *Ann. der Phys.* **17**, 891. [English translation in Einstein, A., Lorentz, H.A., Miakowski, H. and Weyl, H. (1952) *The Principle of Relativity*. Dover, New York.]

Rosser, W.G.V. (1964) *An Introduction to the Theory of Relativity*, p.64. Butterworth, London.
Shankland, R.S. (1963) *Am. J. Phys.* **31**, 47.

4

Special Relativity and the Lorentz Transformations

4.1 POSTULATES OF SPECIAL RELATIVITY

Einstein's two main postulates in the theory of special relativity were

(i) the principle of relativity;

(ii) the principle of the constancy of the speed of light.

According to the *principle of relativity*, the laws of physics, including the laws of relativistic mechanics and of optics, electromagnetism and nuclear physics, are the same in all *inertial reference frames*.

The definition of an inertial reference frame is the same in the theory of special relativity as in Newtonian mechanics (cf. Section 1.2.2). As an example of the scope of the principle of relativity in the theory of special relativity, consider a spaceship coasting along with *uniform velocity* with respect to the fixed stars. If an astronaut performed experiments in mechanics, optics, electromagnetism or nuclear physics inside the spaceship, according to the principle of relativity the laws the astronaut would "derive" on the basis of these experiments would be the same as if they were carried out in any other inertial reference frame such as the laboratory frame. Without looking at anything external to the spaceship, the astronaut could not even say if it was moving, and could not determine its speed relative to the Earth. The astronaut should find no preferred directions in space.

It was shown in Section 2.13 that if the laws of relativistic mechanics, developed in Chapter 2, are to obey the principle of relativity then the appropriate coordinate and time transformations are the Lorentz transformations. Similarly, the Lorentz transformations are the appropriate transformations if the laws of electromagnetism (Maxwell's equations) are to obey the principle of relativity. This will be confirmed in Chapter 8. Thus, as our second postulate, in addition to the principle of relativity, we could postulate that the correct laws of electromagnetism are Maxwell's equations. However, the Lorentz transformations do necessitate a revision of the concepts of absolute space and absolute time, and this aspect of the Lorentz transformations is best developed by taking as our second postulate the principle of the constancy of the speed of light. It was shown in Section 3.3 that this principle follows if it is assumed that Maxwell's equations obey the principle of relativity. The direct and indirect evidence in favour of the principle of the constancy of the speed of light was reviewed in Section 3.4.

According to the principle of the constancy of the speed of light, the speed of light in empty space has the same numerical value in all inertial reference frames. Thus light should travel in straight lines and have the same speed in all directions of empty space in all inertial reference frames. This principle will be taken as *axiomatic* in this chapter. It will be shown that predictions made on the basis of this postulate are in agreement with the experimental results.

In this chapter radar methods will generally be used to determine the positions and times of events. Some people may prefer to use the principle of the constancy of the limiting speed of particles. Instead of radio signals, they can use fictitious space guns to shoot electrons back and forth with speeds extremely close to c. (It is, however, actually possible to use radio signals in practice.)

4.2 RADAR METHODS

As an example of the use of radar methods, we shall consider the determination of the position and velocity of a spaceship moving directly away from the Earth with uniform velocity, as shown in Fig. 4.1. Let a directional antenna transmit a short radio pulse (or radar signal) at a time t_1, measured by a clock at rest in the radar station. This radar signal travels with speed c in empty space. When it reaches the spaceship, a small fraction of the radio waves are reflected in the backward direction, in the direction of the transmitter. These reflected signals can be received and amplified. Suppose that the reflected signal returns to the Earth at a time t_2, measured by the clock at rest in the radar station. According to the principle of the constancy of the speed of light, the speed of the radar signal should be the same, and equal to c, when the signal is going to the spaceship as when the reflected signal is returning. Since the total time for the radar signal to go to the spaceship and back is $t_2 - t_1$, the time the radar signal

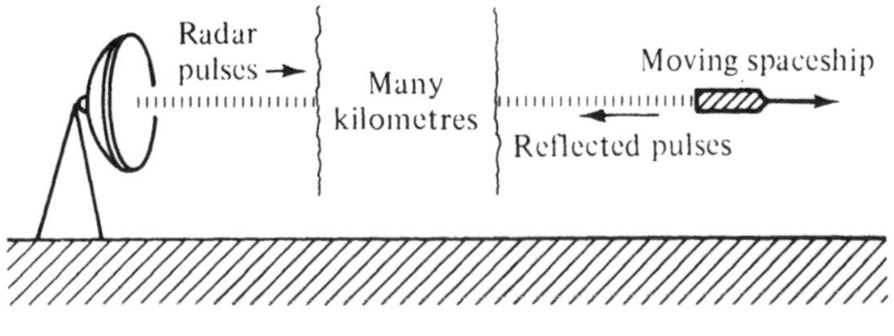

Fig. 4.1 A directional antenna is used to transmit radar pulses, which are reflected by the moving spaceship.

takes to reach the spaceship is $\frac{1}{2}(t_2 - t_1)$ and the time it takes to return is $\frac{1}{2}(t_2 - t_1)$. Since it travels at a speed c, in a time interval of $\frac{1}{2}(t_2 - t_1)$ the radar signal travels a distance $\frac{1}{2}c\,(t_2 - t_1)$. Since it is transmitted at a time t_1, an observer in the radar station will *calculate* that the radar signal reached the spaceship at a time $t_1 + \frac{1}{2}(t_2 - t_1)$; that is, at a time $\frac{1}{2}(t_2 + t_1)$. Hence, if the radar station is at the origin of a coordinate system, and the spaceship is moving along the x axis, the observer in the radar station will calculate that the radar signal was reflected by the spaceship at a time t at a distance x given by

$$x = \frac{1}{2}c\,(t_2 - t_1), \tag{4.1}$$

$$t = \frac{1}{2}(t_2 + t_1), \tag{4.2}$$

By repeating the experiment, the velocity of the spaceship can be determined. For example, let a radar signal be sent from the Earth at 10.00 a.m. Let the reflected signal return to the Earth 2 s later. According to equations (4.1) and (4.2), the coordinates and time of the event of reflection of the signal by the spaceship are

$$x = \frac{1}{2}c\,(t_2 - t_1) = \frac{1}{2} \times 3.00 \times 10^8 \times 2\ \text{m} = 3.00 \times 10^8\ \text{m},$$

$$t = \frac{1}{2}(10.00.00 + 10.00.02)\ \text{a.m.} = 10.00.01\ \text{a.m.}$$

Thus the spaceship was a distance of 3.00×10^8 m from the Earth at a time of 1 s after 10.00 a.m. If a signal transmitted at a time 10.20 a.m. returns to the Earth 4 s later after reflection by the spaceship then according to equations (4.1) and (4.2), the spaceship was 6.00×10^8 m from the Earth at a time 2 s after 10.20 a.m. In the time of 20 min 1 s (1201 s) the spaceship travels a distance of $(6.00 \times 10^8 - 3.00 \times 10^8$ m) or 3.00×10^8 m. Hence, if the spaceship is moving with uniform velocity, its speed is $c/1201$ or about 250 km s^{-1}.

It is convenient to plot the positions of events such as the successive positions of a spaceship on a *displacement–time* curve. For now, it will be assumed that all the events are on the x axis of the laboratory frame Σ, and that the radar station is on the axis at $x = 0$. In the theory of special relativity one is generally concerned with speeds comparable to the speed of light. For example, if an electron leaves the origin $x = 0$ at $t = 0$ and travels at a speed $\frac{1}{2}c$ then after a time $t = 10^{-8}$ s it is at $x = 1.5$ m. With such high-speed particles, it is not usually convenient to plot distance in metres against time in seconds. It is a *convention* to multiply all times by c, the speed of light, before plotting the displacement–time curve. The symbol w will be used for the product ct. Therefore the coordinates of the electron, at time $t = 10^{-8}$ s would be ($x = 1.5$ m, $w = ct = 3$ m). The quantity $w = ct$ has the dimensions of a length. It is also a convention to use $w = ct$ as ordinate and x as abscissa, as shown in Fig. 4.2(a). Such a displacement–time diagram is called a *space–time* diagram. If the time of an event is t it will be shown as ct on space–time diagrams, but will generally be referred to as the time t in the text.

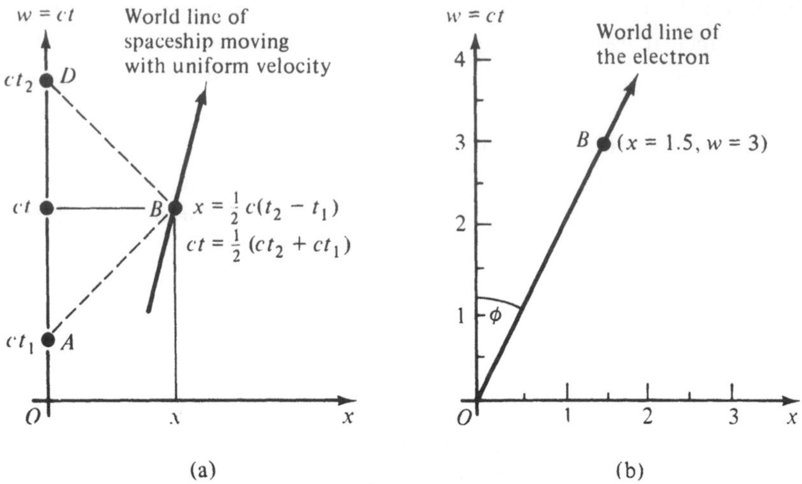

Fig. 4.2 (a) Space–time diagram showing the determination of the position of a spaceship moving with uniform velocity. A radar pulse is transmitted from the radar base having coordinates $x=0$, $w=ct_1$ at the world point A. It is reflected by the spaceship at the world point B at $x=\frac{1}{2}c\,(t_2-t_1)$, $w=ct=\frac{1}{2}(ct_2+ct_1)$, and received back at the radar base at the world point D at $x=0$, $w=ct_2$. (b) The world line of an electron moving with velocity $0.5c$.

The displacement of a spaceship moving with uniform velocity v relative to the laboratory system Σ is shown in Fig. 4.2(a). Such a displacement–time curve is known as a *world line*. The positions (x, $w=ct$) of *events* in such a space–time diagram are known as *world points*. Readers should try to familiarize themselves thoroughly with the definitions introduced in this section. Since light and radio signals go at a speed c, they travel a distance $\Delta x = c\Delta t$ in a time Δt. Hence the slopes of the world lines of light signals going along the $+x$ axis are $c\Delta t/\Delta x = 1$, so that the world lines of light signals going in the positive x direction are at 45° to the x and $w=ct$ axes. Figure 4.2(a) shows the world line of the radar signal transmitted from the radar station at the point $x=0$ of Σ at time t_1 at world point A, which is reflected by the spaceship at world point B and returns to the radar station at $x=0$ at time t_2 at world point D. The world line of the radar signal is at 45° to the x axis for the first half of the journey and at 135° to the x axis for the journey back from the spaceship to the radar station (Fig. 4.2(a)). The x and w coordinates of world point B can be obtained by drawing lines parallel to the $w=ct$ and x axes respectively to cut the x axis and the $w=ct$ axis at the points x and ct respectively, as shown in Fig. 4.2(a). If the spaceship is moving with uniform velocity relative to Σ, its world line is straight. If its velocity is v, in a time Δt the spaceship goes a distance $\Delta x = v\Delta t$, so that the inclination of its world line to the $w=ct$ axis is equal to ϕ, where

$$\tan\phi = \frac{\Delta x}{c\Delta t} = \frac{v\,\Delta t}{c\Delta t} = \frac{v}{c}\,. \tag{4.3}$$

Since for all particles of finite mass, $v < c$, the maximum value of $\tan\phi$ is 1, so that the inclinations of the world lines of all moving particles to the $w = ct$ axis must be less than 45°. The world lines of light (radio) signals are at 45° to the $w = ct$ axis.

If the spaceship is accelerating relative to Σ, its world line is curved.

As a numerical example of a space–time diagram, consider again the example of an electron that leaves the origin of Fig. 4.2(b) at world point O at ($x = 0$, $w = 0$) and travels with uniform velocity $v = 0.5\,c$ along the $+x$ axis. The inclination of the world line of the electron to the $w = ct$ axis is given by

$$\phi = \arctan\left(\frac{v}{c}\right) = \arctan 0.5 = 26.57^\circ$$

as shown in Fig. 4.2(b). In a time of 10^{-8} s the electron travels a distance $x = vt = 0.5 \times 3 \times 10^8 \times 10^{-8}$ m $= 1.5$ m to be at world point B in Fig. 4.2(b), which has coordinates ($x = 1.5$ m, $w = ct = 3$ m).

In this section we have described how one inertial observer with a radar set can determine the positions and times of distant events. We shall now proceed to see how the radar measurements of the coordinates and times of distant events carried out by two inertial observers moving with uniform velocity relative to each other can be related.

4.3 RADAR MEASUREMENTS CARRIED OUT BY TWO ASTRONAUTS IN RELATIVE MOTION

4.3.1 Introduction

Since many people have a mental bias in favour of the laboratory frame, we shall consider two spaceships in outer space moving with different uniform velocities relative to the fixed stars. (If at any stage we want to use the laboratory frame, we can assume that one of the spaceships is at rest on the Earth.) It will be assumed that astronaut John is on one spaceship and astronaut Mary is on the other. Both John and Mary are inertial observers. Let John's spaceship be at rest at the origin of an inertial reference frame Σ, and let Mary's spaceship be at rest at the origin of an inertial reference frame Σ′, as shown in Fig. 4.3(b). Suppose that Mary's spaceship moves with uniform velocity v relative to John's spaceship along the x axis of Σ, which is coincident with the x' axis of Σ′, as shown in Fig. 4.3(b), so that the inertial reference frame Σ′ is moving with uniform velocity v relative to the inertial reference frame Σ.

Suppose that John and Mary construct identical clocks. They must agree to use the same unit of time. It would be sensible for them to use the 1967 definition of the *second* (s), which is still in use, and is

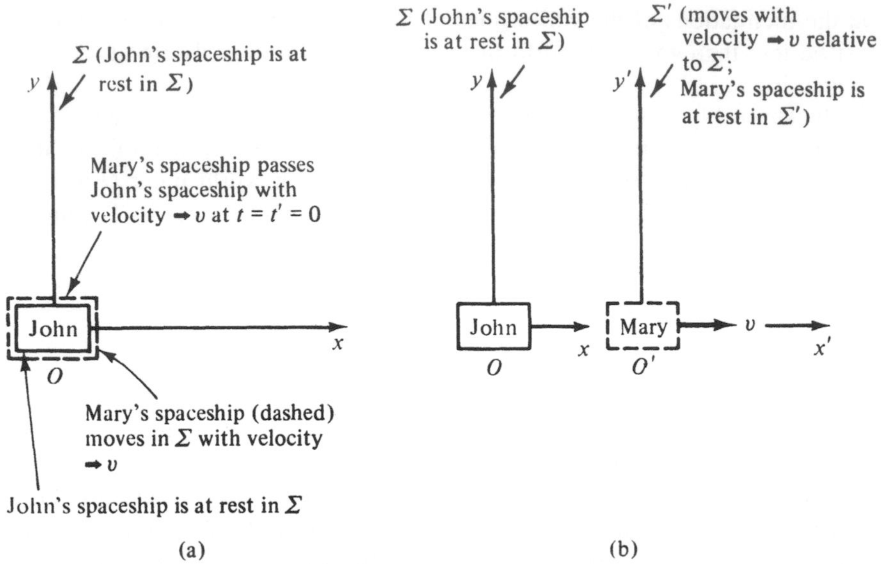

Fig. 4.3 John's spaceship is at rest at the origin of Σ and Mary's spaceship is at rest at the origin of Σ′. Mary's spaceship moves with uniform velocity v relative to John's spaceship along the x axis of Σ, passing John's spaceship at $t = t' = 0$, as shown in (a). The x axis of Σ and the x' axis of Σ′ are coincident.

> The duration of 9 192 631 770 periods of the radiation corresponding to the transition between the two hyperfine levels of the ground state of the caesium-133 atom.

John and Mary must also agree to use the same unit of length. In 1983 the 1960 definition of the *metre*, given in Section 1.1, was replaced by a new definition, in which it is assumed that the velocity of light in empty space is given *exactly* by

$$c = 299\,792\,458 \text{ m s}^{-1}.$$

If John and Mary are to carry out their measurements in a way consistent with the principle of the constancy of the speed of light, they must use the same value for c. It would be sensible for them to agree to use the value of 299 792 458 m s^{-1} (in calculations we shall generally use the *approximate* value of $c = 3 \times 10^8$ m s^{-1}). They can then adopt the 1983 definition of the metre to calibrate their metre rules, which is:

> the distance travelled by light in a vacuum in a time of 1/299 792 458 s.

In Chapter 6, when we come back to relativistic mechanics, we shall need a suitable unit of mass. At present, the kilogram is still defined as the mass of a cylinder of platinum–iridium kept at Sèvres near Paris. Clearly it would not be possible for both John and Mary to take the standard kilogram with them. Pending the definition of the kilogram in terms of some atomic quantity, John and

Mary could agree to assign an exact value to the mass of a proton that is consistent within experimental error with the present definition of the kilogram.

Suppose that John's and Mary's spaceships pass each other when the clocks on their spaceships read $t = 0$ and $t' = 0$ respectively, as shown in Fig. 4.3(a). (The origins of Σ and Σ' therefore coincide at $t = t' = 0$.)

4.3.2 Radar determination of the coordinates of events

Let John send out a radar signal from his spaceship at time t_1 on his clock. Suppose that this signal is reflected at a distant *event* on the x axis of Σ, and that the reflected radar signal returns to John's spaceship at a time t_2 on his clock. If John assumes that the speed of radar signals (or light) is the same and equal to c in all directions, so as to be in agreement with the principle of the constancy of the speed of light, then, using equations (4.1) and (4.2), he will determine the position x and time t of the event to be

$$x = \tfrac{1}{2} c\,(t_2 - t_1) = \tfrac{1}{2}(ct_2 - ct_1), \tag{4.4}$$

$$t = \tfrac{1}{2}(t_2 + t_1), \qquad ct = \tfrac{1}{2}(ct_2 + ct_1). \tag{4.5}$$

Equations (4.4) and (4.5) will be used with different symbols for x, t, t_1 and t_2 throughout this chapter to determine the positions of events relative to the inertial reference frame Σ in which John's spaceship is at rest.

Similarly, let Mary send out a radar signal from her spaceship at time t_1' on her clock. Suppose that this signal is reflected from an *event* on the x' axis of the inertial reference frame Σ', in which Mary's spaceship is at rest, and that the reflected signal reaches her spaceship at time t_2' on her clock. According to the principle of the constancy of the speed of light, which is being taken as axiomatic in this chapter, Mary must use the same value as John for c, the speed of propagation of radar signals in all directions relative to the reference frame Σ' in which her spaceship is at rest. Hence, using equations (4.1) and (4.2), Mary will determine the position x' and the time t' of the event on the x' axis of Σ' to be

$$x' = \tfrac{1}{2} c\,(t_2' - t_1') = \tfrac{1}{2}(ct_2' - ct_1'), \tag{4.6}$$

$$t' = \tfrac{1}{2}(t_2' + t_1'), \qquad ct' = \tfrac{1}{2}(ct_2' + ct_1'). \tag{4.7}$$

These equations will be used throughout this chapter to determine the positions and times of events relative to the inertial reference frame Σ', in which Mary's spaceship is at rest at the origin.

4.3.3 Relative speed of John's and Mary's spaceships

Let John send out a series of radar signals from his spaceship that are reflected back to John's spaceship by Mary's spaceship. Let John use these signals to determine a series of values for the position of Mary's spaceship relative to his own spaceship at various times. John can then calculate the speed at which Mary's spaceship is moving away from his spaceship. Similarly, by transmitting radar signals that are reflected by John's spaceship, Mary can determine the speed at which John's spaceship is moving away from her spaceship. John and Mary must agree on their speed of separation. If they did not determine the same numerical value for this speed, it could only be due to motion in one direction of empty space compared with motion in the opposite direction of empty space. This is contrary to our assumption that there are no preferred directions in empty space, i.e. that space is isotropic. Hence John and Mary must agree on their speed of separation, which will be denoted by v.

4.3.4 The K calculus

A space–time diagram (or displacement–time graph) will be used to represent the positions and times of events relative to the inertial reference frame Σ, in which *John's* spaceship is at rest. As described in Section 4.2, it is conventional to plot

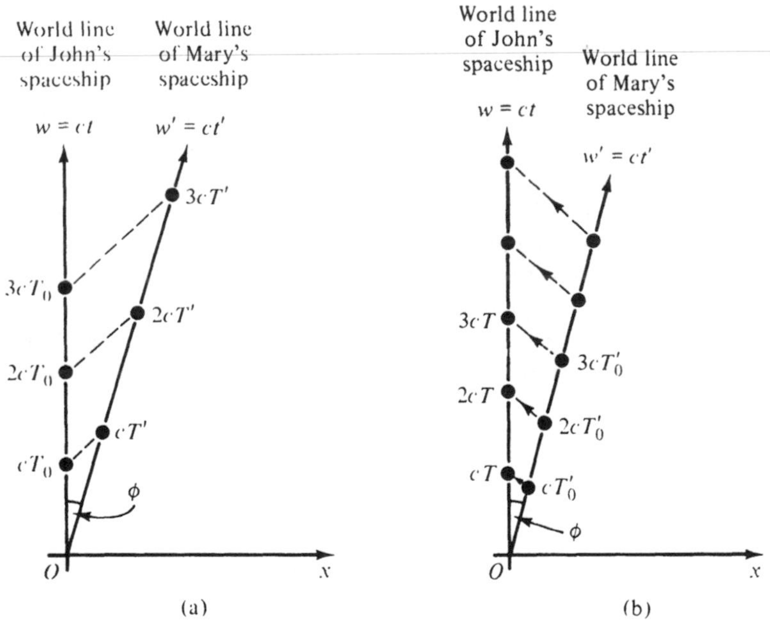

Fig. 4.4 Space–time diagrams relative to the inertial reference frame Σ in which John's spaceship is at rest. (a) John transmits radar signals at times 0, T_0, $2T_0$, $3T_0$ etc. on his clock. These signals reach Mary's spaceship at times 0, T', $2T'$, $3T'$ etc. on her clock, where $T' = KT_0$. (b) Mary transmits radar signals at times 0, T'_0, $2T'_0$,$3T'_0$ etc. on her clock. These signals reach John's spaceship at times 0, T, $2T$, $3T$ etc. on his clock, where $T = KT'_0$.

$w = ct$ against x, as shown in Fig. 4.4(a), where x is the position and t is time of the event measured by John using his radar set. The displacement of Mary's spaceship relative to John's spaceship is shown in Fig. 4.4(a). If her spaceship passes his at $t = t' = 0$ then Mary's world line Ow' is a straight line through the origin. Since Mary's spaceship goes a distance $x = vt = (v/c)\,ct = vw/c$ in a time t relative to John, her world line Ow' is at an angle ϕ to the Ow axis, where $\tan\phi = v/c$.

Throughout this chapter until Section 4.11, it will be assumed that all world points and all world lines are plotted relative to the inertial reference frame Σ *in which John's spaceship is at rest, using* Ox *and* Ow *as rectangular axes. We shall, however, take the liberty of labelling the times of events on Mary's world line* Ow' *as the times measured by Mary's clock.*

Let John send out radio signals at times $t = 0$, T_0, $2T_0$, $3T_0$ etc., as measured by his clock. The time interval between the emission of successive signals is T_0 on John's clock. The world lines of these radio signals are at 45° to the x and $w = ct$ axes in Fig. 4.4(a). Mary's spaceship coincides with John's spaceship when he transmits his first signal at $t = 0$, so that she receives his first signal at $t' = 0$ on her clock. Suppose that Mary receives the radio signal transmitted by John at a time $t = T_0$ on his clock at a time T' measured by her clock. Now T' is not equal to T_0, since Mary is moving away from John during the time interval the radio signal takes to travel from his spaceship to hers. Let

$$T' = KT_0. \tag{4.8}$$

Since Mary is moving away from John, each successive signal transmitted by him has farther to go before reaching her than the previous one. However, if Mary is moving away from John with *uniform* velocity, each successive signal has the same extra distance to travel compared with the previous one. Hence the equally spaced radio signals transmitted by John at times 0, T_0, $2T_0$ etc. on his clock should reach Mary at times 0, T', $2T'$ etc. on her clock, as illustrated in Fig. 4.4(a). This is really an example of the Doppler effect, which will be discussed in Section 4.8 and in detail in Section 5.6. Assume that there is a lamp at rest on John's spaceship. We can treat each wave maximum as a signal transmitted by John at constant time intervals, on his reckoning. If Mary did not receive these signals at constant time intervals on her reckoning, the light coming from John's spaceship would be measured by her to vary in frequency and hence in colour, even though John was moving away from her with uniform velocity v, and he was transmitting the signals at a fixed frequency on his reckoning. Such erratic behaviour is not observed in practice.

As a numerical example, assume that John transmits radio signals at times 0, 1, 2, 3 s on his clock, and the first two signals reach Mary's spaceship at times 0 and 1.5 s on her clock, corresponding to $K = \frac{3}{2}$. If Mary is moving away from John with *uniform* velocity, the successive signals from him should reach her at times 0, 1.5, 3, 4.5, 6 s etc. on her clock. If John only transmitted signals at times 0, 2, 4, 6 s on his clock, Mary would only receive signals at times 0, 3, 6, 9 s measured on her clock. In this case T_0 would be equal to 2 s and T' would be equal to 3 s in equation (4.8), but K would still be equal to $\frac{3}{2}$. This illustrates how, provided that Mary's spaceship continues to move with the same uniform velocity relative to John (Σ), the value of K in equation (4.8) does not depend on the value of the time interval T_0, measured by John in Σ, between the signals transmitted by him.

Let Mary transmit radio signals at times $t' = 0, 1, 2, 3$ s etc. measured by her clock, when she is moving away from John with the same uniform velocity as in the previous numerical example. According to the postulates of special relativity, all directions in space are equivalent and the speed of light is the same in all directions of empty space in all inertial reference frames. There is nothing to differentiate the inertial reference frame Σ' in which Mary is at rest from the inertial reference frame Σ in which John is at rest. John and Mary are just moving apart with uniform velocity relative to each other in outer space. Hence John should receive the radio signals Mary transmits at $t' = 0, 1, 2, 3$ s on her clock at times $t = 0, 1.5, 3, 4.5$ s on his own clock. Thus if Mary measures the interval between John's equally spaced signals to be increased by a factor K when she is going away from him with uniform velocity v John should measure the time interval between Mary's equally spaced signals to be increased by the same factor K, when he is moving away from her with the same uniform velocity v. If Mary transmits radio signals at times $0, T_0', 2T_0', 3T_0'$ etc. measured on her clock, these signals should reach John's spaceship at times $0, T, 2T, 3T$ etc. measured by his clock, where

$$T = KT_0', \tag{4.9}$$

as illustrated in Fig. 4.4(b). Equations (4.8) and (4.9) apply to the time intervals between successive signals. Provided that Mary and John pass each other at $t = t' = 0$, then equations (4.8) and (4.9) can be used to relate the times of single radio signals sent from John to Mary at a time T_0 on his clock, and from Mary to John at a time T_0' on hers.

In summary, provided that Mary's spaceship passes John's spaceship at $t = t' = 0$ and continues to move with uniform velocity v relative to him (Σ), if John transmits a radio signal to Mary at a time T_0 on his clock, and if this signal reaches Mary's spaceship at a time T' on her clock, we have

$$T' = KT_0. \tag{4.10}$$

This is illustrated in Fig. 4.5(a). Conversely, if we know that the radio signal reaches Mary's spaceship from John's at a time T' on her clock, then John must have transmitted the signal at a time T_0 on his clock, where

$$T_0 = T'/K. \tag{4.11}$$

If Mary transmits a radio signal to John's spaceship at a time T_0' on her clock, then the signal should reach him at a time T on his clock given by

$$T = KT_0'. \tag{4.12}$$

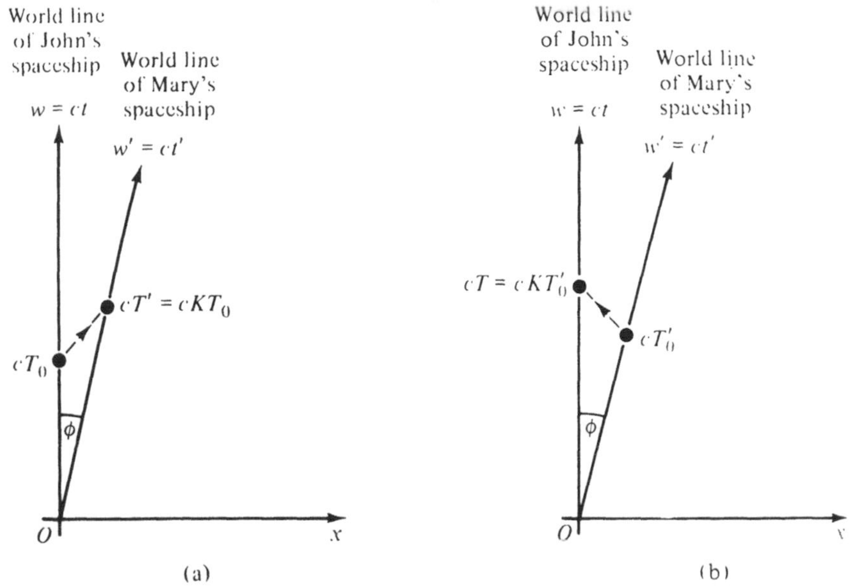

Fig. 4.5 Mary's spaceship passes John's at $t = t' = 0$, so that her world line goes through the origin. (a) If John transmits a radio signals at time T_0 on his clock, it reaches Mary's spaceship at time $T' = KT_0$ on her clock. (b) If Mary transmits a radio signal at a time T_0' on her clock, it reaches John's spaceship at time $T = KT_0'$ on his clock. The space–time diagrams are relative to the inertial reference frame Σ in which John's spaceship is at rest.

This is illustrated in Fig. 4.5(b). Conversely, if we know that Mary's signal reaches John at a time T on his clock then Mary must have transmitted the signal at a time T_0' on her clock, where

$$T_0' = T/K. \tag{4.13}$$

Equations (4.10)–(4.13) are the basic formulae of the K calculus. They hold for all values of T_0 and T_0', provided that Mary passes John at $t = t' = 0$ and they continue to move apart with uniform velocity relative to each other. Readers should familiarize themselves with these equations, and how they can be applied to the conditions illustrated in Figs. 4.5 (a) and (b), *relating the times of transmission and reception of radio or light signals from John to Mary and from Mary to John respectively*. It is assumed that John and Mary coincide at $t = t' = 0$, so that Mary's world line goes through the origin. The value of the constant K will now be determined.

Let John transmit a radar signal at a time T_0 on his clock, as shown in Fig. 4.6, and suppose that this signal reaches Mary at a time T' on her clock where, according to equation (4.10),

$$T' = KT_0. \tag{4.14}$$

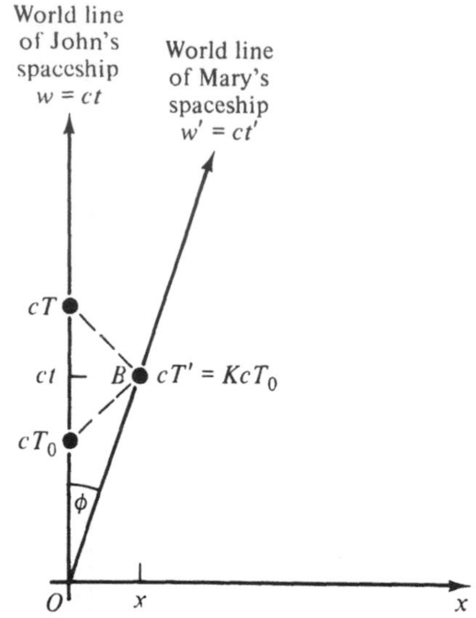

Fig. 4.6 Determination of the constant K. John transmits a radar signal at a time T_0 on his clock, which reaches Mary at a time $T' = KT_0$ on her clock. The reflected signal reaches John's spaceship at time $T = KT' = K^2T_0$ on his clock. He estimates the position of her spaceship as $x = \frac{1}{2}c(T - T_0)$ at $t = \frac{1}{2}(T + T_0)$. From this, and assuming that Mary's spaceship moves at a uniform speed v relative to John's, it follows (see the discussion in the text) that $K = [(1 + v/c)/(1 - v/c)]^{1/2}$.

Let this radar signal be reflected by Mary's spaceship and return to John's at a time T on his clock, as shown in Fig. 4.6. From equation (4.12),

$$T = KT'.$$

Using equation (4.14),

$$T = K^2T_0. \tag{4.15}$$

Using equations (4.4) and (4.5), John estimates that the radar signal was reflected from Mary's spaceship at

$$x = \tfrac{1}{2}c(T - T_0), \qquad t = \tfrac{1}{2}(T + T_0).$$

If Mary's spaceship is moving at uniform speed v relative to John's then, since the two spaceships coincide at $t = 0$, John estimates that Mary's spaceship goes a distance x in a time t. Hence the speed of Mary's spaceship relative to John's is

$$v = \frac{x}{t} = \frac{c(T - T_0)}{T + T_0}.$$

Using equation (4.15),

$$\frac{v}{c}=\frac{K^2T_0-T_0}{K^2T_0+T_0}=\frac{K^2-1}{K^2+1},$$

and rearranging,

$$\frac{v}{c}K^2+\frac{v}{c}=K^2-1,$$

$$K^2\left(1-\frac{v}{c}\right)=1+\frac{v}{c}.$$

So that

$$K=\left(\frac{1+v/c}{1-v/c}\right)^{1/2}. \tag{4.16}$$

Two important expressions involving K can be derived. First, we have

$$\begin{aligned} K+\frac{1}{K}&=\left(\frac{1+v/c}{1-v/c}\right)^{1/2}+\left(\frac{1-v/c}{1+v/c}\right)^{1/2}\\ &=\frac{(1+v/c)+(1-v/c)}{[(1-v/c)\,(1+v/c)]^{1/2}}\\ &=\frac{2}{(1-v^2/c^2)^{1/2}}=2\gamma, \end{aligned} \tag{4.17}$$

where

$$\gamma=(1-v^2/c^2)^{-1/2}. \tag{4.18}$$

Similarly,

$$\begin{aligned} K-\frac{1}{K}&=\frac{2v/c}{(1-v^2/c^2)^{1/2}}\\ &=\frac{2\gamma v}{c}. \end{aligned} \tag{4.19}$$

4.3.5 Time dilatation

In order to give the reader practice in the use of the K calculus, before deriving the Lorentz transformations, we shall apply the K calculus to derive the formula for time dilatation.

Let two successive ticks on Mary's clock be at times t_1' and t_2' respectively, as measured by her clock on her spaceship, which is moving with uniform velocity v relative to John's spaceship. Let John send a radar signal from his spaceship at a time t_3 on his clock, such that it is reflected from Mary's spaceship at the time of the first tick on Mary's clock, and suppose that the reflected signal reaches

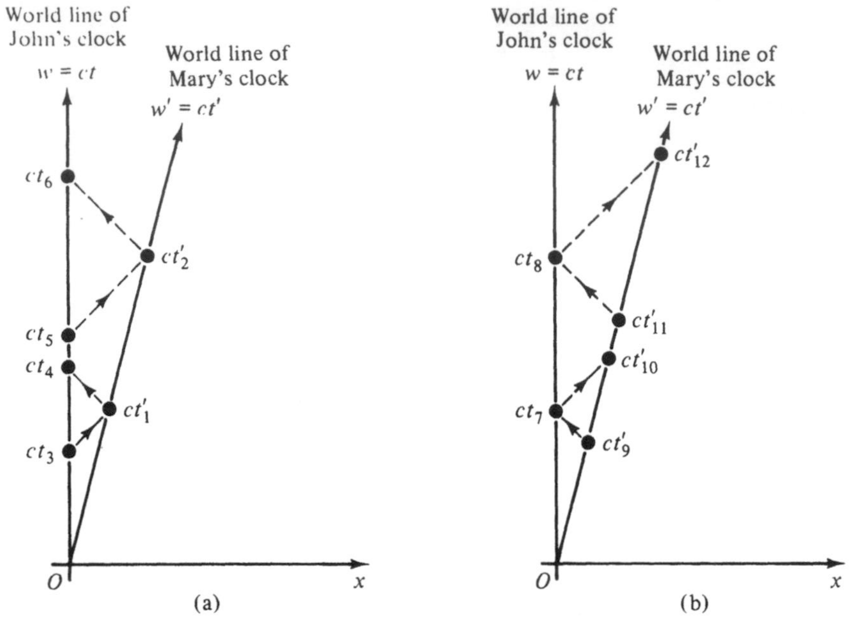

Fig. 4.7 Comparison of the rates of two moving clocks using radar methods. (a) John uses radar methods to determine the times of the ticks at times t_1' and t_2' on Mary's clock. (b) Mary uses radar methods to determine the times of the ticks at times t_7 and t_8 on John's clock.

John's spaceship at a time t_4 on his clock, as illustrated in Fig. 4.7(a). Let John transmit a second radar pulse at a time t_5 on his clock, such that it is reflected from Mary's spaceship at the time of the second tick on Mary's clock, and suppose that the reflected signal returns to John's spaceship at a time t_6 on his clock (Fig. 4.7(a)). Mary estimates the time interval between the two ticks on her clock to be $t_2' - t_1'$. Using equations similar to (4.5), John estimates the times of the two ticks on Mary's clock to be

$$t_1 = \tfrac{1}{2}(t_4 + t_3),$$

$$t_2 = \tfrac{1}{2}(t_6 + t_5).$$

Hence, using radar methods, John estimates the time difference between the two ticks on Mary's clock to be

$$t_2 - t_1 = \tfrac{1}{2}(t_6 + t_5) - \tfrac{1}{2}(t_4 + t_3). \tag{4.20}$$

Using equation (4.11), we have

$$t_3 = \frac{t_1'}{K}, \qquad t_5 = \frac{t_2'}{K}.$$

Using equation (4.12), we have

$$t_4 = Kt_1', \qquad t_6 = Kt_2'.$$

Substituting in equation (4.20),

$$t_2 - t_1 = \tfrac{1}{2}\left(Kt_2' + \frac{t_2'}{K}\right) - \tfrac{1}{2}\left(Kt_1' + \frac{t_1'}{K}\right)$$

$$= \tfrac{1}{2}\left(K + \frac{1}{K}\right)(t_2' - t_1').$$

But from equation (4.17) $K + 1/K = 2\gamma$. Hence

$$t_2 - t_1 = \gamma(t_2' - t_1') = \frac{t_2' - t_1'}{(1 - v^2/c^2)^{1/2}}. \tag{4.21}$$

Equation (4.21) is the expression for time dilatation. Since $\gamma = (1 - v^2/c^2)^{-1/2}$ is always greater than unity, using radar methods John measures the time interval between the two ticks on Mary's clock to be longer than she does. If $t_2' - t_1'$ is one second on Mary's clock then, by John's reckoning, using radar methods, the time between the two ticks on her clock is γ seconds, whereas the time between two successive ticks on John's own identical clock would be one second. Since γ is always greater than unity, *using radar methods*, John *measures* Mary's clock, which is moving with uniform velocity v relative to him, to go at a slower rate than his own clock. For example, if $v = 0.8\,c$ then $\gamma = \frac{5}{3}$, so that, from equation (4.21), the time interval between the two ticks on Mary's clock, which on her measurements in Σ' is 1 s, is measured to be $\frac{5}{3}$ s by John using radar methods in Σ. Hence John concludes that, relative to Σ, there are 36 ticks on Mary's clock per minute, whereas she concludes that there are 60 ticks per minute on her clock.

The time interval between the two events (the two ticks on Mary's clock), as measured by Mary's clock, *which coincides with both events*, is called the *proper time interval* between the two events. John must use radar methods to measure the time interval between the two events. Even if his clock coincided with the first tick on Mary's clock then, owing to their motion relative to each other, John's clock cannot coincide with the second tick on hers. The measurement of the rate of a moving clock, or the time interval between two events on a moving spaceship, necessitates the use of signals to transmit information to the "stationary" observer.

Let two successive ticks on John's clock be at times t_7 and t_8 measured by his clock, when John's and Mary's spaceships are moving apart with uniform speed v relative to each other. John estimates the time interval between the two events to be $t_8 - t_7$. Let Mary transmit radar signals at times t_9' and t_{11}' on her clock, such that they are reflected from John's spaceship at times t_7 and t_8 respectively on his clock, and suppose that the reflected signals return to Mary's spaceship at times t_{10}' and t_{12}' respectively on her clock, as illustrated in Fig. 4.7(b). Using equation (4.7), Mary estimates the times of the two events to be

$$t_7' = \tfrac{1}{2}(t_{10}' + t_9'), \qquad t_8' = \tfrac{1}{2}(t_{12}' + t_{11}').$$

Hence Mary measures the time intervals between the two ticks on John's clock to be $t_8' - t_7'$, where

$$t_8' - t_7' = \tfrac{1}{2}(t_{12}' + t_{11}') - \tfrac{1}{2}(t_{10}' + t_9').$$

Using equations (4.13) and (4.10),

$$t_9' = \frac{t_7}{K}, \qquad t_{10}' = Kt_7, \qquad t_{11}' = \frac{t_8}{K}, \qquad t_{12}' = Kt_8.$$

Hence

$$t_8' - t_7' = \tfrac{1}{2}\left(Kt_8 + \frac{t_8}{K}\right) - \tfrac{1}{2}\left(Kt_7 + \frac{t_7}{K}\right)$$

$$= \tfrac{1}{2}\left(K + \frac{1}{K}\right)(t_8 - t_7). \tag{4.22}$$

Using equation (4.17), equation (4.22) becomes

$$t_8' - t_7' = \gamma(t_8 - t_7) = \frac{t_8 - t_7}{(1 - v^2/c^2)^{1/2}}. \tag{4.23}$$

In this case $t_8 - t_7$ is the proper time interval between the two events on John's spaceship, namely the two ticks on his clock. Using radar methods, Mary measures the time interval between the two ticks on John's clock to be γ times what he does. Thus, using radar methods, Mary measures John's clock to go at a slower rate than her own clock. Time dilatation is perfectly reciprocal, provided that Mary and John continue to move apart with uniform velocity relative to each other. Using radar methods, John measures Mary's clock, which is moving relative to him, to go at a slower rate than his own clock, while, using radar methods, Mary measures John's clock, which is moving relative to her, to go at a slower rate than her own clock. (The case of accelerating clocks will not be considered until Chapter 9.) The use of radar signals illustrates how time dilatation arises when the information is transmitted by signals travelling at finite speeds, and interpreted in a way consistent with the principle of the constancy of the speed of light. An account of experiments that confirm time dilatation will be given in Section 4.6.

4.4 THE LORENTZ TRANSFORMATIONS

It will again be assumed that John's spaceship is at the origin of the inertial reference frame Σ, that Mary's spaceship is at the origin of the inertial reference frame Σ', and that Mary's spaceship is moving with uniform velocity v relative to John's spaceship, such that Σ' is moving with *uniform* velocity v relative to Σ along their common x axis, as shown previously in Fig. 4.3 (b). Let Mary's

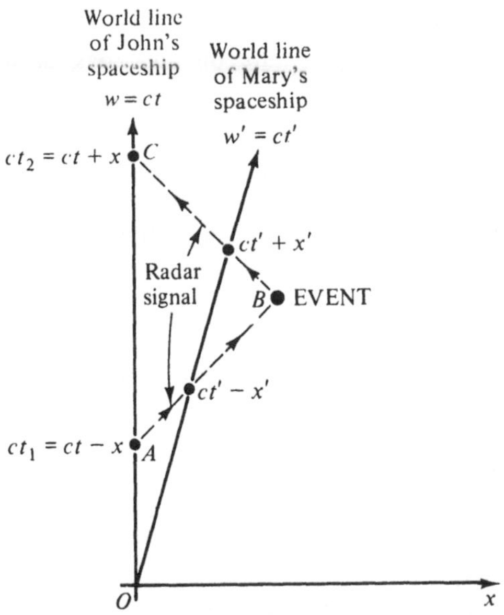

Fig. 4.8 The Lorentz transformations. John transmits a radar signal at a time $t - x/c$ on his clock. The radar signal passes Mary's spaceship at a time $t' - x'/c$ on her clock, and, after reflection at the event, passes her spaceship at a time $t' + x'/c$ on her clock, before reaching John's spaceship at a time $t + x/c$ on his clock.

spaceship pass John's spaceship at $t = t' = 0$, so that the origins of Σ and Σ' coincide at $t = t' = 0$. Suppose that an *event* occurs on the x axis of Σ, at a point farther away from John's spaceship than Mary's spaceship, and that John measures the position and time of this event, using radar methods. Let John transmit the radar signal at a time t_1 on his clock at world point A in Fig. 4.8, and suppose that the signal reflected at the event returns to John's spaceship at a time t_2 measured on his clock at world point C. According to equations (4.4) and (4.5), the x coordinate and time t of the event at world point B in Fig. 4.8, measured by John in Σ using radar methods, are

$$x = \tfrac{1}{2}(ct_2 - ct_1), \tag{4.24}$$

$$ct = \tfrac{1}{2}(ct_2 + ct_1). \tag{4.25}$$

Adding equations (4.24) and (4.25), we find that in Σ

$$ct_2 = ct + x. \tag{4.26}$$

Subtracting equation (4.24) from equation (4.25), we obtain

$$ct_1 = ct - x. \tag{4.27}$$

Hence if John uses x and t to denote the position and time of the event at world point B in Fig. 4.8 then he must have transmitted the radar signal at world point

A at a time $t_1 = t - x/c$ on his clock, and received the signal, reflected at the event at world point B, at a time $t_2 = t + x/c$ on his clock at world point C. As a numerical example, consider an electron moving with velocity $\frac{1}{2}c$ along the x axis of Σ, as shown in the space–time diagram in Fig. 4.2(b). If the position of the electron at world point B having coordinates ($x = 1.5$ m, $w = ct = 3$ m) in Fig. 4.2 (b) is to be measured, using a radar set at rest at the origin, then, according to equation (4.27), the radar signal must be transmitted at a time t_1, given by

$$ct_1 = ct - x = (3 - 1.5)\ \mathrm{m} = 1.5\ \mathrm{m},$$

and, according to equation (4.26), the reflected signal should return to the radar set at a time t_2 given by

$$ct_2 = ct + x = (3 + 1.5)\ \mathrm{m} = 4.5\ \mathrm{m}.$$

Dividing by c gives $t_1 = 0.5 \times 10^{-8}$ s and $t_2 = 1.5 \times 10^{-8}$ s. The reader can check using equations (4.4) and (4.5) that $t_1 = 0.5 \times 10^{-8}$ s and $t_2 = 1.5 \times 10^{-8}$ s do give $x = 1.5$ m and $ct = 3$ m.

Let John's radar signal pass Mary's spaceship at a time $t' - x'/c$ on her clock, and let the signal reflected by the event pass Mary at a time $t' + x'/c$ on her clock, as illustrated in Fig. 4.8, such that using equations (4.6) and (4.7), she estimates the position of the event to be at x' at a time t' relative to Σ'.

Denoting the time from 0 to $t - x/c$ on John's clock by T_0, and the time from 0 to $t' - x'/c$ on Mary's clock by T', since from equation (4.10) $T' = KT_0$, after multiplying both sides by c, we have

$$ct' - x' = K\,(ct - x) = Kct - Kx. \tag{4.28}$$

Denoting the interval between 0 and $t' + x'/c$ on Mary's clock by T_0' and the interval between 0 and $t + x/c$ on John's clock by T, since from equation (4.13) $T_0' = T/K$, we have

$$ct' + x' = \frac{ct + x}{K} = \frac{ct}{K} + \frac{x}{K}. \tag{4.29}$$

Subtracting equation (4.28) from equation (4.29), we obtain

$$2x' = x\left(\frac{1}{K} + K\right) - ct\left(K - \frac{1}{K}\right).$$

From equation (4.17), $K + 1/K = 2\gamma$, and from equation (4.19), $K - 1/K = 2\gamma v/c$. Hence

$$2x' = 2\gamma x - ct\,\frac{2\gamma v}{c},$$

Dividing both sides by 2 gives:

$$x' = \gamma\,(x - vt), \tag{4.30}$$

where

$$\gamma = (1 - v^2/c^2)^{-1/2}. \tag{4.31}$$

Adding equations (4.28) and (4.29) gives

$$2ct' = ct\left(\frac{1}{K} + K\right) - x\left(K - \frac{1}{K}\right).$$

Using equations (4.17) and (4.19), we obtain

$$2ct' = ct\,2\gamma - \frac{x\,2\gamma v}{c}.$$

Dividing both sides by $2c$ gives

$$t' = \gamma\,(t - vx/c^2)\,. \tag{4.32}$$

Equations (4.30) and (4.32) are the *Lorentz transformations* relating the positions x and x' and times t and t' of the event, measured by John and Mary respectively using radar methods. The use of radar methods makes it clear that the Lorentz transformations relate the measured positions and times of events. It is left as an algebraic exercise for the reader to rearrange equations (4.30) and (4.32) to show that

$$x = \gamma\,(x' + vt'), \tag{4.33}$$

$$t = \gamma\,(t' + vx'/c^2). \tag{4.34}$$

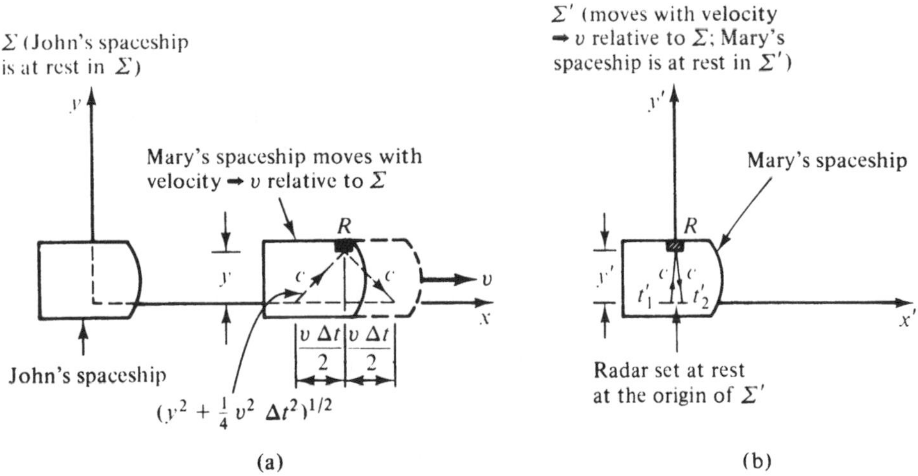

Fig. 4.9 Measurement of a length perpendicular to the direction of relative motion. (a) John's interpretation of Mary's measurements. (b) Mary measures the distance y' between the x' axis and a reflector R on the y' axis of Σ' using radar methods.

These are the *inverse* Lorentz transformations. Note that the inverse transformations can be obtained from equations (4.30) and (4.32) by changing primed into unprimed quantities and unprimed quantities into primed quantities and replacing v by $-v$. This procedure works for all relativistic transformations.

The measurement of lengths in a direction perpendicular to the common x axis, i.e. perpendicular to the direction of motion of Mary's spaceship relative to John's spaceship, will now be discussed. Suppose that Mary places her radar set at $x'=0$ on the x' axis of the inertial reference frame Σ' in which her spaceship is at rest, as shown in Fig. 4.9(b), and uses radar methods to measure the distance y' between her radar set on the x' axis and a reflector R on her spaceship. Let the line joining Mary's radar set and the reflector R be perpendicular to the direction in which John's spaceship is moving with velocity v along the negative x' axis of Σ'. Let Mary transmit a radar signal at a time t_1' on her clock, and suppose that this signal is reflected by R and that the reflected signal returns to her radar set at a time t_2' on her clock. Since the radar signal travels a total distance $2y'$ to the reflector R and back in a time interval $t_2'-t_1'$ relative to the inertial reference frame Σ' in which Mary's spaceship is at rest, she should conclude that in Σ'

$$2y' = c\,(t_2' - t_1'). \tag{4.35}$$

It follows from equation (4.34) that, since Mary transmits the radar signal from $x'=0$ at a time t_1' on her clock, John measures this event to be at a time t_1 in Σ, where

$$t_1 = \gamma\,(t_1' + vx'/c^2) = \gamma t_1'.$$

Similarly, John measures the event, when the reflected radar signal returns to $x'=0$ at a time t_2' on Mary's clock, to be at a time t_2 in Σ, where, according to equation (4.34),

$$t_2 = \gamma\,(t_2' + vx'/c^2) = \gamma t_2'.$$

Hence, using radar methods, John measures the time interval Δt between the ticks at t_1' and t_2' on Mary's clock, when the radar signal left and returned to the x' axis respectively, to be $\Delta t = \gamma\,(t_2' - t_1')$. Using equation (4.35), we have

$$\Delta t = 2\gamma y'/c. \tag{4.36}$$

Relative to the inertial reference frame Σ in which John's spaceship is at rest, Mary's spaceship is moving with uniform velocity v along the $+x$ axis, and, relative to Σ, the radar signal takes the path shown in Fig. 4.9(a). In the time interval Δt measured by John, Mary's spaceship goes a distance $v\Delta t$ relative to John (Σ). By symmetry, Mary's spaceship travels a distance $\frac{1}{2}v\Delta t$ in the time the radar signal takes to go from the transmitter on the x' axis to the reflector R. Let R be at a perpendicular distance y from the x axis, measured relative to Σ, the reference frame in which John's spaceship is at rest. In Fig. 4.9(a) it follows from Pythagoras' theorem that, relative to Σ, the distance from the point of emission of the radar signal to R is $[y^2 + (\frac{1}{2}v\Delta t)^2]^{1/2}$. The total distance travelled by the radar signal relative to Σ is $2\,[y^2 + (\frac{1}{2}\,v\Delta t)^2]^{1/2}$. Since light travels with a speed c during the time Δt it takes to cover this distance, relative to the reference frame Σ shown in Fig. 4.9(a), in which John's spaceship is at rest, we have

$$c\Delta T = 2(y^2 + \tfrac{1}{4}\,v^2\Delta t^2)^{1/2}.$$

Squaring both sides,

$$c^2\Delta t^2 = 4(y^2 + \tfrac{1}{4}\,v^2\,\Delta t^2) = 4y^2 + v^2\Delta t^2.$$

Substituting $2\gamma y'/c$ for Δt from equation (4.36), we have

$$\frac{c^2 4\gamma^2 y'^2}{c^2} = 4y^2 + \frac{v^2 4\gamma^2 y'^2}{c^2}.$$

Rearranging,

$$\gamma^2 y'^2\,(1 - v^2/c^2) = y^2.$$

Since $\gamma^2 = (1 - v^2/c^2)^{-1}$, on taking the square root we have

$$y' = y.$$

Hence John and Mary should agree on the measurements of lengths perpendicular to the x and x' axes, i.e. perpendicular to the directions of their motions relative to each other. Hence, in general, if Σ′ moves with uniform velocity v relative to Σ along a common x axis then

$$y' = y. \tag{4.37}$$

Similarly

$$z' = z.$$

Collecting the Lorentz transformations, we have

$$x' = \gamma\,(x - vt), \qquad x = \gamma\,(x' + vt'), \tag{4.38}$$

$$y' = y, \qquad y = y', \tag{4.39}$$

$$z' = z, \qquad z = z', \tag{4.40}$$

$$t' = \gamma\,(t - vx/c^2), \qquad t = \gamma\,(t' + vx'/c^2). \tag{4.41}$$

The Lorentz transformations relate the coordinates and times of *events* in one inertial coordinate system Σ to the coordinates and time of the *same* event measured relative to another inertial coordinate system Σ′ that is moving relative to Σ along their common x axis and whose origin coincides with the origin of Σ at $t = t' = 0$. The reader should try to become as familiar as possible with these equations. A few simple problems are given at the end of this chapter.

If we put $v/c = \sin\theta$ then

$$\gamma = \frac{1}{(1 - v^2/c^2)^{1/2}} = \frac{1}{(1 - \sin^2\theta)^{1/2}} = \frac{1}{\cos\theta} = \sec\theta.$$

If $v/c = 0.8000$ then one finds $\theta = 53°\ 8'$ and $\sec\theta = 1.6667$. By direct calculation, $\gamma = \frac{5}{3}$.

Note that as $c \to \infty$, the Lorentz transformations become

$$x' = x - vt, \qquad y' = y, \qquad z' = z, \qquad t' = t,$$

in agreement with the Galilean transformations of Chapter 1.

So far in this chapter, radar methods have been used when discussing the determination of the coordinates and times of distant events. This has the advantage of making it clear from the outset that one is dealing with the *measured* times of distant events and that one must allow for the propagation times of signals. It is not always convenient to use radar methods. For example, if an astronaut were in outer space in a spaceship *at rest relative to the Earth*, it would be pointless to use radar methods from a base station on the Earth to measure the times of events inside the spaceship. It is far more convenient for the astronaut to use a clock on the spaceship. The question arises as to how the astronaut can synchronize the clock on the spaceship to keep the same time as a master clock on the Earth. If the astronaut took a synchronized clock from the Earth on the spaceship, one cannot be sure that the clock was not affected by the transportation. Einstein's prescription for synchronizing distant clocks will be illustrated by an imaginary radio conversation between the astronaut and the base station on Earth. The astronaut could ask the base station on Earth to send out a radio signal. On receipt of the radio signal from Earth, the astronaut should send a radio signal back to the Earth without any time delay. The astronaut should also note the time on the clock on the spaceship when the radio signal from Earth reached the spaceship. If the radio signal left the Earth at a time t_1 on the master clock at the base station on the Earth and the signal from the astronaut was received back at a time t_2 on the master clock on the Earth, since the speed of light in empty space is the same in all directions, the controller at the base station will estimate that the radio signal reached the astronaut at a time $\frac{1}{2}(t_1 + t_2)$. This information can be transmitted to the astronaut, who can then set the clock on the spaceship to agree with the master clock on the Earth.

Clocks synchronized in this way, and distributed throughout space, could be used to determine the times of distant events when and where they occur. In the laboratory it is convenient to send messages from events as electrical signals along electric cables. It does not matter what method is used to measure the times of distant events, provided it is done in a way consistent with the theory of special relativity. In future, when applying the Lorentz transformations, we shall *not* always specify precisely how the measurements are carried out, but we shall assume that they are carried out and interpreted in a way consistent with the theory of special relativity.

Some of the conclusions arising from the Lorentz transformations, in particular (i) non-absolute simultaneity (ii) time dilatation and (iii) length contraction, will now be considered.

4.5 NON-ABSOLUTE SIMULTANEITY

Let two events occur at two separated points x_1 and x_2 on the x axis of the inertial reference frame Σ, and be measured to occur at the same time t in Σ. According to the Lorentz transformations, these would be recorded at times t_1' and t_2' by synchronized clocks at rest in Σ', where t_1' and t_2' are given by

$$t_1' = \gamma(t - vx_1/c^2), \qquad t_2' = \gamma(t - vx_2/c^2).$$

Since x_1 is not equal to x_2, t_1' cannot be equal to t_2', so that, according to the Lorentz transformations, two spatially separated events that are simultaneous in Σ would not be measured to be simultaneous in Σ'. Similarly, if two events occur simultaneously at two spatially separated points x_1' and x_2' in Σ' then, from the Lorentz transformations, they would not be simultaneous in Σ. Thus, according to the theory of special relativity, the simultaneity of spatially separated events is not an absolute property, as it was assumed to be in Newtonian mechanics. This is an important result since, if inertial observers in different inertial reference frames do not agree on simultaneity, they will not agree on the measured time interval between two events. It will be shown in Section 4.7 that non-absolute simultaneity plays an important role in the phenomenon of length contraction. An example due to Einstein is given in Appendix B, which illustrates how the non-absolute simultaneity of the measured times of events arises if the principle of the constancy of the speed of light is used to interpret the measurements.

4.6 TIME DILATATION

4.6.1 Application of the Lorentz transformations

The example of time dilatation, discussed in Section 4.3.5 using radar methods and the K calculus and shown in Fig. 4.7(a), will now be solved using the Lorentz transformations. Consider the two ticks at times t_1' and t_2' on Mary's clock, as illustrated in Fig. 4.7(a). In Σ' the position of Mary's clock is $x' = 0$. Using the Lorentz transformations, the times of the two ticks on Mary's clock measured in Σ are, since $x' = 0$,

$$t_1 = \gamma(t_1' + vx'/c^2) = \gamma t_1',$$

$$t_2 = \gamma(t_2' + vx'/c^2) = \gamma t_2'.$$

Hence

$$t_2 - t_1 = \gamma (t_2' - t_1') = \frac{t_2' - t_1'}{(1 - v^2/c^2)^{1/2}}. \tag{4.42}$$

Equation (4.42) is in agreement with equation (4.21). The time interval $t_2' - t_1'$ is the proper time interval measured by *one* clock (Mary's) that coincides with both events. John must use two synchronized clocks (or radar methods) to measure $t_2 - t_1$, since, if a clock at rest relative to Σ coincides with Mary's clock at a time t_1' on her clock, it does not coincide with it at the later time t_2' on her clock, since the latter is moving relative to Σ.

Since John's clock is at $x = 0$ in Σ, relative to the reference frame Σ' in which Mary's spaceship is at rest, the ticks at times t_7 and t_8 on John's clock (cf. Fig. 4.7(b)) are at times

$$t_7' = \gamma (t_7 - vx/c^2) = \gamma t_7,$$

$$t_8' = \gamma (t_8 - vx/c^2) = \gamma t_8$$

relative to Σ'. Subtracting,

$$t_8' - t_7' = \gamma (t_8 - t_7). \tag{4.43}$$

This is in agreement with equation (4.23), showing that time dilatation is reciprocal. The time interval $t_8 - t_7$ is the proper time interval in this case, since it is measured by one clock (John's) that coincides with both events. Equations (4.42) and (4.43) show that proper time intervals in either Σ or Σ' are measured to be longer in the other reference frame. Some of the experimental evidence in favour of time dilatation will now be discussed.

4.6.2 Experimental verification of time dilatation

According to the law of radioactive decay, if N_0 radioactive atoms are at rest at $t = 0$ the *average* number N left after a time t is given by

$$N = N_0 \exp(-\lambda t),$$

where λ is the decay constant. In a time equal to the half-life $T_{1/2} = 0.693/\lambda$, on average half of the radioactive atoms will have decayed. In a further time $T_{1/2}$, on average half of the remaining radioactive atoms will have decayed, etc. It is shown in Problem 4.5 that the average or mean time T_0 that a radioactive atom lives before it decays is equal to $1/\lambda$. Hence the law of radioactive decay can be rewritten in the form

$$N = N_0 \exp\left(-\frac{t}{T_0}\right),$$

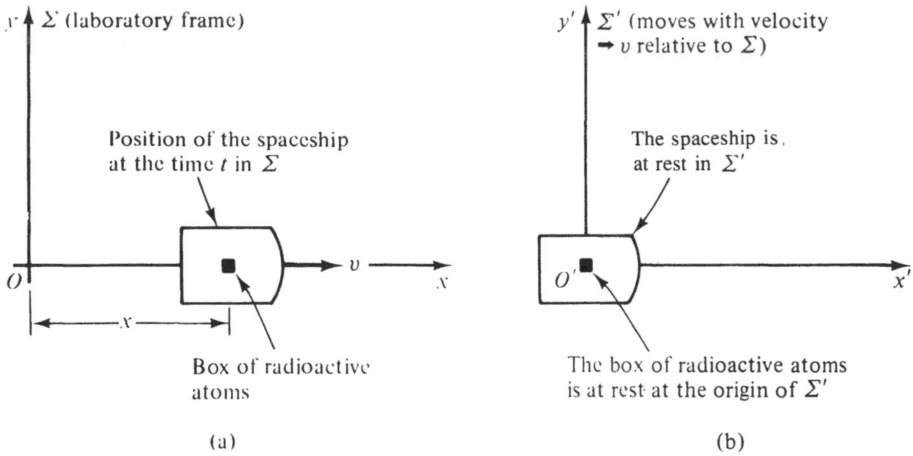

Fig. 4.10 (a) There is a box of radioactive atoms inside a spaceship moving with uniform velocity v along the x axis of Σ. (b) The spaceship and box of radioactive atoms are at rest in Σ'.

where T_0 is the *average* or *mean life* of the radioactive atoms. The half-life $T_{1/2} = 0.693\, T_0$. In high-energy physics it is conventional to use the mean life T_0, rather than λ and $T_{1/2}$, and we shall follow this convention.

Consider a box of radioactive atoms inside a spaceship at rest at the origin of the inertial reference frame Σ' that is moving with uniform velocity v relative to the laboratory frame Σ, as shown in Fig. 4.10(a,b). Let the spaceship pass the origin of the laboratory frame Σ at a time $t = 0$ in Σ and $t' = 0$ in Σ'. Let there be a total of N_0 radioactive atoms in the box on the spaceship at time $t' = 0$ in Σ'. According to the law of radioactive decay, the number N left in the spaceship after a time t', relative to the spaceship (Σ'), is

$$N = N_0 \exp\left(-\frac{t'}{T_0}\right), \tag{4.44}$$

where T_0 is the average time that a radioactive particle lives, measured in the inertial reference frame in which it is at rest (the spaceship (Σ') in this case). Relative to the laboratory frame Σ, the radioactive atoms are moving with uniform velocity v along the x axis, as shown in Fig. 4.10(a). Corresponding to $x' = 0$, $t' = t'$ relative to the spaceship (Σ'), from the Lorentz transformations, in the laboratory frame Σ we have

$$x = \gamma(x' + vt') = \gamma v t'; \qquad \text{or} \quad t' = \frac{x}{\gamma v}, \tag{4.45}$$

$$t = \gamma(t' + vx'/c^2) = \gamma t'; \qquad \text{or} \quad t' = \frac{t}{\gamma}. \tag{4.46}$$

Substituting for t' in equation (4.44), using equations (4.46) and (4.45), we find

$$N = N_0 \exp\left(-\frac{t}{\gamma T_0}\right) \tag{4.47}$$

$$= N_0 \exp\left(-\frac{x}{\gamma v T_0}\right). \tag{4.48}$$

Hence, relative to the laboratory frame Σ, the radioactive particles live for an average time γT_0, and travel an average distance $\gamma v T_0$ before decaying. Equations (4.47) and (4.48) have been confirmed by experiments with charged pions (denoted by π) and charged muons (denoted by μ).

Pions are produced in collisions between high-energy nucleons (see Section 6.6.3). The mass of the charged pion is 139.6 MeV/c^2. The charged pion is unstable, decaying into a muon and a neutrino. The mean lifetime of a charged pion is 2.55×10^{-8} s in the inertial reference frame in which it is at rest. The mass of a muon is 105.7 MeV/c^2, it decays into an electron plus two neutrinos, and its mean life is 2.2×10^{-6} s in the inertial reference frame in which it is at rest. Since the mean lifetime of charged pions, when they are at rest in the laboratory frame Σ, is $T_0 = 2.55 \times 10^{-8}$ s, and the maximum or limiting speed of particles is the speed of light $c = 3 \times 10^8$ m s^{-1}, if there were no time dilatation then charged pions would only go an average distance of $3 \times 10^8 \times 2.55 \times 10^{-8}$ m $= 7.6$ m before decaying, even if they travelled at the speed of light. In 100 m the intensity of a beam of charged pions would decrease to exp $(-100/7.6\text{ m}) \approx 2 \times 10^{-6}$ of the original intensity, if there were no time dilatation. Allowing for time dilatation, charged pions of total energy 4.5 GeV, $u = 0.9995c$, $\gamma \approx 32$, should live a mean time $\gamma T_0 = 8.1 \times 10^{-7}$ s and travel a mean distance $\gamma v T_0 = 244$ m before decaying. If allowance is made for time dilatation then, according to equation (4.48), in a distance of 100 m the intensity of a beam of pions, each of energy 4.5 Gev should on average decrease to exp $(-x/\gamma v T_0) =$ exp $(-100/244) = 0.66$ of the original intensity, whereas, if there were no time dilatation, the beam intensity would decrease to about 2×10^{-6} of the original intensity. Experiments at CERN have confirmed that charged pions of energy 4.5 GeV do travel a mean distance of about 250 m before decaying. In fact, at CERN the pion detectors are placed 100 m or more from the production point, where, but for time dilatation, the number of undecayed pions left would be extremely small.

Another example of time dilatation concerns cosmic-ray muons. The primary cosmic-ray protons and α particles produce charged pions in high-energy nuclear reactions near the top of the Earth's atmosphere at altitudes of about 60 km. These charged pions decay quickly into muons. The mean lifetime of a muon is $T_0 = 2.2 \times 10^{-6}$ s in the reference frame in which the muon is at rest. Even if a cosmic-ray muon travelled at the speed of light, if there were no time dilatation, the muon would only travel a mean distance of $3 \times 10^8 \times 2.2 \times 10^{-6}$ m $= 0.66$ km before decaying, and, if ionization loss is neglected, the fraction reaching sea level would be exp $(-60/0.66) =$ exp $(-91) \approx 10^{-40} \approx 0$. In practice a significant fraction of cosmic-ray muons reach sea level. The mean (total) energy of the muons reaching sea level is $E_\mu = 3.17$ GeV. Since $m_\mu c^2 = 105.7$ MeV and since $u = v$, it follows from equation (2.50) that for a muon of total energy $E_\mu = 3.17$ GeV, $\alpha = \gamma = E_\mu / m_\mu c^2 \approx 30$. Allowing for time dilatation, such muons would travel a mean distance $\gamma v T_0 \approx 30 \times 0.9995 \times 3 \times 10^8 \times 2.2 \times 10^{-6}$ m ≈ 20 km in a vacuum before decaying. According to equation (4.48), if ionization loss is neglected, a fraction equal to $\exp(-x/\gamma v T_0) \approx \exp(-60/20) = 0.05$ of muons of energy 3.17 GeV, produced at an altitude of 60 km, should reach sea level. This result accounts for the significant number of cosmic-ray muons that do reach sea level.

These results on pions and muons confirm equation (4.48), which was derived using the Lorentz transformations and was interpreted in terms of time dilatation relative to the laboratory frame. These results give us confidence in the validity of the Lorentz transformations.

4.7 THE LORENTZ LENGTH CONTRACTION

In order to measure the length of a *stationary* object, in principle, one could lay the object on a ruler and then look successively at the positions of each end of the object. If the object were moving then both ends would have to be measured at the same time. This point is brought out vividly by the example illustrated in Fig. 4.11(a). If a fish is swimming with velocity v relative to the laboratory frame (which will be identified with Σ), one would not measure the position of the fish's tail at the point B and then wait until the fish had swum until its mouth was opposite the point C before measuring the position of the other end of the fish. Not even the most fervent angler would stoop to that but would insist that one should measure both ends of the moving fish at the same time in the laboratory frame (Σ). The reference frame moving with the fish will be identified with Σ'.

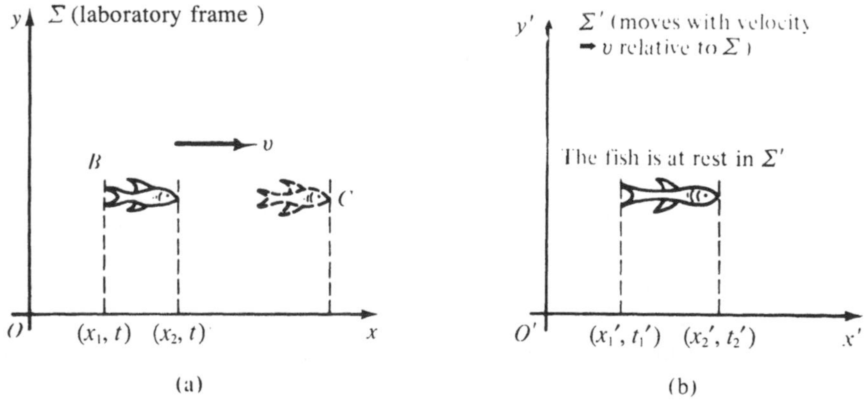

Fig. 4.11 (a) Measurement of the length of a fish that is swimming with uniform velocity v relative to Σ. (b) In the inertial frame Σ', the fish is at rest.

Let the positions of the ends of the fish be measured at the *same* time t in Σ and recorded to be at the points x_1 and x_2 as shown in Fig. 4.11(a). According to the Lorentz transformations, the fish would record the events associated with the measurement of its length at the following positions and times in Σ':

$$x_1' = \gamma(x_1 - vt), \qquad t_1' = \gamma(t - vx_1/c^2), \tag{4.49}$$

$$x_2' = \gamma(x_2 - vt), \qquad t_2' = \gamma(t - vx_2/c^2). \tag{4.50}$$

Since the fish does not move in Σ', its extremities can be measured at any time in Σ', so that, even though t_1' is not equal to t_2', the length of the fish measured in Σ' is $x_2' - x_1'$. For example, an inertial observer in Σ' can put a ruler, which is at rest in Σ', next to the fish, measure the position of the front of the fish, have a lunch break and then return later to measure the position of the other end of the fish. Neither the fish nor the ruler will have moved relative to Σ' in the intervening period. Subtracting equations (4.49) from equations (4.50) gives

$$x_2' - x_1' = \gamma(x_2 - x_1)$$

or

$$x_2 - x_1 = (x_2' - x_1')/\gamma.$$

Therefore

$$l = l_0(1 - v^2/c^2)^{1/2}, \tag{4.51}$$

where $l_0 = x_2' - x_1'$ is the length of the fish measured in the coordinate system in which it is at rest, and is called the *proper* length of the fish, while $l = x_2 - x_1$ is the length of the fish measured in an inertial reference frame, relative to which the fish is moving with uniform velocity v. This is the phenomenon of *length contraction*. It is also widely known as *the Lorentz contraction* or the *Lorentz–FitzGerald contraction*. Since, from the Lorentz transformations $y = y'$, the dimensions of the fish measured in a direction perpendicular to its direction of motion are the same in Σ and Σ'.

Since x_2 is greater than x_1 in Fig. 4.11(a), it follows from equations (4.49) and (4.50) that t_2' is less than t_1', so that in Σ' the event at (x_2', t_2') is measured to occur before the event at (x_1', t_1'). Since t_2' is not equal to t_1', from the fish's point of view, according to its clocks, the measurements were not carried out at the same time. For example, let two observers, at rest in the laboratory frame Σ, carry out the actual experiment of measuring the length of the fish by noting, at the same time t, the positions of the front and rear of the fish on a ruler, *which is stationary in the laboratory frame* Σ. From the fish's point of view, relative to the reference frame Σ' in which it was at rest, one observer measured the front of its mouth at a time t_2', as shown in Fig. 4.12(a). Their ruler was then moved with velocity $-v$ in Σ', until, at the later time t_1' in Σ', the second observer measured the position of the fish's tail as shown in Fig. 4.12(b). From the fish's point of view it is no wonder that the laboratory observers measured its length

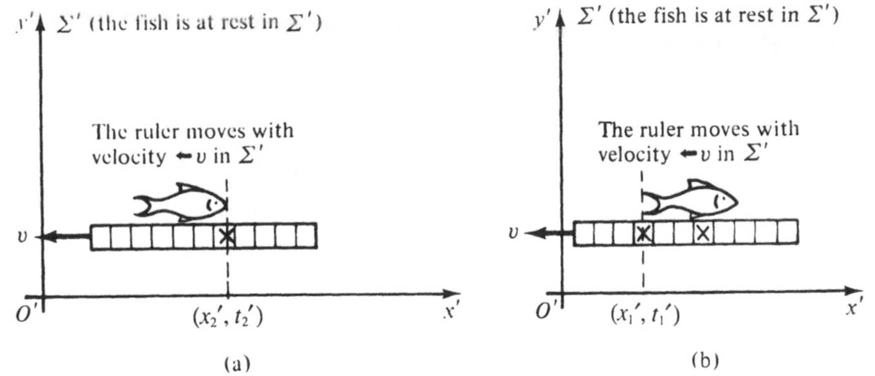

Fig. 4.12 Measurement of the length of a moving fish. According to the fish, the positions of its extremities were measured at different times t_1' and t_2' using a moving ruler.

to be too short—they moved the ruler. It can be seen that the measurement of length involves the question of the simultaneity of spatially separated events. A simple example, due to Einstein, of how the non-absolute simultaneity of spatially separated events follows from the principle of the constancy of the speed of light is given in Appendix B.

Since, from the Lorentz transformations, $y = y'$ and $z = z'$, the dimensions of moving bodies perpendicular to their direction of motion are not measured to be Lorentz-contracted.

The phenomenon of length contraction is reciprocal. If the fish measured the length of an object that was stationary in the laboratory at the same time t' in Σ' and found the extremities of the object to be at the points x_4' and x_3' then, from the Lorentz transformations, one would have

$$x_4 - x_3 = \gamma(x_4' + vt') - \gamma(x_3' + vt')$$

or

$$x_4' - x_3' = (x_4 - x_3)/\gamma. \tag{4.52}$$

Hence, in general, if an object is moving relative to an observer, the length of the object in the direction of motion is *measured* by that observer to be less than the length measured by an observer at rest relative to the object. Note that what was said was that a moving object is *measured* to contract in the direction of motion, if the positions of the extremities of the moving body are measured *at the same time* in the inertial reference frame in which the body is moving with uniform velocity v. If one looks at, or photographs, a moving body, one does not generally see the precise Lorentz contraction, given by equation (4.51). What one sees or photographs depends on the light actually reaching the eye (or camera) at that instant, and the light from different parts of the object will have left the object at different times depending on the distances of the various parts of the object from the observer. This effect is discussed in detail in Section 5.8.

Let a body of proper volume V_0 move with uniform velocity v relative to Σ. If it is divided into thin rods parallel to v, since the measured length of each one of these is reduced in length by an amount given by equation (4.51), while their cross-sectional area is unchanged, the measured total volume of the body relative to Σ should be

$$V = V_0 (1 - v^2/c^2)^{1/2}. \tag{4.53}$$

So far in our discussions, time dilatation and length contraction have been treated as independent phenomena. An example will now be given illustrating the close connection between time dilatation and length contraction.

Example 1

The diameter of our galaxy is 10^5 light years. How long does it take a proton (in the proton's rest frame) to cross the galaxy if its energy is 10^{19} eV? Take the rest mass of the proton to be 938 MeV/c^2. Ignore the effects of galactic magnetic fields, which in practice would deflect such a proton. What are the dimensions of the Galaxy in the proton's rest frame ?

Solution To simplify the discussion, assume that the cross-section of the Galaxy is circular in the (x, y) plane of the inertial reference frame Σ, in which the Galaxy is at rest, as shown in Fig. 4.13(a). Let the proton travel along the x axis and enter the Galaxy at Event 1 at (x_1, t_1) in Σ and let it leave at Event 2 at (x_2, t_2) in Σ (Fig. 4.13(a)). Relative to Σ, the proton covers the distance $d = x_2 - x_1$ at a velocity v, so that the time $t_2 - t_1$ that the proton takes to cross the Galaxy, measured in Σ, is

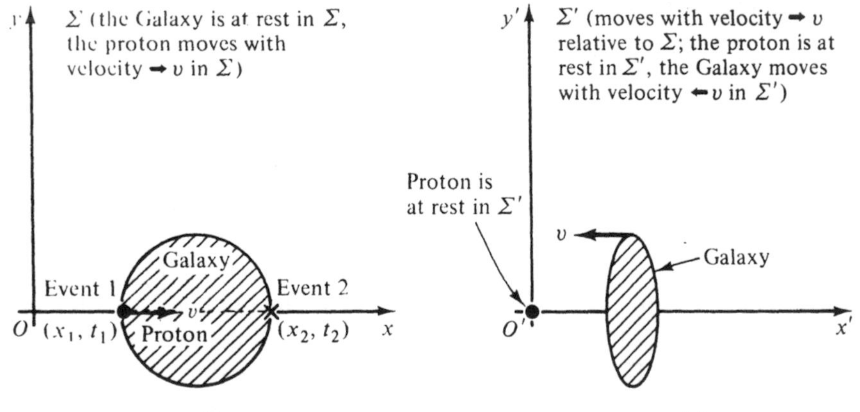

Fig. 4.13 (a) The Galaxy is at rest in Σ. The proton moves with uniform velocity v along the x axis of Σ. (b) The proton is at rest at the origin O' of Σ'. The Galaxy moves to the left with velocity v and is measured to be length-contracted in Σ'.

$$t_2 - t_1 = d/v,$$

where $d = 10^5 \times 365 \times 24 \times 3600 \times 3 \times 10^8\ \text{m} = 9.46 \times 10^{20}\ \text{m}$.

According to equation (2.50), for a proton of (total) energy 10^{19} eV we have, since $u = v$,

$$\alpha = \gamma = \frac{E}{mc^2} = \frac{10^{19}}{938 \times 10^6} = 1.066 \times 10^{10}.$$

Therefore

$$\frac{v}{c} = \left(1 - \frac{1}{\gamma^2}\right)^{1/2} \approx 1 - \frac{1}{2\gamma^2} = 1 - 4.4 \times 10^{-21}.$$

Hence the time $t_2 - t_1$ that it takes the proton to cross the Galaxy, measured in Σ, is

$$t_2 - t_1 = \frac{d}{v} = \frac{9.46 \times 10^{20}\ \text{m}}{(1 - 4.4 \times 10^{-21})\, c}$$

$$= 3.15 \times 10^{12}\ \text{s}.$$

Since the speed of the proton is very close to the speed of light, $t_2 - t_1$ is only fractionally greater than 10^5 years.

It will now be assumed that the proton is at rest at the origin of the inertial reference frame Σ', which is moving with uniform velocity v relative to Σ, as shown in Fig. 4.13(b). In Σ' the Galaxy moves to the left with uniform velocity v (Fig. 4.13(b)). Let the Galaxy start to pass the proton at Event 1 at $(x_1' = 0, t_1')$ in Σ', and let the other end of the Galaxy pass the proton at Event 2 at $(x_2' = 0, t_2')$ in Σ'. The time that the Galaxy takes to pass the proton, measured in Σ', is $t_2' - t_1'$. From the Lorentz transformations, since $x_2' = x_1' = 0$, we have

$$t_1 = \gamma\,(t_1' + v x_1'/c^2) = \gamma t_1',$$

$$t_2 = \gamma\,(t_2' + v x_2'/c^2) = \gamma t_2'.$$

Subtracting and dividing by γ, we have

$$t_2' - t_1' = (t_2 - t_1)\,/\gamma.$$

This is the normal formula for *time dilatation*. Putting $t_2 - t_1 = 3.15 \times 10^{12}$ s and $\gamma = 1.066 \times 10^{10}$, we obtain $t_2' - t_1' = 3.15 \times 10^{12}\ \text{s}/1.066 \times 10^{10} = 296$ s. The time interval $t_2' - t_1' = 296$ s is the *proper* time interval between events 1 and 2, which can be measured by one clock at rest with the proton at the origin of Σ'. On the other hand, the time interval $t_2 - t_1 = 3.15 \times 10^{12}$ s measured in Σ must be measured by two clocks, which are at rest at x_1 and x_2 in Σ in Fig. 4.13(a), and which must be synchronized in a way consistent with the theory of special relativity—for example by using radar methods.

An inertial observer at rest at the origin of Σ' in Fig. 4.13(b) would say that the Galaxy moved past the stationary proton for a time $t_2' - t_1'$ travelling at a speed $-v$. The observer in Σ' would conclude that the diameter d' of the moving Galaxy, measured in the direction in which the Galaxy is moving relative to Σ', is

$$d' = v\,(t_2' - t_1') = \frac{v\,(t_2 - t_1)}{\gamma} = \frac{v}{\gamma}\frac{d}{v}$$

$$= \frac{d}{\gamma}.$$

This is the standard expression for the Lorentz length contraction. In this example the proper length is the diameter d of the Galaxy measured in the inertial reference frame Σ in which the Galaxy is at rest. Substituting numerical values, we have

$$d' = \frac{d}{\gamma} = \frac{9.46 \times 10^{20}\,\text{m}}{1.066 \times 10^{10}} = 8.87 \times 10^{10}\,\text{m}.$$

This value for d' is less than the distance from the Sun to the Earth, which is 1.5×10^{11} m. This result shows that the diameter of the Galaxy, which is 10^5 light years in its rest frame, is measured, in the proton's rest frame to be less, in the direction in which the Galaxy is moving relative to the proton, than the dimensions of the solar system. Since, from the Lorentz transformations, $y = y'$, the measured dimensions of the Galaxy are the same in the y' direction in Σ' as in the y direction in Σ. The circular cross-section of the Galaxy, in the (x, y) plane of Σ, is measured to be of elliptical shape in the (x', y') plane of Σ', if the measurements are carried out at the same time in Σ'. This is illustrated in Figs. 4.13(a,b). (See Problem 4.12).

Example 2

A spaceship is moving directly away from the Earth with uniform velocity. A radar pulse is transmitted from the Earth at 12.00 GMT. The pulse reflected from the rear of the spaceship is received back at the radar base at 12.02 GMT. The pulse reflected from the front of the spaceship is received back at the radar base 2 μs later. A second radar pulse is transmitted at 12.04 GMT and, after reflection at the rear end of the spaceship, is received back at base at 12.18 GMT. Find the speed and proper length of the spaceship.

Solution If it returns to the radar base at 12.02 GMT, the radar pulse transmitted at 12.00 GMT reaches the rear of the spaceship at 12.01 GMT. The distance travelled by the radar signal in the 60 s between 12.00 and 12.01 GMT is $60c$, or 1.8×10^{10} m. Thus at 12.01 GMT the rear of the spaceship is 1.8×10^{10} m from the Earth. Similarly, the radar signal transmitted at 12.04 GMT reaches the rear of the spaceship at 12.11 GMT, when it is $7 \times 60c$ or 12.6×10^{10} m from the Earth. In the 10×60 s between 12.01 and 12.11 GMT the spaceship travels a distance of $6 \times 60c$. Hence its speed is $0.6c$ in the laboratory frame Σ.

The first radar signal is reflected from the rear of the spaceship at 12.01 GMT and from the front 1 μs later. If v is the speed of the spaceship then in 1 μs the front of the spaceship moves a distance $10^{-6}\,v$, so that, after passing the rear of the spaceship, the radar pulse covers a distance $l + 10^{-6}\,v$ before reflection at the front end 1 μs later, where l is the length of the *moving* spaceship in Σ. In 1 μs the radar signal travels a distance $10^{-6}c$. Hence

$$10^{-6}\,c = l + 10^{-6}\,v.$$

Since $v = 0.6c$, we have

$$l = 10^{-6} \times 3 \times 10^{8}\,(1 - 0.6)\ \text{m} = 120\,\text{m}.$$

This is the length of the moving spaceship. From equation (4.51) for the Lorentz contraction,

$$l = l_0\,(1 - v^2/c^2)^{1/2}.$$

Therefore

$$l_0 = l\,(1 - v^2/c^2)^{-1/2} = 120 \times [1 - (0.6)^2\,]^{-1/2}\ \text{m}$$

$$= 150\,\text{m}.$$

Hence the proper length of the spaceship is 150 m. This example illustrates how one could determine the length of a moving body using radar methods.

4.8 THE DOPPLER EFFECT (*K* - CALCULUS)

In 1842 Doppler suggested that the motion of a light source affects the wavelength and frequency of the light reaching the observer. The *K*-calculus method is really an example of the Doppler effect. Consider the example shown previously in Fig. 4.4(b) and again in Fig. 4.14(a). Mary's spaceship passes John's spaceship at $t = t' = 0$ and moves away with uniform velocity v relative to it. Let Mary transmit radio signals at times $t' = 0$, T_0', $2T_0'$ etc. on her clock. She measures the frequency of the transmission of these signals from her spaceship to be $\nu_0 = 1/T_0'$. Let these signals reach John's spaceship at times $t = 0$, T, $2T$, $3T$ etc. on his clock (Fig. 4.14(a)). According to equation (4.12)

$$T = KT_0'.$$

Hence John measures the frequency of signals reaching him to be

$$\nu = \frac{1}{T} = \frac{1}{KT_0'} = \frac{\nu_0}{K},$$

$$= \nu_0 \left(\frac{1 - v/c}{1 + v/c}\right)^{1/2}, \tag{4.54}$$

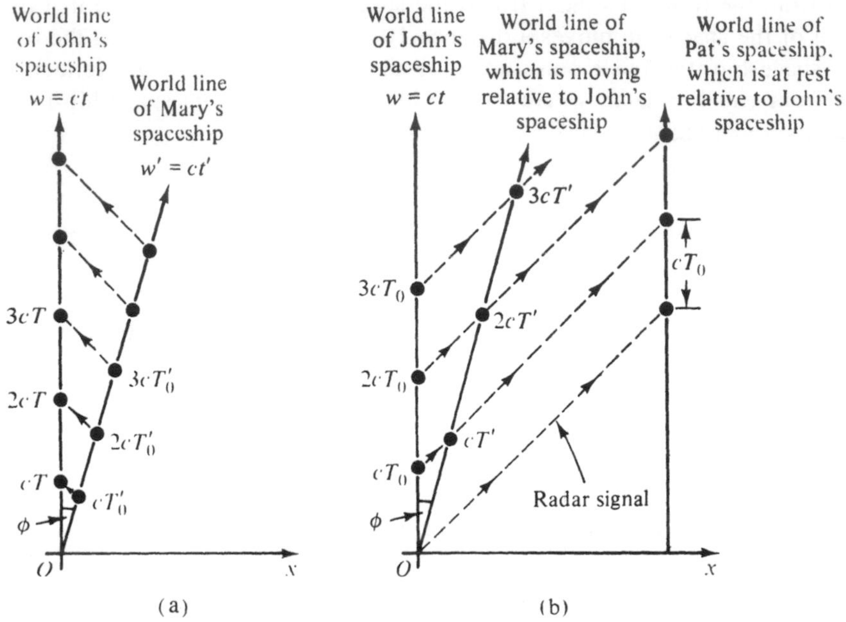

Fig. 4.14 (a) Mary transmits radio signals at times 0, T_0', $2T_0'$, $3T_0'$ etc. on her clock. These signals reach John's spaceship at times 0, T, $2T$, $3T$ etc. on his clock, where $T = KT_0'$. (b) John transmits radio signals from his spaceship at times 0, T_0, $2T_0$ etc. on his clock. The world lines of these signals are shown dashed. Mary is in a spaceship moving away from John's spaceship with uniform velocity v. John's radio signals reach Mary's spaceship at times 0, T', $2T'$ etc. on her clock. Pat's spaceship is at rest relative to John's spaceship. Pat receives John's radio signals at time intervals of T_0 on Pat's clock.

where K is given by equation (4.16) and ν_0 is the frequency of the signals measured in the inertial reference frame in which the source of the signals is at rest. One can treat each successive maximum of a monochromatic light wave as a signal. Hence equation (4.54) should hold for light. In this case the light source on Mary's spaceship is moving *away* from the observer (John). Since the speed of light is invariant, $\lambda_0\nu_0 = c$ and $\lambda\nu = c$. Hence

$$\lambda = \lambda_0 \left(\frac{1 + v/c}{1 - v/c} \right)^{1/2} . \tag{4.55}$$

Thus when the light source is moving away from the observer, the frequency of the light reaching the observer decreases, and the wavelength increases. For example, if $v = 0.6c$ then the observed frequency is

$$\nu = \nu_0 \left(\frac{1 - 0.6}{1 + 0.6} \right)^{1/2} = \nu_0 \left(\frac{0.4}{1.6} \right)^{1/2} = \tfrac{1}{2}\nu_0 .$$

The observed wavelength λ is $\lambda = 2\lambda_0$. Hence a yellow light source of wavelength $\lambda_0 = 600$ nm (6000 Å) would have a measured wavelength of 1200 nm (12 000 Å) if it moved directly away from the observer with a speed of $v = 0.6c$. This wavelength is in the infrared region.

If $v << c$ then, expanding equation (4.54) using the binomial theorem, ignoring terms of order v^2/c^2, we have, when the source is moving away from the observer,

$$\nu = \nu_0\,(1 - v/c)^{1/2}\,(1 + v/c)^{-1/2}$$

$$\approx \nu_0\left(1 - \frac{1}{2}\frac{v}{c}\right)\left(1 - \frac{1}{2}\frac{v}{c}\right),$$

and thus

$$\nu \approx \nu_0\,(1 - v/c). \tag{4.56}$$

Similarly, if $v << c$, when the source is moving away from the observer, we have

$$\lambda \approx \lambda_0\,(1 + v/c). \tag{4.57}$$

This is in agreement with the classical non-relativistic result.

Let astronaut Pat be in a spaceship that is at *rest* relative to John's spaceship (Σ). Let Pat be farther away from John's spaceship than Mary. Pat's world line is shown in Fig. 4.14(b). It is parallel to the ct axis (John's world line). Consider the radio signals transmitted from John's spaceship at times 0, T_0, $2T_0$, etc. on John's clock. Since Pat is at rest relative to John, the time between successive signals reaching Pat should be T_0 on Pat's clock. Let Mary transmit radio signals whenever she receives one from John at times 0, T', $2T'$ etc. on her clock, where, from equation (4.10), $T' = KT_0$. These signals should also reach Pat at intervals of T_0 on Pat's clock, just as if there had been no intermediate stage when Mary received and retransmitted radar signals without time delay. Hence if Mary transmits signals of frequency $\nu_0 = 1/T'$ measured by her clock, the frequency of the signals reaching Pat must be $\nu = 1/T_0$, when Mary is approaching Pat. Since, from equation (4.11), $T_0 = T'/K$, substituting for K from equation (4.16) gives

$$\nu = \frac{1}{T_0} = \frac{K}{T'} = K\nu_0$$

$$= \nu_0\left(\frac{1 + v/c}{1 - v/c}\right)^{1/2}. \tag{4.58}$$

In this case the light source on Mary's spaceship is approaching the observer, who is Pat in this case. Note that if the frequency of light from a receding light source decreases from ν_0 to ν_0/K, the frequency of light from the same light source, when the light source is approaching the observer with the same uniform velocity, increases from ν_0 to $K\nu_0$.

According to equation (4.58), when a light source is approaching an observer the frequency of the light increases and the wavelength decreases. For example, if $v = 0.6c$ then $\nu = 2\nu_0$ and $\lambda = \frac{1}{2}\lambda_0$; thus in this case, for yellow light of wavelength $\lambda_0 = 600$ nm, $\lambda = 300$ nm and the light reaching the observer would be in the near-ultraviolet. As v tends to c, the frequency of light from a source moving towards an observer goes farther and farther into the ultraviolet, reaching the X-ray and γ-ray regions.

The full relativistic theory of the Doppler effect will be developed in Section 5.6 using the wave theory of light and in Section 6.7.2 using the photon theory of light.

4.9 DISCUSSION OF SPECIAL RELATIVITY

It has been shown how Newtonian mechanics and the concept of absolute time must be abandoned in favour of the theory of special relativity. It has also been shown that the predictions of special relativity about time dilatation are in agreement with experiments carried out in the laboratory. In principle the theory of special relativity is always better than Newtonian mechanics, but in the mechanical phenomena in our normal daily lives it is an unnecessary over-elaboration to use special relativity, since Newtonian mechanics and the concept of absolute time are satisfactory *approximations*. Special relativity is important in electricity and optics, and absolutely essential in high-energy nuclear and particle physics.

It is often asked whether length contraction is real. What the principle of relativity says is that the laws of physics are the same in all inertial reference frames moving relative to each other, but the measures of particular quantities may be different in different inertial reference frames. For example, if a ball rolls on the deck of a ship that is moving with uniform velocity relative to the Earth, the speed of the ball relative to the ship is different to its speed relative to the Earth. Is this change real? According to the theory of special relativity, not only will the velocity of the ball relative to the ship and the Earth be different, but so will the measures of the length of the ship, the times of events etc. in the two inertial reference frames. The laws of physics are the same in both systems, but the numerical values of physical quantities are different.

Events happen independently of the choice of any particular inertial reference frame. As an example, consider a game of tennis played on the deck of a ship moving with uniform velocity relative to the Earth. An inertial observer at rest in the laboratory frame (Σ) and an inertial observer at rest on the ship (Σ') would measure different speeds for the tennis ball. In the context of special relativity, they would also measure different dimensions for the tennis court in the direction of motion of the ship and they would measure different times for the lengths of the rallies, but they would agree on whether a tennis ball landed in or out of court, since this is an event which is independent of the choice of any particular inertial reference frame. The inertial observers in Σ and Σ' should use the same laws of physics in Σ and Σ' and should get consistent results in both Σ and Σ'. The Lorentz transformations relate the results obtained in Σ and Σ'.

It has been assumed throughout this chapter that Σ' (Mary's spaceship) always moves with *uniform* velocity v relative to Σ (John's spaceship). The theory of special relativity is only applicable to inertial reference frames.

★ 4.10 INTERVALS BETWEEN EVENTS

Time intervals and lengths of moving bodies are invariants in Newtonian mechanics. This is not true in special relativity. However, a certain combination of the time interval between two events and the spatial separation of the two events is an invariant.

Consider two events, one at (x, y, z) at a time t and the other at $(x + \delta x, y + \delta y, z + \delta z)$ at a time $t + \delta t$ relative to the inertial reference frame Σ, where, in general the increments δx, δy, δz, and δt are not infinitesimals but have finite values. From the Lorentz transformations, in an inertial frame Σ' moving with uniform velocity v relative to Σ,

$$x' = \gamma (x - vt),$$
$$y' = y, \qquad z' = z,$$
$$t' = \gamma (t - vx/c^2)$$

and

$$x' + \delta x' = \gamma [x + \delta x - v (t + \delta t)],$$
$$y' + \delta y' = y + \delta y, \qquad z' + \delta z' = z + \delta z,$$
$$t' + \delta t' = \gamma [t + \delta t - v (x + \delta x) /c^2].$$

Subtraction gives

$$\delta x' = \gamma (\delta x - v \delta t),$$
$$\delta y' = \delta y, \qquad \delta z' = \delta z,$$
$$\delta t' = \gamma [\delta t - v \delta x/c^2].$$

Notice the Lorentz transformations can be used to transform the increments δx, δy, δz and δt. Consider

$$\delta x'^2 + \delta y'^2 + \delta z'^2 - c^2 \delta t'^2 = \gamma^2 (\delta x - v \delta t)^2 + \delta y^2 + \delta z^2 - c^2 \gamma^2 (\delta t - v \delta x/c^2)^2$$
$$= \gamma^2 [\delta x^2 (1 - v^2/c^2) - c^2 \delta t^2 (1 - v^2/c^2)] + \delta y^2 + \delta z^2.$$

Since

$$\gamma^2 = (1 - v^2/c^2)^{-1},$$

we have

$$\delta x'^2 + \delta y'^2 + \delta z'^2 - c^2 \delta t'^2 = \delta x^2 + \delta y^2 + \delta z^2 - c^2 \delta t^2,$$

so that

$$\delta s^2 = \delta x^2 + \delta y^2 + \delta z^2 - c^2 \delta t^2$$
$$= \delta x^2 + \delta y^2 + \delta z^2 - \delta w^2 \qquad (4.59)$$

is an invariant; that is, it has the same numerical value in both Σ and Σ'. The quantity δs is called *the interval* between the events.† According to the theory of special relativity, if two events occur, two observers, one at rest in Σ and one at rest in Σ', will record different distance and time separations between the two events, but they will record the same *interval*. If δs^2 is an invariant then it cannot change sign on transformation from one inertial reference frame to another.

If δs^2 is positive, that is if $\delta x^2 + \delta y^2 + \delta z^2 > c^2\,\delta t^2$, then the interval δs between the two events is called a *spacelike* interval. If δs^2 is negative, that is if $\delta x^2 + \delta y^2 + \delta z^2 < c^2\,\delta t^2$ then δs is called a *timelike* interval.

If two events occur on the path of a light signal, the distance $(\delta x^2 + \delta y^2 + \delta z^2)^{1/2}$ between them is equal to the product of the speed of light c and the time interval δt between the two events, which is equal to $c\,\delta t$. In this special case δs^2 is zero, so the interval δs between two events on the path of a light signal is zero and is consequently called a *null* interval.

To simplify the discussion, in the rest of this section it will be assumed that δy and δz are always zero.

Timelike intervals Consider the equation

$$\delta x' = \gamma\,(\delta x - v\,\delta t).$$

The distance between the two events is zero in Σ', that is $\delta x' = 0$, if

$$\delta x = v\,\delta t,$$

or

$$\frac{v}{c} = \frac{\delta x}{c\,\delta t}. \tag{4.61}$$

If δs is timelike then δs^2 is negative, $c^2\,\delta t^2 > \delta x^2$ and $c\,|\,\delta t\,| > |\,\delta x\,|$, and so v/c is less than unity in equation (4.61). Thus if δs is a timelike interval, it is always possible to find a reference frame in which $\delta x'$ is zero—that is, a reference frame in which the two events are measured to occur at the same point of space. In this reference frame the time interval between the two events is the proper time interval between them. For example, let one event occur at $x = 0$, $t = 0$ and the other at $x = 3 \times (3 \times 10^8)$ m $= 3c$, $t = 5$ s (or $w = ct = 5c$), relative to the inertial reference frame Σ. From equation (4.59),

$$\delta s^2 = \delta x^2 - c^2\,\delta t^2 = \delta x^2 - \delta w^2 = 9c^2 - 25c^2 = -16c^2.$$

---

† Some authors prefer to define the interval δs using the relation

$$\delta s^2 = c^2\,\delta t^2 - \delta x^2 - \delta y^2 - \delta z^2. \tag{4.60}$$

In this case δs^2 is positive for a timelike interval and negative for a spacelike interval. Equation (4.59) has been chosen here so as to fit in with our initial complex-number treatment of 4-vectors in Chapter 7. The reader should always check what definition is being used for the interval by other authors.

Hence δs is a timelike interval. Substitution in equation (4.61) shows that $\delta x'$ is zero in a reference frame moving with velocity $v = \frac{3}{5}c$ relative to Σ. In this reference frame the proper time interval between the two events, denoted by $\delta\tau$, is

$$\delta\tau = \delta t' = \gamma(\delta t - v\,\delta x/c^2) = 1.25\,(5 - \tfrac{3}{5}c \times 3c/c^2)\text{ s} = 4\text{ s}.$$

If δs were spacelike then $\delta x^2 > c^2\,\delta t^2$, $|\delta x| > c\,|\delta t|$, and in equation (4.61) v/c would have to be greater than unity. Hence it is impossible to find a reference frame in which $\delta x'$ is zero if the interval δs is spacelike.

Spacelike intervals In the equation

$$\delta t' = \gamma\left(\delta t - \frac{v\,\delta x}{c^2}\right) = \gamma\,\delta t\left(1 - \frac{v\,\delta x}{c^2\,\delta t}\right) \tag{4.62}$$

the time interval $\delta t'$ is zero if

$$\delta t = \frac{v\,\delta x}{c^2},$$

or

$$\frac{v}{c} = \frac{c\,\delta t}{\delta x}. \tag{4.63}$$

If δs is spacelike then δs^2 is positive, $\delta x^2 > c^2\,\delta t^2$ and $|\delta x| > c\,|\delta t|$. Hence if δs is spacelike, it is possible to satisfy equation (4.63) with v less than c and find an inertial reference frame in which $\delta t'$ is zero—that is, a reference frame in which the events are simultaneous. For example, if one event occurs at $x = 0$, $t = 0$ and the other at $x = 5c$, $t = 3$ s (or $w = ct = 3c$) relative to Σ then

$$\delta s^2 = \delta x^2 - c^2\,\delta t^2 = (5c)^2 - c^2 3^2 = +16c^2,$$

so that δs is spacelike. If in equation (4.62) we put $\delta x = 5c$, $\delta t = 3$ s and $v = \frac{3}{5}c$, we find that $\delta t'$ is zero. If $v/c > c\,|\delta t|/|\delta x|$, but v/c remains less than unity, then $\delta t'$ is opposite in sign to δt, and the time order of the two events is reversed. For example, if for $\delta t = 3$ s, $\delta x = 5c$ we put $v = \frac{4}{5}c$ in equation (4.62), we find that $\delta t' = -\frac{5}{3}$ s so that in a reference frame Σ' moving with velocity $\frac{4}{5}c$ relative to Σ, the event that is at $x = 0$, $t = 0$ relative to Σ is measured to occur $1\frac{2}{3}$ s after the event that is at $x = 5c$, $t = 3$ s relative to Σ.

At first sight it seems to go against our common sense to find that the measured temporal order of events can be different in different inertial reference frames. This reversal of the temporal order of events can only occur if δs^2 is spacelike; that is, if $\delta x^2 > c^2\,\delta t^2$ or $|\delta x| > c\,|\delta t|$. If $|\delta x| > c\,|\delta t|$, it is impossible to send a light signal the distance $|\delta x|$ between the events in the time interval $|\delta t|$. Since energy and momentum cannot be transmitted at a speed exceeding the speed of light, if δs^2 is positive then what happens at one event cannot influence what happens at the other. Since, when δs^2 is positive, one event cannot influence what happens at the other, it does not matter which event is measured to occur first. There can be no causal connection between the events. For example, one cannot influence what is going on at present, or in the next few years, on a star a thousand light years away, since, if one sent a radio signal now, it would not reach the star for a thousand years. One cannot influence what is going on on the Moon in one second's time, since a radio signal would take 1.28 s to reach there. When δs^2 is timelike, it is possible to send a light or radio signal from one event to the other. For example, one could influence what happens on the Moon in a month's time by sending a spaceship to get there in less than a month. Since $v < c$, if δs^2 is negative and $|\delta x| < c|\delta t|$ then in equation (4.62) $v\delta x/c^2\delta t < 1$ and $\delta t'$ always has the same sign as δt, so that, when δs^2 is timelike and causal connection between the two events is possible, there is a definite time order for the two events, which is the same in all inertial reference frames. These conclusions can be illustrated graphically.

Consider first events along the x axis of the inertial reference frame Σ. It will be assumed that one event is here and now at the point $x = 0$ at the time $t = 0$ at the origin of Σ. A space–time diagram will be used to plot the positions and times of other events; see Fig. 4.15. The world lines of light rays passing the origin $x = 0$ at $t = 0$ are given by $x = \pm ct$ and are at 45° to the x and $w = ct$ axes, as shown in Fig. 4.15. Events in the region labelled FUTURE in the figure can be reached from the origin with speeds less than the speed of light, so that the reader could influence all events in this part of the figure. The world line of the reader's future

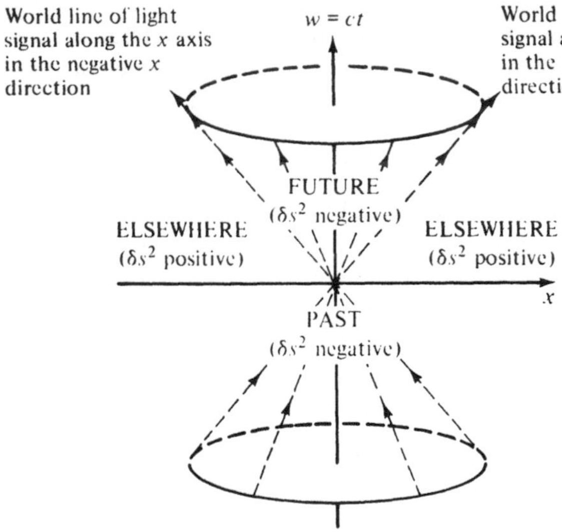

Fig. 4.15 One event takes place at the origin at $x = 0$, $t = 0$. If another event takes place in either of the regions marked PAST or FUTURE then δs^2 is negative, light signals can go from one of the events to the other, and one of the events can influence what happens at the other event. If the other event is in the ELSEWHERE regions then δs^2 is positive and one event cannot influence what happens at the other.

is also in this region. Events in the region labelled PAST could have sent signals to reach the reader at the origin at or before $t = 0$ and could influence what happens at $x = 0$, $t = 0$. The world line of the reader's past history is in the PAST region. Events in the regions labelled ELSEWHERE cannot send light signals to reach $x = 0$ by $t = 0$, neither can light signals be sent from the origin at $x = 0$, $t = 0$ to reach these events to have any influence on them. The intervals between the event at the origin, $x = 0$ at $t = 0$, and events in either the PAST or FUTURE regions are timelike ($\delta s^2 < 0$), and the intervals between the event at the origin and events in the ELSEWHERE regions are spacelike ($\delta s^2 > 0$). For events on the world lines of light signals, δs^2 is zero.

For motion in two dimensions we must use the coordinates (x, y, $w = ct$). In this case light signals, which leave the origin of Σ at $t = 0$, lie on the surface of a cone whose axis is along the $w = ct$ axis, that is of semi-angle 45° and whose apex is at the origin in Fig. 4.15. Since $\delta s^2 = 0$ for events on the world lines of light signals, the surface of the cone is called the *null cone* or the *light cone*. One can visualize the light cone by assuming that the y axis is perpendicular to the paper in Fig. 4.15. The surface of the light cone is generated by rotating the dashed world lines, which are at 45° to the x axis, around the $w = ct$ axis as sketched in the figure. The FUTURE region, where $\delta s^2 < 0$, is inside the forward cone ($w > 0$). The PAST region is inside the cone bounded by light rays, which reach the origin at $x = 0$, $y = 0$ at the time $w = ct = 0$. The ELSEWHERE region is outside the light cones in the region where $\delta s^2 > 0$. For motion in three dimensions, one must use the four-dimensional space (x, y, z, $w = ct$).

The invariance of the interval defined by equation (4.59) under a Lorentz transformation is a very important result, which is often used as the starting point for more advanced treatments of special relativity, for example that in Landau and Lifshitz (1975). The invariance of timelike intervals will be used extensively in our development of 4-vector methods in Chapter 7.

★ 4.11 MINKOWSKI DIAGRAMS

In this chapter, in particular in Sections 4.2–4.4, when space–time diagrams were drawn, a rectangular coordinate system with axes Ox and Ow, where $w = ct$, was used to plot the positions and times of events relative to Σ, the inertial reference frame in which John's spaceship was at rest at the origin. It will now be shown how the *same* world points and world lines can be used to represent the positions and times of the events relative to the inertial reference frame Σ' that moves with uniform velocity v relative to Σ and in which Mary's spaceship is at rest at the origin, provided that oblique axes are used to represent the positions and times of events relative to Σ' and the scales for x' and $w' = ct'$ are chosen appropriately.

In Fig. 4.16(a) rectangular axes Ox, Ow are used to represent the positions and times of events relative to Σ. On the Ow', or time, axis of the oblique coordinate system that will be used to represent the positions and times of events relative to Σ', we must have $x' = 0$. From the Lorentz transformations, if $x' = 0$ then

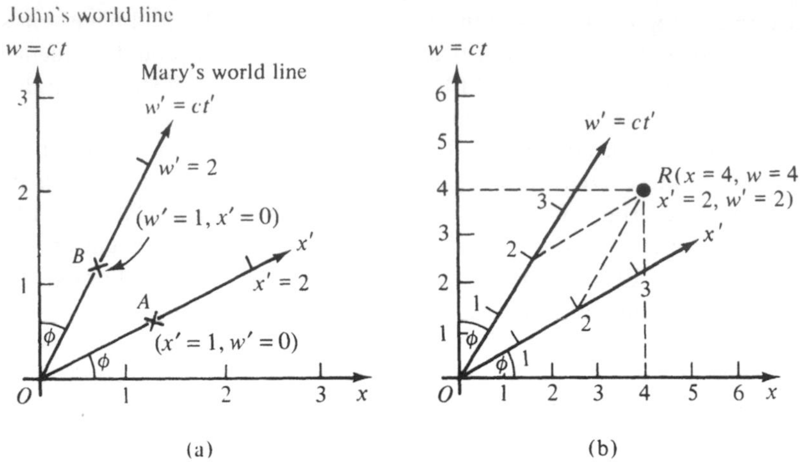

Fig. 4.16 (a) A Minkowski diagram. The rectangular axes Ox, Ow are used to represent the coordinates and times of events relative to Σ, the inertial reference frame in which John's spaceship is at rest at the origin. The oblique axes Ox', Ow' can be used to represent the positions of the events relative to the inertial frame Σ' in which Mary's spaceship is at rest at the origin. (b) A Minkowski diagram for the case when $v = 0.6c$. The event $x = 4$, $w = 4$ is shown. Relative to the oblique axes Ox', Ow' this event has coordinates $x' = 2$, $w' = 2$.

$$x' = \gamma(x - vt) = 0,$$

so that, relative to the Ox, Ow axes of Σ, the Ow' axis of Σ' is given by

$$x = vt = \frac{v}{c}ct = \frac{v}{c}w.$$

The line $x = (v/c)w$ is represented by Ow' in Fig. 4.16(a). It is at an angle $\phi = \arctan(v/c)$ to the Ow axis in the figure. This is the world line of Mary's spaceship relative to the Ox, Ow rectangular axes. The angle ϕ varies depending on the velocity of Mary's spaceship relative to John's (Σ).

Along the Ox' axis of the oblique coordinate system that will be used to represent the positions and times of events relative to Mary's spaceship (Σ'), $w' = ct' = 0$. From the Lorentz transformations, if $t' = 0$ then

$$t' = \gamma(t - vx/c^2) = 0$$

so that, relative to the Ox, Ow axes of Σ, the Ox' axis of Σ' is given by

$$w = ct = \frac{vx}{c}.$$

The line $w' = 0$ is represented by Ox' in Fig. 4.16(a). It makes an angle ϕ with the Ox axis, where $\phi = \arctan(v/c)$, as shown in the figure. The lines Ox' and Ow' will now be used as oblique axes to represent the coordinates and times of events relative to Σ'.

The unit of length along the Ox' axis is the distance from the origin O to the point A at $x' = 1$, $w' = ct' = 0$ in Fig. 4.16(a). Relative to the rectangular axes Ox, Ow, from the Lorentz transformations, the point A has coordinates

$$x = \gamma(x' + vt') = \gamma x' = \gamma,$$
$$w = ct = \gamma(ct' + vx'/c) = \gamma v/c.$$

Using the scales appropriate to the rectangular axes Ox, Ow, the distance of the point A from the origin of Σ is

$$g_1 = (\gamma^2 + \gamma^2 v^2/c^2)^{1/2} = \gamma(1 + v^2/c^2)^{1/2}$$
$$= \left(\frac{1 + v^2/c^2}{1 - v^2/c^2}\right)^{1/2}. \tag{4.64}$$

Similarly, the unit of w' along Ow' is the distance from the origin to the point B at $w' = ct' = 1$, $x' = 0$ in Fig. 4.16(a). From the Lorentz transformations, for the point B in Σ we have

$$x = \gamma(x' + vt') = \gamma(x' + vw'/c) = \gamma v/c,$$
$$w = ct = \gamma(ct' + vx'/c) = \gamma.$$

Using the scales appropriate to the Ox, Ow rectangular axes, the distance from O to B is given by

$$g_2 = (\gamma^2 + \gamma^2 v^2/c^2)^{1/2} = \gamma(1 + v^2/c^2)^{1/2} = \left(\frac{1 + v^2/c^2}{1 - v^2/c^2}\right)^{1/2}. \tag{4.65}$$

If we now use Ox' and Ow' in Fig. 4.16(a) as oblique axes and change our scales of x' and w' according to equations (4.64) and (4.65) then we can use the *same* world points and world lines to represent the positions and times of events relative to Σ', using Ox' and Ow' as oblique axes, and relative to Σ, using the rectangular Ox, Ow axes. A proof is given by Rosser (1964). We shall merely use a simple numerical example to illustrate the invariance of world points.

The reader should make a large-scale drawing of Fig. 4.16(b). Start by drawing rectangular axes Ox, Ow to represent the positions and times of events relative to Σ. Let Σ' move with velocity $v = \frac{3}{5}c$ relative to Σ. Hence $\phi = \arctan \frac{3}{5}$. Draw the Ox' and Ow' axes. For example, join the origin to the point $x = 5$, $w = 3$ to get the line Ox'. If $v/c = \frac{3}{5}$ then

$$g_1 = g_2 = \left(\frac{1 + v^2/c^2}{1 - v^2/c^2}\right)^{1/2} = \left(\frac{1 + 3^2/5^2}{1 - 3^2/5^2}\right)^{1/2} = 1.46.$$

Mark out units of $x' = 1, 2, 3, 4$ etc. along Ox', using the distance from O to $x' = 1$ as 1.46 on the scale used originally for the rectangular axes Ox, Ow. For example, if the unit of x is 1 cm and the unit of $w = ct$ is 1 cm in the Ox, Ow rectangular axes then the distance from O to $x' = 1$ along Ox' is 1.46 cm. Similarly, mark out $w' = 1, 2, 3, 4$ etc. on the Ow' axis, making the distance from O to

$w' = 1$ equal to 1.46 times the length from O to $w = 1$. Plot the world point R at $x = 4$, $w = ct = 4$ relative to Σ, using the Ox, Ow axes, as shown in Fig. 4.16(b). To determine the coordinates of this world point relative to the Ox', Ow' oblique axes, draw lines through R parallel to the Ow' and Ox' axes, as shown in Fig. 4.16(b). It can be seen from the graph that these lines cut the oblique axes at $x' = 2$ and $w' = 2$ respectively. If the position of the world point R need not be changed, we should expect the coordinates of the event at $x = 4$, $w = 4$ relative to Σ to be at $x' = 2$, $w' = 2$ relative to Σ'. For $v = \frac{3}{5}c$

$$\gamma = (1 - 3^2/5^2)^{-1/2} = \tfrac{5}{4}.$$

Using the Lorentz transformations, for $x = 4$, $w = ct = 4$,

$$x' = \gamma\left(x - \frac{v}{c}ct\right) = \tfrac{5}{4}\left(4 - \frac{3}{5}\times 4\right) = 2,$$

$$t' = \gamma(t - vx/c^2),$$

$$w' = ct' = \gamma\left(ct - \frac{v}{c}x\right) = \tfrac{5}{4}\left(4 - \frac{3}{5}\times 4\right) = 2.$$

This shows that, with the use of the oblique axes Ox' and Ow' and the choice of lengths of units of x' and w' given by equations (4.64) and (4.65) respectively, the positions of world points and world lines need not be changed in Fig. 4.16(b), and the *same* world points and world lines can be used to represent the positions and times of events relative to Σ using the Ox, Ow axes, and relative to Σ' using the oblique Ox', Ow' axes. This is true whatever the value of v, provided that it is less than c. Therefore the *same* world points and world lines can be used to represent the positions of events relative to all inertial reference frames moving along the x axis of Σ and whose origins coincide with the origin of Σ at $t = t' = 0$. Thus John and Mary can use the same world points on the same graph, if John uses the rectangular axes Ox, Ow and Mary uses the oblique axes Ox', Ow' provided that Mary uses different scales to John for x' and w', namely the scales given by equations (4.64) and (4.65). (We could have used rectangular axes to represent the positions and times of events relative to the inertial reference frame Σ' in which Mary's spaceship is at rest, but we would then have had to use oblique axes to represent the world points relative to the reference frame Σ in which John's spaceship is at rest.)

Diagrams such as Figs. 4.16(a,b) are known as *Minkowski diagrams*, in honour of H. Minkowski, who first developed the method in 1908.

If a particle moves in the (x, y) plane of Σ, one needs three dimensions (x, y, $w = ct$) to represent its displacement at various times. For motion in three dimensions, one needs four-dimensional space–time diagrams, using (x, y, z, $w = ct$). Since $y' = y$ and $z' = z$, there is no need to change the units of y' and z' and the Oy and Oy' and the Oz and Oz' axes are coincident. Minkowski suggested that the four-dimensional space (x, y, z, $w = ct$) represents "the world". In this world the world points and world lines are invariants.

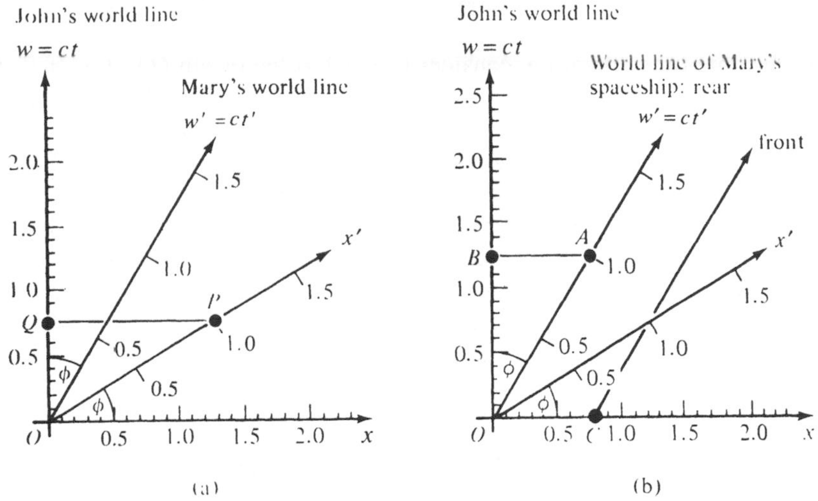

Fig. 4.17 (a) Minkowski diagram to illustrate non-absolute simultaneity. (b) Minkowski diagram to illustrate time dilatation and length contraction.

Minkowski diagrams can be used to illustrate non-absolute simultaneity, time dilatation and length contraction. We shall use the same conditions as shown in Fig. 4.16(b), and shown again on an enlarged scale in Figs. 4.17(a,b). The reader should make large-scale diagrams to check our conclusions. It will be recalled that Mary's spaceship (Σ') moves with uniform velocity $v = 0.6c$ relative to John's spaceship (Σ), such that $\tan\phi = 0.6$, $\gamma = \frac{5}{4}$ and $g_1 = g_2 = 1.46$. (Note the different lengths of the units of x and x' and of w and w' in Figs. 4.17(a,b), consistent with $g_1 = g_2 = 1.46$.) The reader should remember that the world points are fixed on a Minkowski diagrams for all inertial reference frames.

Consider two events, one at the origin O at $x' = 0$, $w' = ct' = 0$ relative to Σ' and the other at world point P in Fig. 4.17(a) at $x' = 1$, $w' = ct' = 0$ on the Ox' axis of the oblique axes (Σ'). These events are simultaneous at $t' = 0$ relative to Σ'. The event at the origin O is at $w = ct = 0$ relative to Σ. In order to determine the time of the event at the world point P relative to the rectangular axes (Σ), draw a line through P parallel to the Ox axis to cut the Ow axis at the world point Q. From Fig. 4.17(a), it can be seen that at Q, graphically $w = ct = 0.75$. Thus, relative to Σ, the event at P does not occur at the same time as the event at O at the origin, though the events at O and P are simultaneous at $t' = 0$, relative to Σ'. From the Lorentz transformations for $x' = 1$, $t' = 0$,

$$t = \gamma(t' + vx'/c^2) = \gamma v/c^2,$$

$$w = ct = \gamma v/c = \tfrac{5}{4} \times \tfrac{3}{5} = 0.75.$$

Thus the graphical method using a Minkowski diagram gives the same result as the Lorentz transformations. It is left as an exercise for the reader to show that events at $x = 0$ and $x = 1$ at the time $w = ct = 0$ relative to Σ are not simultaneous relative to Σ'.

To illustrate time dilatation, consider Event 1 at the origin O in Fig. 4.17(b) at $x' = 0$, $w' = ct' = 0$ relative to the oblique axes (Σ') and Event 2 at world point A at $x' = 0$, $w' = ct' = 1$ relative to Σ'. The time of the event at the origin O is $w = ct = 0$ relative to the rectangular axes (Σ). To determine the time of Event 2 at A, relative to the rectangular axes (Σ), draw a line through A parallel to the Ox axis to cut the Ow axis at world point B. It can be seen from Fig. 4.17(b) that at B, $w = ct = 1.25$. Thus the time between the two events at world points O and A is $\delta w = c\,\delta t = 1.25$ relative to Σ and $\delta w' = c\,\delta t' = 1$ in the inertial reference frame Σ', in which the events occur at the same point $x' = 0$, illustrating time dilatation. From the Lorentz transformations, for $x' = 0$,

$$w = ct = \gamma\,(ct' + vx'/c) = \gamma ct'.$$

For Events 1 and 2, $t' = 0$ and $t' = 1$ respectively. Hence in Σ

$$\delta w = c\,\delta t = \gamma c\,\delta t' = \tfrac{5}{4} \times 1 = \tfrac{5}{4}$$

Thus the graphical method using a Minkowski diagram and the Lorentz transformations give the same result.

As an example of length contraction, assume that Mary's spaceship is of unit length, that its rear is at $x' = 0$ and its front at $x' = 1$ in Σ'. The world lines of the front and rear are shown in Fig. 4.14(b). To measure the length of Mary's spaceship relative to Σ, John must measure the distance between the world lines of its front and rear at the same time relative to Σ, say at $t = 0$. Thus, relative to the inertial reference frame Σ in which John's spaceship is at rest, the length of Mary's spaceship, measured at the time $t = 0$ in Σ, is equal to the distance OC between the world lines of the front and rear of Mary's spaceship, which is equal to 0.8 in Fig. 4.17(b). From the Lorentz transformations, if $x' = 1$ in Σ', at $t = 0$ in Σ we have

$$x' = \gamma\,(x - vt) = \gamma x,$$

$$x = \frac{x'}{\gamma} = \frac{1}{1.25} = 0.8.$$

Thus the graphical method using a Minkowski diagram again gives the same result as the Lorentz transformations.

It is left as an exercise for the reader to show graphically that time dilatation and length contraction are reciprocal, by showing that the event at $x = 0$, $w = ct = 1$ in Σ is at $w' = ct' = 1.25$ relative to Σ', and that if John's spaceship is of unit length in Σ then, relative to Σ', its length is 0.8. A reader interested in a fuller account of Minkowski diagrams is referred to Rosser (1964).

According to the theory of special relativity, the numerical values of lengths and time intervals are not absolute, but depend on the inertial reference frame chosen to represent the motions. However, it was shown in Section 4.10 that the interval between "events" is absolute—that is it has the same numerical value in all inertial frames of reference. The interval δs was defined by the relation

$$\delta s^2 = \delta x^2 + \delta y^2 + \delta z^2 - c^2\,\delta t^2.$$

Thus, although the spatial and temporal intervals between two events are not absolute, according to the theory of special relativity a certain combination of the two, namely the interval between them, is absolute. So are the "world points" and "world lines". This is probably the way in which one should interpret the opening remarks that Minkowski made in his address delivered to the Eightieth Assembly of German Natural Scientists and Physicians at Cologne, on 21 September 1908. In this address he introduced the geometrical interpretation of space and time for the first time. Minkowski said:

> The views of space and time which I wish to lay before you have sprung from the soil of experimental physics, and therein lies their strength. They are radical. Henceforth space by itself, and time by itself, are doomed to fade away into mere shadows, and only a kind of union of the two will preserve an independent reality.

In the 4-dimensional (x, y, z, ct) "world" the kinematics of particles can be developed using 4-dimensional methods. There is an alternative approach in which the four variables (x, y, z, ict) are used, where $i = \sqrt{(-1)}$. This approach is mathematically simpler and will be developed in Chapter 7.

REFERENCES

Landau, L.D. and Lifshitz, E.M. (1975) *The Classical Theory of Fields*. Pergamon Press, Oxford.

Rosser, W.G.V. (1964) *An Introduction to the Theory of Relativity*, Chap.6, Butterworth, London.

PROBLEMS

Assume that the velocity of light $c = 3.00 \times 10^8$ m s^{-1}.

4.1 Assume that the inertial reference frame Σ' moves with uniform velocity $0.6c$ relative to Σ along the x axis of Σ. Let the origins of Σ and Σ' coincide at $t = t' = 0$. Use the Lorentz transformations to do the following.

(a) Find the coordinates and times relative to Σ' of the events that have the following coordinates and times relative to Σ:

(i) $x = 4$ m, $t = 6$ s;

(ii) $x = 7 \times 10^8$ m, $t = 2$ s;

(iii) $x = 6 \times 10^{10}$ m, $t = 3$ s.

(b) Find the coordinates and times relative to Σ of the events that have the following coordinates and times relative to Σ':

(i) $x' = 10$ m, $t' = 4$ s;

(ii) $x' = 9 \times 10^9$ m, $t' = 4$ s;

(iii) $x' = 10^{11}$ m, $t' = 50$ s.

4.2 By what amount is the Earth shortened along its diameter (as measured by an observer at rest relative to the Sun) owing to its orbital motion around the Sun? (Take the velocity of the Earth to be 30 km s^{-1} and its radius to be 6371 km.)

4.3 A spaceship is moving at such a speed in the laboratory frame that its measured length is half its proper length. How fast is the spaceship moving relative to the laboratory system?

4.4 A rod of proper length L_0 is at rest in an inertial frame Σ'. The rod is inclined at an angle θ' to the x' axis in Σ'. What is the length of the rod in Σ, and what is its inclination to the x axis in Σ, if Σ' moves with uniform velocity v relative to Σ along the common x axis ?

4.5 Show that the average lifetime T_0 of a radioactive atom is $1/\lambda$, where λ is the decay constant. [*Hint*: If $\mathrm{d}N$ is the number of radioactive atoms that decay in the time interval between t and $t + \mathrm{d}t$ then the average (mean) life is $N_0^{-1}\int t\,\mathrm{d}N$. It follows from the definition of the decay constant λ that $\mathrm{d}N = \lambda N\,\mathrm{d}t = \lambda N_0 e^{-\lambda t}\mathrm{d}t$. Substitute for $\mathrm{d}N$ and integrate by parts between the limits $t = 0$ and $t = \infty$.]

4.6 If the mean lifetime of a muon (μ) when it is at rest is 2.2×10^{-6} s, calculate the average distance it will travel *in vacuum* before decay, if its velocity, is (a) $0.9c$; (b) $0.99c$; (*c*) $0.999c$. [*Hint*: Use equation (4.48).]

4.7 Calculate the average distance that a pion (π) will travel before it undergoes spontaneous decay, if its velocity is (a) $0.75c$; (b) $0.9c$; (c) $0.99c$; (d) $0.999c$; (e) $0.9995c$. The mean lifetime of a stationary pion is 2.55×10^{-8} s.

4.8 A beam of unstable particles, each of velocity $0.8c$ in the laboratory frame, passes through two particle detectors that are 20 m apart. It is found that, on average, when there are 1000 counts in the first detector there are only 100 counts in the second detector. Assuming that this decrease is due entirely to the decay of the particles in flight, estimate the mean lifetime of the unstable particles, measured in the inertial reference frame in which they are at rest.

4.9 What is the Doppler shift in the wavelength of the H_α (656.3 nm) line from a star moving directly away from the Earth with a velocity of 300 km s^{-1} ?

4.10 Calculate the wavelength (using the relativistic formula) for light of wavelength 500 nm when the source is approaching the observer with velocity (a) $0.1c$; (b) $0.9c$.

4.11 A physicist is arrested for driving through the red lights at a road junction. At the trial the physicist claims that the car was going so fast that the red light appeared green to the driver. "Plea accepted", said the judge, "but I fine you a pound for each kilometre per hour your speed exceeded the speed limit of 45 kilometres per hour". Calculate the fine, taking the wavelength of green light to be 530 nm and that of red light to be 630 nm.

4.12 Show that the circle $x^2 + y^2 = R^2$ in Σ is measured to be the ellipse

$$\frac{x'^2}{R^2/\gamma^2} + \frac{y'^2}{R^2} = 1$$

in the inertial reference frame Σ' that is moving along the x axis of Σ with uniform velocity v, if the measurements are carried out in Σ' at the time $t' = 0$.

4.13 A cosmic-ray muon is created at an altitude of 60 km in the Earth's atmosphere. The muon moves vertically downwards with a constant speed $0.9995c$ relative to the Earth. It lives for a time of 2.2×10^{-6} s in the inertial reference frame in which it is at rest.

(a) How long does the muon live in the laboratory frame ?

(b) How far does the muon travel relative to the Earth before it decays ?

(c) Through what length of the Earth's atmosphere does the muon pass in its lifetime, measured in the inertial reference frame in which the muon is at rest? (This is an example of length contraction.)

4.14 One event occurs at the origin of an inertial frame Σ at the time $t = 0$. Another event occurs at $x = 3c$, $y = z = 0$ at a time $t = 6$ s relative to Σ.

(a) What is the velocity (relative to Σ) of the inertial frame Σ' in which the two events are recorded at the same point in space ?

(b) What is the time interval between the events relative to Σ'?

4.15 One event occurs at the origin of the inertial frame Σ at $t = 0$. A second event occurs at the point $x = 5c$, $y = z = 0$ at a time $t = 4$ s relative to Σ. Find the velocities, relative to Σ, of the inertial reference frames in which (a) the events are simultaneous; (b) Event 2 precedes Event 1 by 1 s; (c) Event 1 precedes Event 2 by 1 s.

4.16 (a) An event occurs at $x = 3c$, $t = 4$ s in the inertial reference frame Σ. Use a Minkowski diagram to determine the coordinates and time of the event relative to a reference frame Σ' moving with uniform velocity $0.6c$ relative to Σ. Find the velocity relative to Σ of the reference frame in which the event occurs at $x' = 0$

(b) If an event is at $x = 4c$, $t = 3$ s in Σ, use a Minkowski diagram to find its coordinates relative to a reference frame moving with velocity $0.6c$ relative to Σ. Find the velocity relative to Σ of the reference frame in which the event occurs at $t' = 0$.

Check your answers using the Lorentz transformations.

4.17 In Σ events occur at the origin and at the point $(x, 0, 0)$ simultaneously at $t = 0$. The time interval between the events in the inertial reference frame Σ' that is moving with uniform velocity v relative to Σ along the common x axis is T. Show that the distance between the events in Σ' is $(x^2 + c^2T^2)^{1/2}$ and that $v = c\,(1 + x^2/c^2T^2)^{-1/2}$. [*Hint*: Use the invariance of the interval between the events to determine the spatial separation in Σ'.]

4.18 Show that if $v/c = \tanh\alpha$ then the Lorentz transformations can be written in the form

$$x' = x\cosh\alpha - ct\sinh\alpha,$$

$$y' = y, \qquad z' = z,$$

$$ct' = -x\sinh\alpha + ct\cosh\alpha.$$

[*Hint*: $\gamma = (1 - \tanh^2\alpha)^{-1/2} = \cosh\alpha$.]

5

Relativistic Kinematics and Optics

5.1 INTRODUCTION

In Chapter 4 we discussed how the coordinates and times of events are transformed from one inertial reference frame to another. In practice one is often interested in how other quantities, such as the velocity, momentum and energy of a particle, change if one changes one's standard of rest; that is, if one transforms from one inertial reference frame to another moving relative to it. In this chapter the relativistic velocity transformations will be derived and some of their applications will be discussed. It will be assumed throughout this chapter that the inertial reference frame Σ' moves with uniform velocity v relative to Σ along their common x axis and that the origins of Σ and Σ' coincide at $t = t' = 0$.

5.2 THE VELOCITY TRANSFORMATIONS

Consider a particle that is moving in three dimensions with velocity $\boldsymbol{u}$ in Σ, as shown in Fig. 5.1(a) (this is not a space–time diagram, since the $w = ct$ axis is not shown). Let the position of the particle be determined at Event 1 at (x, y, z) at a time t in Σ, and then a very short time $\mathrm{d}t$ later at Event 2 at $(x + \mathrm{d}x, y + \mathrm{d}y, z + \mathrm{d}z)$ at the time $t + \mathrm{d}t$ in Σ (Fig. 5.1(a)). The components of the velocity $\boldsymbol{u}$ of the particle in Σ are given in the limit as $\mathrm{d}t$ tends to zero by

$$u_x = \frac{\mathrm{d}x}{\mathrm{d}t}, \qquad u_y = \frac{\mathrm{d}y}{\mathrm{d}t}, \qquad u_z = \frac{\mathrm{d}z}{\mathrm{d}t}\,. \tag{5.1}$$

The magnitude of $\boldsymbol{u}$ is

$$u = (u_x^2 + u_y^2 + u_z^2)^{1/2}. \tag{5.2}$$

The two successive measurements of the position of the particle are events, since they involve measurements such as the passing of the particle past marks on a scale, or the passage of a high-energy atomic particle through two particle detec-

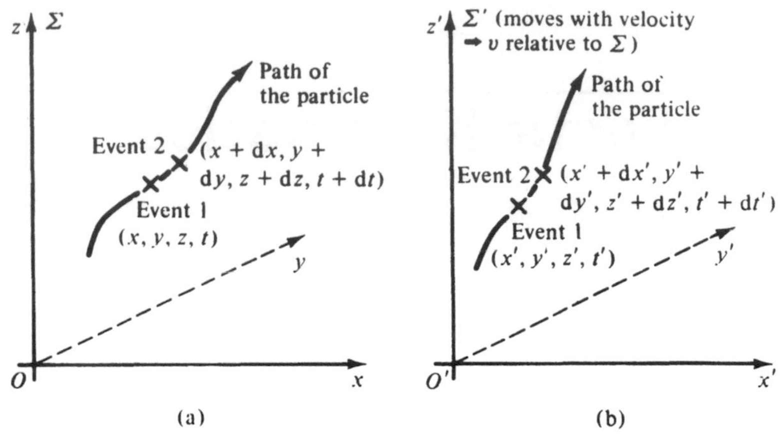

Fig. 5.1 Velocity of a particle in Σ and Σ′.

tors. The coordinates and times of these events relative to Σ′ can be calculated using the Lorentz transformations. For Event 1 at (x, y, z, t) in Σ we have in Σ′

$$\left.\begin{aligned} x' &= \gamma(x - vt), \\ y' &= y, \qquad z' = z, \\ t' &= \gamma(t - vx/c^2). \end{aligned}\right\} \tag{5.3}$$

For Event 2 at $(x + dx, y + dy, z + dz, t + dt)$ in Σ we have in Σ′

$$\left.\begin{aligned} x' + dx' &= \gamma[x + dx - v(t + dt)], \\ y' + dy' &= y + dy, \qquad z' + dz' = z + dz, \\ t' + dt' &= \gamma[t + dt - v(x + dx)/c^2]. \end{aligned}\right\} \tag{5.4}$$

Subtracting (5.3) from (5.4), we have

$$\left.\begin{aligned} dx' &= \gamma(dx - v\,dt), \\ dy' &= dy, \qquad dz' = dz, \\ dt' &= \gamma(dt - v\,dx/c^2). \end{aligned}\right\} \tag{5.5}$$

The velocity of the particle, measured in Σ′, has components:

$$u_x' = \frac{dx'}{dt'} = \frac{\gamma(dx - v\,dt)}{\gamma(dt - v\,dx/c^2)}$$

$$= \frac{dx/dt - v}{1 - v\,(dx/dt)/c^2} = \frac{u_x - v}{1 - vu_x/c^2}, \tag{5.6}$$

$$u_y' = \frac{dy'}{dt'} = \frac{dy}{\gamma(dt - v\,dx/c^2)}$$

$$= \frac{dy/dt}{\gamma[1 - v\,(dx/dt)/c^2]} = \frac{u_y}{\gamma(1 - vu_x/c^2)}, \tag{5.7}$$

$$u_z' = \frac{dz'}{dt'}$$

$$= \frac{u_z}{\gamma(1 - vu_x/c^2)} . \quad (5.8)$$

These are the *relativistic velocity transformations*. The factor $\gamma = (1 - v^2/c^2)^{-1/2}$ depends only on the velocity v with which Σ' is moving relative to Σ and on c, the velocity of light in empty space.

The inverse relations can be obtained by changing unprimed quantities into primed quantities, primed into unprimed, and replacing v by $-v$, giving

$$\left.\begin{aligned} u_x &= \frac{u_x' + v}{1 + vu_x'/c^2} \\ u_y &= \frac{u_y'}{\gamma(1 + vu_x'/c^2)} \\ u_z &= \frac{u_z'}{\gamma(1 + vu_x'/c^2)} \end{aligned}\right\} \quad (5.9)$$

Note that if $u << c$ and $v << c$ then, from equations (5.6)–(5.8), we have

$$u_x' \approx u_x - v, \qquad u_y' \approx u_y, \qquad u_z' \approx u_z.$$

This is in agreement with the Galilean velocity transformations of Newtonian mechanics.

5.3 NUMERICAL EXAMPLES OF THE RELATIVISTIC VELOCITY TRANSFORMATIONS

As a typical example from Newtonian mechanics, consider a ship moving with a uniform velocity of 18 m s^{-1} relative to the Earth. Let a ball be rolled at a speed of $u_x' = 2$ m s^{-1} relative to the ship, in the direction of motion of the ship. From equations (5.9), the speed of the ball relative to the Earth is

$$u_x = \frac{u_x' + v}{1 + vu_x'/c^2} = \frac{(18 + 2)\ \text{m s}^{-1}}{1 + 2 \times 18 / (3 \times 10^8)^2} = \frac{20\ \text{m s}^{-1}}{1 + 4 \times 10^{-16}}.$$

Expanding the denominator using the binomial theorem, we have

$$u_x \approx 20\,(1 + 4 \times 10^{-16})^{-1}\ \text{m s}^{-1} = (20 - 80 \times 10^{-16})\ \text{m s}^{-1}$$

$$= 19.999\,999\,999\,999\,992\ \text{m s}^{-1}.$$

According to the Galilean velocity transformation, equation (1.21),

$$u_x = u_x' + v = (18 + 2)\ \text{m s}^{-1} = 20\ \text{m s}^{-1}.$$

This example illustrates how for normal everyday macroscopic phenomena the deviations from Newtonian mechanics are generally negligible.

Let a radioactive atom move with a uniform velocity $v = 0.2c$ along the x axis of the laboratory frame Σ. Let it emit a β particle of velocity $0.95c$ relative to the inertial reference frame Σ' in which the radioactive atom is at rest. If the β particle is emitted along the x' axis of Σ', such that $u_x' = 0.95c$ then, from the relativistic velocity transformation, equation (5.9), its speed in the laboratory frame Σ is

$$u_x = \frac{u_x' + v}{1 + vu_x'/c^2} = \frac{0.95c + 0.2c}{1 + 0.2 \times 0.95}$$

$$= 0.966c.$$

According to the Galilean transformations,

$$u_x = u_x' + v = 0.95c + 0.2c = 1.15c.$$

Thus the deviations from the Galilean transformations are important in high-energy physics.

If the β particle of velocity $0.95c$ were emitted in a direction parallel to the y' axis of the inertial reference frame Σ' in which the decaying radioactive atom was at rest, such that $u_x' = 0$, $u_y' = 0.95c$ and $u_z' = 0$ in Σ', then, with $v = 0.2c$, $\gamma = 1.021$, according to equations (5.9) in Σ we would have

$$u_x = \frac{u_x' + v}{1 + vu_x'/c^2} = \frac{0 + 0.2c}{1 + 0.2 \times 0} = 0.2c,$$

$$u_y = \frac{u_y'}{\gamma(1 + vu_x'/c^2)} = \frac{0.95c}{1.021} = 0.931c.$$

$$u_z = 0,$$

$$u = (u_x^2 + u_y^2 + u_z^2)^{1/2} = 0.952c.$$

The direction of motion of the particle in Σ is at an angle θ to the x axis of Σ, where

$$\theta = \arctan \frac{u_y}{u_x} = \arctan \frac{0.931c}{0.2c} = \arctan 4.655$$

$$= 77.9°.$$

The reader can check that one cannot obtain velocities exceeding c by adding a number of velocities that are themselves less than c. For example, if $u_x' = 0.99c$, $u_y' = u_z' = 0$ in Σ' and $v = 0.99c$ then in Σ we have

$$u_x = \frac{u_x' + v}{1 + vu_x'/c^2} = \frac{0.99c + 0.99c}{1 + 0.99 \times 0.99} = 0.99995c.$$

If $u_x' = c$ in Σ' then in Σ

$$u_x = \frac{u_x' + v}{1 + vu_x'/c^2} = \frac{c + v}{1 + vc/c^2} = \frac{c\,(c + v)}{c + v} = c.$$

Thus the speed of light in empty space has the same numerical value in Σ and Σ′, illustrating that the velocity transformations are consistent with the principle of the constancy of the speed of light, as of course they should be, since the Lorentz transformations were derived from this principle in Chapter 4. It is consistent with the relativistic velocity transformations to use the 1981 definition of the metre and to assume that $c = 299\,792\,458\ \text{m s}^{-1}$ in both Σ and Σ′. However, if this value for c is used in Σ then, according to the Galilean velocity transformations, the speed of light in Σ′ should vary between $c - v$ and $c + v$. Hence it is not consistent with the Galilean transformations and Newtonian mechanics to assume that c has the same value in both Σ and Σ′. However, it has been shown that the Galilean transformations and Newtonian mechanics are excellent *low-velocity approximations* to the Lorentz transformations and relativistic mechanics even when the 1981 definition of the metre is used in both Σ and Σ′.

5.4 FIZEAU'S EXPERIMENT

In 1851 Fizeau showed that the speed of light in moving water depended on the speed of the water relative to the laboratory. The experiment was improved by Michelson and Morley in 1886. Let Σ be the laboratory frame, in which the water is moving with uniform velocity v in the positive x direction, and let the direction of the light be in the positive x direction also (Fig. 5.2(a)). In the inertial frame Σ′, moving with uniform velocity v relative to Σ, the water is at rest, (Fig. 5.2(b)). Let the refractive index of stationary water be n' measured in Σ′. The velocity of the light in Σ′ has the components:

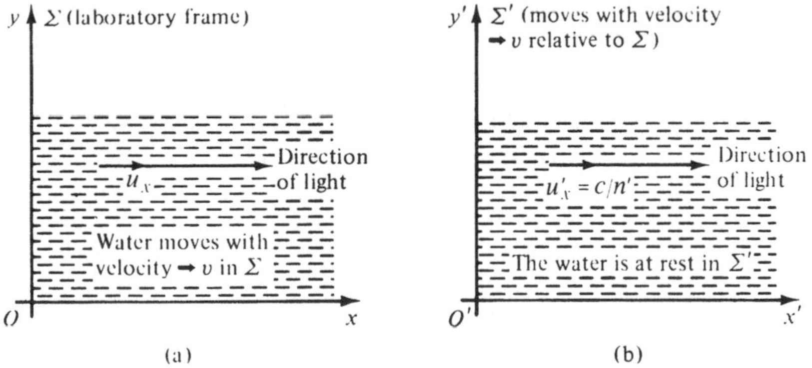

Fig. 5.2 In Σ′ the water is at rest, and the speed of light is the same and equal to c/n' in all directions in Σ′, as shown in (b). In the laboratory frame Σ the water is moving with velocity v, as shown in (a). According to the velocity transformations of special relativity, the speed of light parallel to the x axis is equal to $c/n' + v\,(1 - 1/n'^2)$ in Σ.

$$u_x' = \frac{c}{n'}, \qquad u_y' = u_z' = 0.$$

Applying equations (5.9), we have

$$u_x = \frac{u_x' + v}{1 + vu_x'/c^2} = \frac{c/n' + v}{1 + vc/n'c^2} = \frac{c}{n'}\left(1 + \frac{n'v}{c}\right)\left(1 + \frac{v}{n'c}\right)^{-1},$$

$$u_y = u_z = 0.$$

Expanding $(1 + v/n'c)^{-1}$ by the binomial theorem and neglecting terms of order v^2/c^2, we find

$$u_x \approx \frac{c}{n'}\left(1 + \frac{n'v}{c}\right)\left(1 - \frac{v}{n'c}\right)$$

$$\approx \frac{c}{n'} + v\left(1 - \frac{1}{n'^2}\right). \tag{5.10}$$

If the Galilean transformations were correct, one would have

$$u_x = u_x' + v = \frac{c}{n'} + v. \tag{5.11}$$

In 1886 Michelson and Morley showed that the velocity of light in water that is moving with velocity v in the laboratory frame Σ was equal to $c/n' + fv$ in Σ when the light was propagating in the direction of motion of the water. Their experimental value for f was 0.434 ± 0.020. Putting $n' = 1.33$ in equation (5.10), we find that, according to the relativistic transformations,

$$f = 1 - \frac{1}{n'^2} = 1 - \frac{1}{1.33^2} = 0.435.$$

This is in good agreement with the experimental results of Michelson and Morley. According to the Galilean velocity transformations, which give equation (5.11), f should be equal to 1. The reason why the Galilean transformations are inaccurate in this case is that, since $n' = 1.33$, one of the velocities, namely c/n' is comparable to c. If n' were, say, 1000 then $1 - 1/n'^2$ would be $1 - 10^{-6}$ and equation (5.11) would be a satisfactory approximation to equation (5.10).

By measuring the speed of light in moving water, it is possible to determine the speed of the water. This does not contravene the principle of relativity. If an astronaut inside a spaceship moving with uniform velocity relative to the fixed stars repeated Fizeau's experiment, all the astronaut could determine would be the speed of the water relative to the spaceship. The astronaut could not determine the speed of the spaceship relative to the Earth by this experiment.

5.5 TRANSFORMATION OF $(1 - u^2/c^2)^{1/2}$ AND $(1 - u'^2/c^2)^{1/2}$

The quantities

$$\alpha = \frac{1}{(1 - u^2/c^2)^{1/2}}, \tag{5.12a}$$

$$\alpha' = \frac{1}{(1 - u'^2/c^2)^{1/2}}, \tag{5.12b}$$

where u and u' are the speeds of the particles in Σ and Σ' respectively, appear frequently in relativistic mechanics. For example, according to equation (2.1), the relativistic momentum of a particle of mass m moving with velocity $\boldsymbol{u}$ in Σ is $\boldsymbol{p} = \alpha m \boldsymbol{u}$. If the velocity of the particle in Σ' is $\boldsymbol{u}'$, its momentum in Σ' is $\alpha' m \boldsymbol{u}'$. The transformations for α and α' will now be derived.

The square of the total velocity of a particle relative to Σ' is

$$u'^2 = u_x'^2 + u_y'^2 + u_z'^2.$$

Using equations (5.6), (5.7) and (5.8), we have

$$u'^2 = \frac{(u_x - v)^2 + (1 - v^2/c^2)\, u_y^2 + (1 - v^2/c^2) u_z^2}{(1 - vu_x/c^2)^2}.$$

Remembering that

$$u^2 = u_x^2 + u_y^2 + u_z^2,$$

or

$$u_y^2 + u_z^2 = u^2 - u_x^2.$$

we find

$$u'^2 = \frac{(u_x - v)^2 + (u^2 - u_x^2)\,(1 - v^2/c^2)}{(1 - vu_x/c^2)^2}.$$

Thus

$$1 - \frac{u'^2}{c^2} = 1 - \frac{\left(\frac{u_x}{c} - \frac{v}{c}\right)^2 + \left(\frac{u^2}{c^2} - \frac{u_x^2}{c^2}\right)\left(1 - \frac{v^2}{c^2}\right)}{(1 - vu_x/c^2)^2}$$

$$= \frac{1 - \frac{2vu_x}{c^2} + \frac{v^2u_x^2}{c^4} - \frac{u_x^2}{c^2} + \frac{2vu_x}{c^2} - \frac{v^2}{c^2} - \frac{u^2}{c^2} + \frac{u_x^2}{c^2} + \frac{v^2u^2}{c^4} - \frac{v^2u_x^2}{c^4}}{(1 - vu_x/c^2)^2}$$

$$= \frac{1 - v^2/c^2 - u^2/c^2 + v^2u^2/c^4}{(1 - vu_x/c^2)^2} = \frac{(1 - v^2/c^2)(1 - u^2/c^2)}{(1 - vu_x/c^2)^2}.$$

Taking the square root, we have

$$(1 - u'^2/c^2)^{1/2} = \frac{[(1 - v^2/c^2)(1 - u^2/c^2)]^{1/2}}{1 - vu_x/c^2}, \tag{5.13}$$

Using equations (5.12a, b) and $\gamma = (1 - v^2/c^2)^{-1/2}$, we have

$$\alpha' = \gamma\alpha\,(1 - vu_x/c^2). \tag{5.14}$$

Similarly, the inverse relations are

$$(1 - u^2/c^2)^{1/2} = \frac{[(1 - v^2/c^2)(1 - u'^2/c^2)]^{1/2}}{1 + vu_x'/c^2}, \tag{5.15}$$

$$\alpha = \gamma\alpha'\,(1 + vu_x'/c^2). \tag{5.16}$$

As a numerical example, consider a particle having total speed $u' = 0.5c$, with components $u_x' = 0.4c$, $u_y' = 0.3c$ and $u_z' = 0$ in Σ'. The value of α' is $(1 - 0.5^2)^{-1/2} = 1.1547$. Let the velocity of Σ' relative to Σ be $v = 0.6c$, such that $\gamma = (1 - 0.6^2)^{-1/2} = 1.25$. According to equation (5.16), the value of α in Σ is

$$\alpha = \gamma\alpha'\,(1 + vu_x'/c^2) = 1.25 \times 1.1547\,(1 + 0.6 \times 0.4) = 1.790.$$

As an exercise, the reader can check that, according to the relativistic velocity transformations,

$$u_x = 0.8065c, \qquad u_y = 0.1935c, \qquad u_z = 0, \qquad u = 0.8294;$$

and therefore

$$\alpha = (1 - u^2/c^2)^{-1/2} = (1 - 0.8294^2)^{-1/2} = 1.790.$$

5.6 THE DOPPLER EFFECT (WAVE THEORY)

In Section 4.8 we applied the K-calculus method to study the Doppler effect in the special cases when the light source was moving directly towards and directly away from the observer. We now go on to consider the general case, using the wave theory of light. Consider a monochromatic light source, such as a sodium lamp, that is at rest at the origin O' of the inertial reference frame Σ' that is moving with uniform velocity v relative to the laboratory frame Σ along their common x axis, as shown in Fig. 5.3(b). The light source is moving with uniform velocity v along the x axis of the laboratory frame Σ, as shown in Fig. 5.3(a). The light is detected by an inertial observer at rest in Σ at the point P in Fig. 5.3(a). Let a wave crest (positive maximum) be emitted from the light source, which is at rest at O' in Fig. 5.3(b), at Event 1 at the point ($x' = 0$, $y' = 0$, $z' = 0$) at a time t_1' measured in Σ'. Let the next wave crest (positive maximum) be

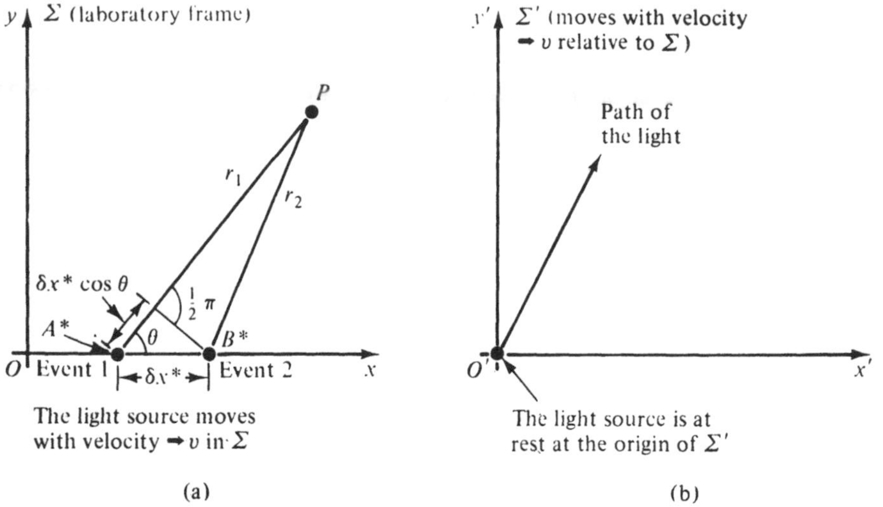

Fig. 5.3 (a) The light source is moving with uniform velocity v along the x axis of Σ. The light is detected at P. (b) The light source is at rest at the origin O' in Σ'.

emitted from O' at Event 2 at the point $(x'=0,\ y'=0,\ z'=0)$ at a time $t_2'=t_1'+\delta t'$ in Σ'. If ν_0 is the frequency of the light measured by an inertial observer at rest in the inertial reference frame Σ' in which the source is at rest then

$$\delta t' = t_2' - t_1' = \frac{1}{\nu_0}. \tag{5.17}$$

Let Event 1 occur when the light source is at position A^*, with coordinates $(x_1^*, 0, 0,)$ at a time t_1^* in the laboratory frame Σ, as shown in Fig. 5.3(a). The position of the light source at the instant the light is emitted will be called the retarded position of the source, and the time the light is emitted will be called the retarded time. Hence A^* is the *retarded position* of the source, when the first wave crest is emitted by the light source in Event 1 at the *retarded time* t_1^* in Σ. From the Lorentz transformations, since $x_1'=0$ for Event 1 in Σ', the coordinates and time of Event 1 at A^*, measured in Σ are

$$x_1^* = \gamma(x_1' + vt_1') = \gamma v t_1',$$

$$t_1^* = \gamma(t_1' + vx_1'/c^2) = \gamma t_1'. \tag{5.18}$$

Let Event 2, when the second wave crest is emitted, occur at the retarded position B^* in Σ, as shown in Fig. 5.3(a). From the Lorentz transformations, since $x_2'=0$ for Event 2 in Σ', the coordinates and time of Event 2 at B^*, measured in Σ, are

$$x_2^* = \gamma v t_2', \qquad t_2^* = \gamma t_2'. \tag{5.19}$$

It follows from equations (5.17)–(5.19) that

$$\delta x^* = x_2^* - x_1^* = \gamma v\,(t_2' - t_1') = \gamma v\,\delta t'. \tag{5.20}$$

$$t_2^* - t_1^* = \gamma\,(t_2' - t_1') = \gamma\,\delta t'. \tag{5.21}$$

Equation (5.21) is the normal expression for time dilatation. The time interval $t_2' - t_1'$ between the two events, namely the emission of two successive wave crests, measured in the inertial frame Σ' in which the light source is at rest, is the proper time interval between the two events, which can, in principle, be measured by one clock at rest by the light source that is at rest in Σ'. On the other hand, in Σ, $t_2^* - t_1^*$ must be measured by two spatially separated synchronized clocks at rest at A^* and B^* in Σ in Fig. 5.3(a).

The light from Event 1 at A^* at t_1^* in Σ must travel a distance r_1 from A^* to P in Fig. 5.3(a) at speed c to reach P at the *time of observation* T_1, which is equal to $t_1^* + r_1/c$ in Σ. Light from Event 2 at B^* must travel a distance r_2 from B^* to P in Fig. 5.3(a) at speed c to reach P at a time $T_2 = t_2^* + r_2/c$ in Σ. Hence the time interval $T = T_2 - T_1$ between the *receipt* of the two successive wave crests at the point P in Fig. 5.3(a) is

$$\delta T = T_2 - T_1 = t_2^* + r_2/c - (t_1^* + r_1/c),$$

and therefore

$$\delta T = t_2^* - t_1^* + (r_2 - r_1)/c = t_2^* - t_1^* - (r_1 - r_2)/c. \tag{5.22}$$

In practice the distance $\delta x^* = x_2^* - x_1^* = \gamma v\,\delta t'$ between Events 1 and 2, at the retarded positions A^* and B^* respectively in Fig. 5.3(a), is very much less than the distances r_1 and r_2. For example, if $v = 0.6c$ and $\nu_0 = 5.09 \times 10^{14}$ Hz, corresponding to the yellow sodium spectral line of wavelength 589 nm, using equations (5.20) and (5.17), we have

$$\delta x^* = \frac{\gamma v}{\nu_0} = \frac{1}{(1 - 0.6^2)^{1/2}}\,\frac{0.6 \times 3 \times 10^8}{5.09 \times 10^{14}} = 4.4 \times 10^{-7}\ \text{m},$$

which is much less than r_1 and r_2 in normal laboratory experiments. It can be seen from Fig. 5.3(a) that if $\delta x^* << r_1$ and $\delta x^* << r_2$ then, to an excellent approximation,

$$r_1 - r_2 = \delta x^* \cos\theta.$$

Using equation (5.20) to substitute for δx^*, we have

$$r_1 - r_2 = \gamma v\,\delta t' \cos\theta. \tag{5.23}$$

Substituting for $r_1 - r_2$ from equation (5.23) and for $t_2^* - t_1^*$ from equation (5.21) into equation (5.22), we obtain

$$\delta T = \gamma\,\delta t' - \frac{\gamma v}{c}\,\delta t' \cos\theta.$$

Putting $\delta t' = 1/\nu_0$ and $\delta T = 1/\nu$, where ν is the frequency of the light measured at the point P in Fig. 5.3(a) by an inertial observer at rest at the point P in the laboratory frame Σ, and rearranging, we have

$$\nu = \frac{\nu_0}{\gamma[1 - (\upsilon/c)\cos\theta]} . \tag{5.24}$$

In equation (5.24), ν_0 is the frequency of the light measured in the inertial reference frame Σ' in which the source of light is at rest, ν is the frequency measured at the point P in Fig. 5.3(a) in the laboratory frame Σ, in which the light source is moving with velocity $\boldsymbol{\upsilon}$ at its retarded position, and θ is the angle between $\boldsymbol{\upsilon}$ and the vector $\boldsymbol{r}$ from the retarded position of the source to the point of observation P in Σ. Equation (5.24) reduces to equations (4.58) and (4.54) when $\theta = 0$ and $\theta = \pi$ respectively.

The transformations for the wavelength can be obtained from equation (5.24) using $\lambda_0\nu_0 = \lambda\nu = c$, where λ_0 and λ are the wavelengths of the light measured in Σ' and Σ respectively. We obtain

$$\lambda = \lambda_0\gamma\left(1 - \frac{\upsilon}{c}\cos\theta\right). \tag{5.25}$$

5.7 THE ABERRATION OF LIGHT

Consider again a monochromatic light source, such as a sodium lamp, that is at rest at the origin O' of the inertial reference frame Σ', shown previously in Fig. 5.3(b) and again in Fig. 5.4(b). The light source is moving with uniform velocity υ along the x axis of the laboratory frame Σ, as shown in Fig. 5.4(a). Consider a short pulse of light that is emitted in the (x', y') plane of Σ' at an angle θ' to the x' axis of Σ', as shown in Fig. 5.4(b). Since the light pulse moves outwards from the origin O' of Σ' with a velocity c, the components of its velocity in Σ' are

$$u_x' = c\cos\theta', \qquad u_y' = c\sin\theta', \qquad u_z' = 0.$$

Using the inverse relativistic velocity transformations, given by equations (5.9), we find that the components of the velocity of the light pulse in Σ are

$$u_x = \frac{u_x' + \upsilon}{1 + \upsilon u_x'/c^2} = \frac{c\cos\theta' + \upsilon}{1 + (\upsilon/c)\cos\theta'} , \tag{5.26}$$

$$u_y = \frac{u_y'}{\gamma(1 + \upsilon u_x'/c^2)} = \frac{c\sin\theta'}{\gamma[1 + (\upsilon/c)\cos\theta']} , \tag{5.27}$$

$$u_z = 0. \tag{5.28}$$

Dividing equation (5.27) by equation (5.26), we obtain in Σ

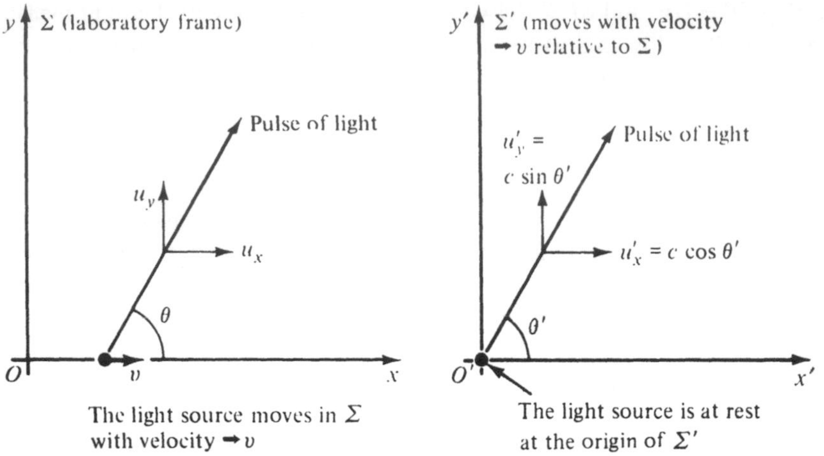

Fig. 5.4 The aberration of light. (a) The light source is moving along the x axis of Σ. (b) The light source is at rest at the origin O' in Σ′.

$$\tan\theta = \frac{u_y}{u_x} = \frac{\sin\theta'}{\gamma(\cos\theta' + v/c)}, \tag{5.29}$$

where θ is the inclination of the path of the pulse of light to the x axis of Σ, as shown in Fig. 5.4(a)..

Consider the special case when the pulse of light is emitted at $\theta' = 90°$ along the y' axis of Σ′. It follows from equation (5.29) that, since for $\theta' = 90°$ $\sin\theta' = 1$ and $\cos\theta' = 0$,

$$\tan\theta = \frac{1}{\gamma(0 + v/c)} = \frac{c}{\gamma v}. \tag{5.30}$$

For example, if $v/c = 0.1, 0.6, 0.9$ and 0.9999 in Σ, and if $\theta' = 90°$ in Σ′, then the values of θ, measured in the laboratory frame Σ, are 84°, 53°, 26° and 0.81° respectively. It can be seen that as the velocity of the lamp is increased in Σ more and more of the light emitted is concentrated in the forward direction in Σ. When $v = 0.9999c$, most of the light is concentrated into a cone of semi-angle about 0.8°. This effect is sometimes called the "headlight effect". In addition to the concentration of light in the forward direction, the frequency of the light is changed by the Doppler effect. For example, according to equation (5.25), for $v = 0.9999c$ and $\lambda_0 = 589$ nm for the yellow sodium spectral line, the wavelength of the light emitted in the forward direction ($\theta = 0$) in the laboratory frame Σ is 4.17 nm, which is in the ultraviolet. The wavelength of the light emitted in the backward direction ($\theta = 180°$) is 8.3×10^4 nm, which is in the infrared. A speed of $v = 0.9999c$ may seem unrealistic for a sodium lamp, but, according to equation (2.50), this corresponds to an electron of (total) energy 36 MeV, which will emit electromagnetic radiation (*Bremsstrahlung*), when it is accelerated or decelerated.

★ 5.8 VISUAL APPEARANCE OF A MOVING OBJECT

One should always say that a moving body is *measured* to be Lorentz-contracted. What one sees (or photographs) depends on the light actually reaching the eye (or camera) at that given instant of time. For example, if one photographs the light coming from the stars in the heavens, some of the light will have left some of the stars millions of years ago, whereas the light from the Moon left 1.28 s ago. In the millions of years since the light was emitted by the distant stars, the stars will have moved. The position of a light source, when the light was emitted, will be called the *retarded position* of the light source, and the time the light was emitted will be called the *retarded time*. The photograph of the stars will represent the retarded positions of the stars at their appropriate retarded times.

Two examples of the photograph of a moving rod will now be considered.

(a) A rod that is parallel to its direction of motion

Consider a very short rod AB that is at rest in the (x', y') plane of the inertial reference frame Σ', as shown in Fig. 5.5(b). The rod lies parallel to the x' axis of Σ' at a perpendicular distance y_0' from the x' axis of Σ'. The coordinates of the ends A and B of the rod in Σ' are $(x_1', y_0', 0)$ and $(x_2', y_0', 0)$ respectively, where $x_1' > x_2'$. The proper length l_0 of the rod, measured in the inertial reference frame Σ' in which it is at rest, is

$$l_0 = x_1' - x_2'. \tag{5.31}$$

In the inertial reference frame Σ, shown in Fig. 5.5(a), the rod is moving with uniform velocity v parallel to the x axis of Σ, at a perpendicular distance $y_0 = y_0'$ from the x axis of Σ. A camera at rest at the origin O of the inertial

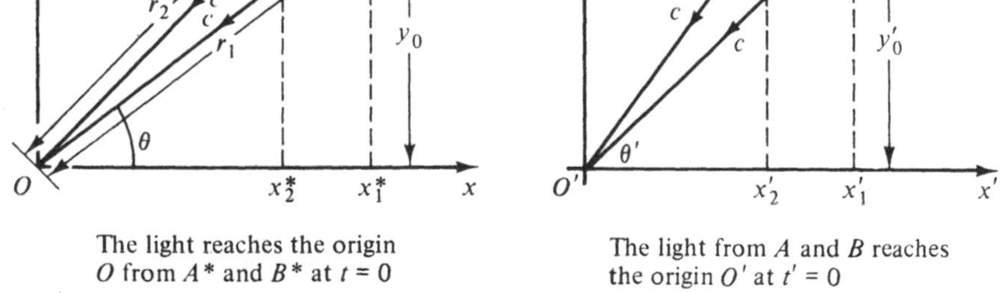

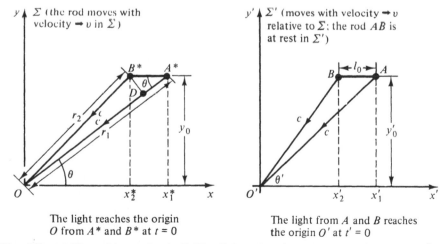

Fig. 5.5 (a) The rod is moving in Σ. The light pulses from the retarded positions A^* and B^* reach the origin O at $t = 0$ in Σ. (b) The rod AB is at rest in Σ'. The light pulses from A and B reach the origin O' at $t' = 0$ in Σ'.

reference frame Σ in Fig. 5.5(a) takes a photograph of the moving rod AB at the time $t=0$, when the origins of Σ and Σ' coincide, which is at the time $t'=0$ in Σ'. Let the light from the end A of the rod leave A when it is at the retarded position A^* in Fig. 5.5(a) at Event 1 at $(x_1^*, y_0, 0)$ in Σ. If the distance from A^* to the origin O is r_1, it takes the light from A^* a time r_1/c in Σ to reach the origin O at $t=0$. Hence the appropriate retarded time, when the light that reaches the origin of Σ at $t=0$ was emitted from A^*, is $t_1^* = -r_1/c$. From the Lorentz transformations, the x_1' coordinate of Event 1, measured in Σ', is

$$x_1' = \gamma(x_1^* - vt_1^*) = \gamma(x_1^* + vr_1/c). \tag{5.32}$$

Let the light from the end B of the moving rod, which reaches the origin of Σ at $t=0$, leave B when it is at the retarded position B^* in Fig. 5.5(a) at Event 2 at $(x_2^*, y_0, 0)$ in Σ. If the distance from B^* to the origin O is r_2, the retarded time, when the light left B^* to reach the origin of Σ at $t=0$, is $t_2^* = -r_2/c$ in Σ. From the Lorentz transformations, the x_2' coordinate of Event 2 in Σ' is

$$x_2' = \gamma(x_2^* - vt_2^*) = \gamma(x_2^* + vr_2/c). \tag{5.33}$$

Subtracting equation (5.33) from equation (5.32) and using equation (5.31) to substitute for $x_1' - x_2'$, we have

$$x_1' - x_2' = l_0 = \gamma(x_1^* - x_2^*) + \frac{\gamma v}{c}(r_1 - r_2). \tag{5.34}$$

The difference in the path lengths from A^* to O and from B^* to O in Fig. 5.5(a) is $r_1 - r_2$, which is equal to the distance from D to A^* in Fig. 5.5(a). Strictly speaking the line from B^* to D is an arc of a circle of radius r_2, but, provided that the length of the rod AB is infinitesimal, the arc from D to B^* can be treated as a straight line perpendicular to the line OA^*, so that to a very good approximation

$$r_1 - r_2 = (DA^*) = (A^*B^*)\cos\theta = (x_1^* - x_2^*)\cos\theta. \tag{5.35}$$

Substituting for $r_1 - r_2$ in equation (5.34) and rearranging, we obtain

$$x_1^* - x_2^* = \frac{l_0}{\gamma[1 + (v/c)\cos\theta]}, \tag{5.36}$$

where $x_1^* - x_2^*$ is the distance between the retarded positions A^* and B^* in Fig. 5.5(a). When the photograph is taken, at the time $t=0$, using the camera at rest at the origin of Σ, the light pulses from the ends A and B of the moving rod come from the retarded positions A^* and B^* respectively in Σ. Hence the visual appearance (image) of the moving rod AB on the photograph would be the same as the photograph of a rod that was at rest in Σ, that extended from A^* to B^* in Fig. 5.5(a), and that was of length $x_1^* - x_2^*$ given by equation (5.36). Events 1 and 2 corresponding to the retarded positions A^* and B^* of the ends of the rod are at times t_1^* and t_2^* respectively, and are not simultaneous. It follows from equation (4.51) that if the positions of the ends of the moving rod were *measured* at the same time in Σ then the *measured* length of the rod would be l_0/γ, the normal formula for the *Lorentz length contraction*. It follows from equation (5.36) that if θ were less than $\frac{1}{2}\pi$, which is true for the case shown in Fig. 5.5(a), then $x_1^* - x_2^*$ would be less than l_0/γ. According to equation (5.36), if θ were greater than $\frac{1}{2}\pi$ in Fig. 5.5(a) then $x_1^* - x_2^*$ would be greater than l_0/γ. It is only for $\theta = \frac{1}{2}\pi$ that $x_1^* - x_2^*$ is equal to l_0/γ.

According to equation (5.36), if the rod AB in Fig. 5.5(b) were of finite length (for example if it were a ruler) then the photographic images of the various sections of rod would change according to equation (5.36). For example, the separations of the lines on the photograph of the moving ruler would be closer together on the right-hand side of the photograph of the moving ruler compared with the left-hand side.

(b) A rod that is perpendicular to the direction of motion

Consider a straight rod AB of finite proper length l_0 that is at rest in the (x', y') plane of the inertial reference frame Σ', lying parallel to the y' axis of Σ' at a perpendicular distance a from the y' axis, as shown in Fig. 5.6(b). The rod moves with uniform velocity v in the $+x$ direction in Σ, as shown in Fig. 5.6(a). A photograph of the moving rod is taken by a camera at rest at the origin O of Σ at the time $t = 0$, when the origins of Σ and Σ' coincide. Consider the light coming from the point D on the rod, which is at $(x' = a,\ y',\ z' = 0)$ in Σ'. The distance from D to the origin of Σ' is $(a^2 + y'^2)^{1/2}$. The light from D, which reaches the origin O' of Σ' in Fig. 5.6(b) at the time $t' = 0$ in Σ', must leave the point D at $t' = -(a^2 + y'^2)^{1/2}/c$ in Σ'. From the Lorentz transformations, the x coordinate of this event in Σ, corresponding to $(x' = a,\ y',\ z' = 0,\ t' = -(a^2 + y'^2)^{1/2}/c)$ in Σ' is, with $y' = y$,

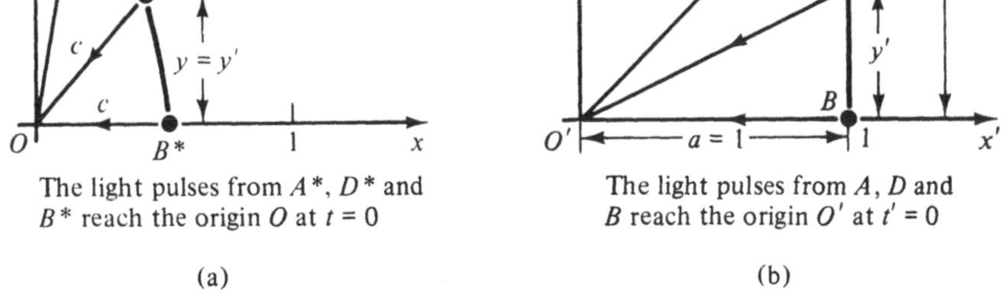

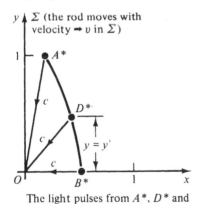

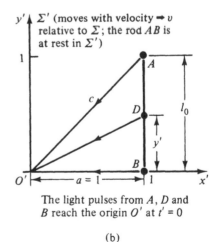

Fig. 5.6 (a) The rod AB is moving with uniform velocity v in Σ. The light pulses from the retarded positions A^*, D^* and B^* reach the origin O of Σ at the same time $t = 0$ in Σ. (b) The rod AB is at rest in Σ'.

$$x = \gamma(x' + vt') = \gamma\left[a - \frac{v}{c}(a^2 + y^2)^{1/2}\right]. \tag{5.37}$$

Equation (5.37) gives the retarded position D^* of the point D on the moving rod AB in Fig. 5.6(a), where, relative to Σ, the light that reaches the origin O of Σ at $t = 0$ leaves the point D on the moving rod.

As a numerical example, put $a = 1$ m, $l_0 = 1$ m and $v = 0.6c$, such that $\gamma = 1.25$. Equation (5.37) becomes

$$x = 1.25\,[1 - 0.6\,(1 + y^2)^{1/2}]. \tag{5.38}$$

The values of x for $y = 0$, 0.5 and 1.0 are 0.5, 0.411 and 0.189 respectively. These points lie on the curve A^*B^* in Fig. 5.6(a). The photographic image of the

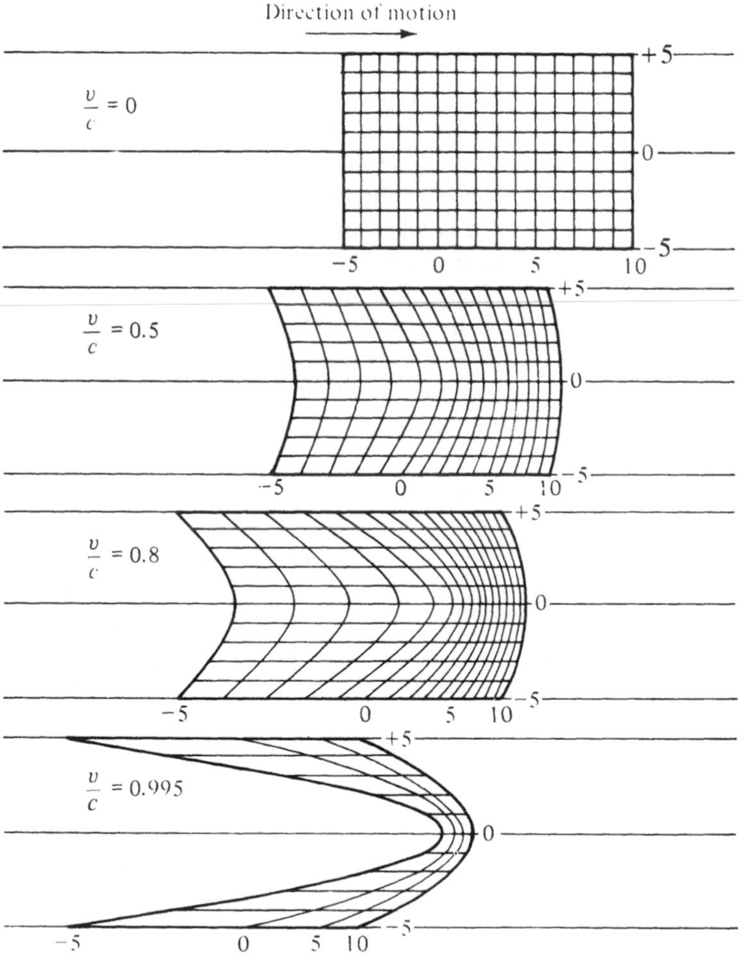

Fig. 5.7 Photograph of a plane grid moving at relativistic speeds. The observer is unit distance in front of the origin. For each view the direction in which the observer sees the origin is perpendicular to the motion. (By courtesy of *Am. J. Phys.*)

moving rod AB in Fig. 5.6(a) is the same on the photograph, taken by the camera at rest at the origin of Σ at $t = 0$, as the photographic image of a curved rod, which is at rest in Σ and whose shape follows the curve given by equation (5.38). In the case shown in Fig. 5.6(a) the light from the end A of the rod must always leave before the light from B, whatever the position of the rod, so that A^* is always to the left of B^*.

To illustrate both cases (a) and (b), consider the expected photographic image of a moving grid on a photograph taken by a camera at unit distance directly above the origin of the grid at (0, 0) (Fig. 5.7). The reader can check that the shapes of the vertical lines and the spacings between the vertical lines fit in with the ideas developed when cases (a) and (b) were discussed. (See Scott and Viner (1965).)

REFERENCE

Scott, G.D. and Viner, M.R. (1965) *Am. J. Phys.* **33**, 534.

PROBLEMS

Assume that the velocity of light $c = 3.00 \times 10^8$ m s^{-1}.

5.1 A train is passing through a station at a speed of 20 m s^{-1}. A marble is rolled along the floor of one of the compartments with a velocity of 10 m s^{-1} relative to the train. Calculate the speed of the marble relative to an observer standing on the platform

(a) if the marble rolls in the direction of motion of the train;

(b) if it rolls perpendicular to the direction of motion of the train as measured by a passenger on the train.

5.2 A radioactive nucleus is moving with a velocity $0.1c$ relative to the laboratory when it emits a β particle with a velocity $0.9c$ relative to the coordinate system in which the decaying nucleus is at rest. Calculate the velocity and direction of the β particle relative to the laboratory if, relative to the radioactive nucleus, it is emitted

(a) in the direction of motion of the nucleus relative to the laboratory;

(b) perpendicular to the direction of motion.

5.3 A radioactive nucleus that is at rest in Σ', but is moving with velocity $\frac{1}{6}c$ along the x axis of Σ, emits a β particle of velocity $0.8c$ (relative to Σ') at an angle of 45° to the x' axis of Σ'. What is the velocity of the β particle relative to an observer at rest in an inertial reference frame Σ'' going at a velocity of $\frac{1}{2}c$ along the negative x axis of Σ? [*Hint*: Find the velocity of Σ' relative to Σ'', and then apply the velocity transformations between Σ' and Σ''.]

5.4 Particles 1 and 2 leave the origin of the inertial frame Σ at the same time in Σ. Particle 1 moves with a velocity of $0.8c$ along the x axis of Σ, and particle 2 moves with a velocity of $0.8\,c$ along the y axis of Σ. What is the speed of particle 2 relative to particle 1? [*Hint*: Determine the total speed of particle 2 in the inertial frame in which particle 1 is at rest.]

5.5 An observer moving along the x axis of Σ with velocity v observes a body of proper volume V_0 moving with velocity u along the x axis of Σ. Show that the observer measures the volume to be

$$V_0\left[\frac{(c^2-v^2)(c^2-u^2)}{(c^2-uv)^2}\right]^{1/2}.$$

[*Hint*: The volume is $V_0(1-u'^2/c^2)^{1/2}$. Use equation (5.13) to transform $(1-u'^2/c^2)^{1/2}$.]

5.6 An observer at rest at the origin of an inertial frame Σ sees two particles travelling with equal speeds $0.8c$. One goes along the positive x axis, the other along the negative x axis.

(a) What is the relative velocity of the particles in Σ? How can this exceed the velocity of light?

(b) With what speed are the particles transferring energy and momentum in Σ?

(c) What is the velocity of one particle relative to the other?

5.7 In a cathode-ray tube the speed at which the electron beam sweeps across the fluorescent screen can exceed the velocity of light in empty space. How can you reconcile this with the limiting character of the velocity of light in empty space? [*Hint*: What is the speed of individual electrons?]

5.8 In Σ the acceleration $\boldsymbol{a}$ of a particle has components $a_x = du_x/dt$ etc., while in Σ' its acceleration $\boldsymbol{a}'$ has components $a_x' = du_x'/dt'$ etc. Now

$$a_x' = \frac{du_x'}{dt'} = \frac{du_x'/dt}{dt'/dt} = \frac{d}{dt}\left(\frac{u_x - v}{1 - vu_x/c^2}\right) \Big/ \frac{d}{dt}[\gamma(t - vx/c^2)].$$

Carry out the differentiations, remembering that v and γ are constants, to show that

$$a_x' = \frac{(1-v^2/c^2)^{3/2}}{\left(1-\dfrac{vu_x}{c^2}\right)^3}a_x. \tag{5.39}$$

Also show that

$$a_y' = \frac{1-\dfrac{v^2}{c^2}}{\left(1-\dfrac{vu_x}{c^2}\right)^2}\left(a_y + \frac{vu_y/c^2}{1-vu_x/c^2}a_x\right), \tag{5.40}$$

with a similar expression for a_z'. Show that if $u \ll c$ and $v \ll c$ then $\boldsymbol{a}' \approx \boldsymbol{a}$.

5.9 A particle is at rest at the origin of an inertial frame Σ at time $t = 0$. It moves such that it has a constant acceleration g relative to the inertial frame in which it is instantaneously at rest. Find the velocity and position of the particle relative to Σ after a time t measured in Σ.

Outline of solution From equation (5.39) with $v = u$

$$g = a_x' = \frac{(1 - u^2/c^2)^{3/2}}{(1 - u^2/c^2)^3} a_x$$

$$= \frac{d}{dt}\left[\frac{u}{(1 - u^2/c^2)^{1/2}}\right].$$

Then proceed as in Section 2.5 except that now $g = qE/m$. Show that, corresponding to equation (2.28),

$$u = \frac{gt}{\left(1 + \frac{g^2t^2}{c^2}\right)^{1/2}},$$

or

$$t = \frac{u}{g\,(1 - u^2/c^2)^{1/2}}.$$

Corresponding to equation (2.35),

$$x = \frac{c^2}{g}\left[\left(1 + \frac{g^2t^2}{c^2}\right)^{1/2} - 1\right].$$

Show that, corresponding to equation (2.36),

$$\left(x + \frac{c^2}{g}\right)^2 - c^2t^2 = \frac{c^4}{g^2},$$

or

$$gx^2 + 2c^2x - gc^2t^2 = 0.$$

This is the equation for a hyperbola. This type of motion is consequently known as hyperbolic motion.

★ **5.10** Prove that light signals sent out from the origin of Σ after a time $t = c/g$ in Σ will never catch the particle discussed in the previous problem. [*Hint*: Show that the line $x = c\,(t - c/g)$ is an asymptote to the hyperbola $(x + c^2/g)^2 - c^2t^2 = c^4/g^2$.]

5.11 A light source is moving past an observer stationary in the laboratory frame with a velocity of $0.6c$. The distance of closest approach is 10 m from the observer. Find the distance (measured in the laboratory system) of the source from the observer when the source emits light that shows no Doppler shift when it reaches the observer—that is, when the frequency of the light reaching the observer, measured in the laboratory system, is equal to the frequency of the light measured in the rest frame of the source.

5.12 Assume that light is emitted isotropically in the coordinate system in which a sodium lamp is at rest. If the lamp moves with velocity $\frac{1}{4}c$ relative to the laboratory, calculate the half-angle of the cone containing half the emitted photons. [*Hint*: Consider light emitted at 90° relative to the direction of motion of the lamp in the reference system in which the lamp is at rest. Calculate its direction in the laboratory frame.]

5.13 When excited ^{57}Fe nuclei decay, they sometimes emit photons of energy 14.4 keV, corresponding to a frequency of 3.46×10^{18} Hz. If the ^{57}Fe nucleus moves with velocity $\frac{1}{5}c$ when it decays via the 14.4 keV level, calculate the frequencies of radiation emitted at angles of 0°, 60°, 90°, 120° and 180° (in the laboratory frame) relative to the direction of motion of the nucleus. [*Hint*: Use equation (5.24).]

★ **5.14** A plane mirror is moving in the direction of its normal with uniform velocity v in the positive x direction in the inertial frame Σ. Let a ray of light of wavelength λ_1 strike the moving mirror at an angle of incidence α_1, and let the ray be reflected at an angle α_2 to the normal with wavelength λ_2. Show that if the mirror is moving towards the incident light

$$\frac{\sin \alpha_1}{\cos \alpha_1 + v/c} = \frac{\sin \alpha_2}{\cos \alpha_2 - v/c},$$

$$\frac{\lambda_1}{\lambda_2} = \frac{c + v \cos \alpha_1}{c - v \cos \alpha_2}.$$

[*Hint*: The angle of reflection is equal to the angle of incidence in the reference frame in which the mirror is at rest. The wavelength of the light is unchanged on reflection in this reference frame. Use the formulae for aberration and for the Doppler effect.]

6

Relativistic Mechanics and Conservation Laws

6.1 RÉSUMÉ OF CHAPTER 2

An account was given in Chapter 2 of experiments on the motions of high-speed charged particles, such as electrons and protons, in electric and magnetic fields that confirmed that the relativistic momentum $\boldsymbol{p}$ of a particle of mass m moving with velocity $\boldsymbol{u}$ is given by

$$\boldsymbol{p} = \alpha m \boldsymbol{u} \tag{6.1}$$

where

$$\alpha = \frac{1}{(1 - u^2/c^2)^{1/2}} . \tag{6.2}$$

Note that α depends only on u, the speed of the particle, and on c, the speed of light in empty space. The invariant mass m can be measured (using e.g. a mass spectrometer) whatever the speed of the particle, provided that the laws of relativistic mechanics are applied (see Section 2.12). The mass m can also be measured using Newtonian mechanics in the zero-velocity limit.

According to equation (2.24), when a charge of magnitude q is moving with velocity $\boldsymbol{u}$ in an electric field of intensity $\boldsymbol{E}$ and a magnetic field of magnetic induction $\boldsymbol{B}$, the expression for the Lorentz force becomes

$$\boldsymbol{f} = \frac{\mathrm{d}\boldsymbol{p}}{\mathrm{d}t} = \frac{\mathrm{d}}{\mathrm{d}t}(\alpha m \boldsymbol{u})$$

$$= \frac{\mathrm{d}}{\mathrm{d}t}\left[\frac{m\boldsymbol{u}}{(1 - u^2/c^2)^{1/2}}\right] = q\boldsymbol{E} + q\boldsymbol{u} \times \boldsymbol{B}. \tag{6.3}$$

It was shown in Chapter 2 that, provided the charge is not emitting electromagnetic radiation, equation (6.3) is in excellent agreement with the experimental results and can be used to design proton accelerators capable of accelerating protons up to energies of more than 1 TeV.

Using equation (6.1), it was shown in Section 2.2 that the kinetic energy T of a particle of mass m moving with velocity $\boldsymbol{u}$ is

$$T = mc^2 (\alpha - 1) . \tag{6.4}$$

According to equation (2.49), the (total) energy E of the particle is defined by

$$E = T + mc^2. \tag{6.5}$$

It follows from equations (6.4) and (6.5) that

$$E = \alpha mc^2. \tag{6.6}$$

According to equation (2.55), the momentum p and the (total) energy E of a particle are related by

$$E^2 = c^2p^2 + m^2c^4. \tag{6.7}$$

Dividing equation (6.1) by equation (6.6), we have

$$\boldsymbol{u} = \frac{c^2\boldsymbol{p}}{E} . \tag{6.8}$$

It was shown in Section 2.5 that it follows from equation (6.3) that there is a limiting speed for electrons accelerated in an electric field, which is equal to the speed of light in empty space. It was shown in Section 2.13 that the Lorentz transformations follow if it is assumed that this experimentally confirmed result holds in all inertial reference frames, illustrating that the Lorentz transformations are the appropriate coordinate and time transformations, if the above equations for the relativistic mechanics of a charged particle moving in an electric and a magnetic field obey the principle of relativity.

In this chapter we shall extend the theory of relativistic mechanics to the interactions of particles. In Newtonian mechanics linear momentum is conserved in both elastic and inelastic collisions. We shall start by assuming that the relativistic momentum, defined by equation (6.1), which is the same as equation (2.1), is conserved in all inertial reference frames in the interactions of "point" particles. It will be shown in Section 6.2 that this leads to the general form of the law of conservation of energy in relativistic mechanics.

Historical comment. It is not necessary to develop the relativistic mechanics of a particle before deriving the Lorentz transformations. The approach used in Chapter 2 was adopted to convince the reader from the outset that, when the speeds of particles are comparable to the speed of light, Newtonian mechanics is completely inadequate and the Galilean transformations must be replaced by the Lorentz transformations. This approach was designed to prepare the reader psychologically for the changes in the concepts of space and time necessitated by the Lorentz transformations. In the more traditional approach to special relativity the experimental evidence in favour of the principle of the constancy of the speed of light is based mainly on optical experiments such as that of

Michelson and Morley. The Lorentz transformations and the relativistic velocity transformations can then be derived before even mentioning relativistic mechanics, which is then developed by seeing how the momentum of a particle must be redefined if, in the case of an idealized elastic collision, the laws of conservation of momentum and of energy are to obey the principle of relativity, when the coordinates and time are transformed using the Lorentz transformations. This approach leads to equation (6.1) (see Rosser (1964).)

6.2 THE LAW OF CONSERVATION OF ENERGY

Consider the general interaction of two particles labelled 1 and 2, of masses m_1 and m_2, that before the event are moving with velocities $\boldsymbol{u}_1$ and $\boldsymbol{u}_2$ in the inertial reference frame Σ and velocities $\boldsymbol{u}_1'$ and $\boldsymbol{u}_2'$ in the inertial frame Σ', as shown in Figs. 6.1(a,b). Σ' is moving with uniform velocity v relative to Σ along their common x axis. Suppose that the two particles present after the interaction have masses m_3 and m_4 and move with velocities $\boldsymbol{u}_3$ and $\boldsymbol{u}_4$ respectively in Σ and velocities $\boldsymbol{u}_3'$ and $\boldsymbol{u}_4'$ in Σ'. It will be assumed that all the particles are small enough for the event to be considered as taking place at one point of space at one instant of time.

It will be assumed that the total linear momentum, determined using equation (6.1), is conserved in the event in *both* Σ and Σ'. Applying the law of conservation of momentum in the $+x$ direction in Σ, using equation (6.1) to define momentum, we have

$$\alpha_1 m_1 u_{1x} + \alpha_2 m_2 u_{2x} = \alpha_3 m_3 u_{3x} + \alpha_4 m_4 u_{4x}, \tag{6.9}$$

where u_{1x} is the x component of $\boldsymbol{u}_1$ etc. and $\alpha_1 = (1 - u_1^2/c^2)^{-1/2}$ etc. It will be assumed, by analogy with equation (6.1) that the momentum, measured in Σ', of

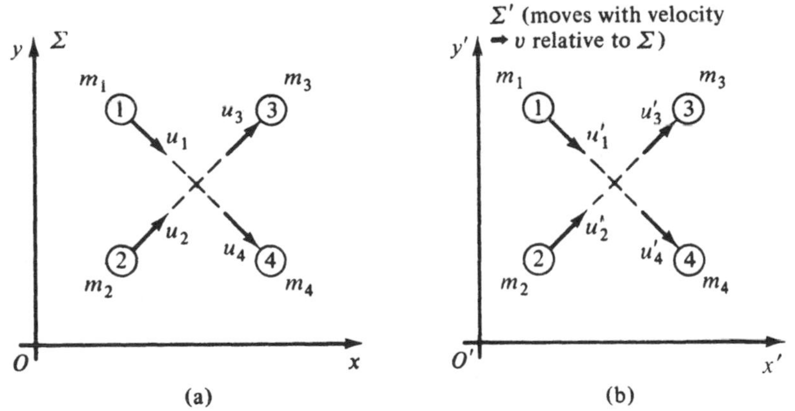

Fig. 6.1 General interaction of two particles of masses m_1 and m_2. After the event there are two particles of masses m_3 and m_4. (a) The experimental conditions in Σ. (b) The experimental conditions in Σ'.

a particle of mass m moving with velocity $\boldsymbol{u}'$ in Σ' is $\alpha' m\boldsymbol{u}'$, where $\alpha' = (1 - u'^2/c^2)^{-1/2}$. Applying the law of conservation of momentum in the $+x'$ direction of Σ', we have

$$\alpha_1' m_1 u_{1x}' + \alpha_2' m_2 u_{2x}' = \alpha_3' m_3 u_{3x}' + \alpha_4' m_4 u_{4x}'. \tag{6.10}$$

where u_{1x}' is the x' component of $\boldsymbol{u}_1'$ etc. and $\alpha_1' = (1 - u_1'^2/c^2)^{-1/2}$ etc. The same values are used for the invariant masses m_1, m_2, m_3 and m_4 in both Σ and Σ'. Both equations (6.9) and (6.10) must be valid if momentum is to be conserved in both Σ and Σ'.

It follows from the relativistic velocity transformations, equations (5.6) and (5.14), that

$$u_{1x}' = \frac{u_{1x} - v}{1 - vu_{1x}/c^2}$$

$$\alpha_1' = \gamma\alpha_1 (1 - vu_{1x}/c^2),$$

and therefore

$$\alpha_1' u_{1x}' = \gamma\alpha_1 (u_{1x} - v),$$

where $\gamma = (1 - v^2/c^2)^{-1/2}$ and v is the uniform velocity at which Σ' is moving relative to Σ along their common x axis. Similar expressions hold for $\alpha_2' u_{2x}'$, $\alpha_3' u_{3x}'$ and $\alpha_4' u_{4x}'$. Substituting for $\alpha_1' u_{1x}'$ etc. into equation (6.10), which expresses the law of conservation of momentum in the $+x'$ direction of Σ', we have

$$\gamma m_1\alpha_1 (u_{1x} - v) + \gamma m_2\alpha_2 (u_{2x} - v) = \gamma m_3\alpha_3 (u_{3x} - v) + \gamma m_4\alpha_4 (u_{4x} - v).$$

Cancelling γ and rearranging, we obtain

$$\alpha_1 m_1 u_{1x} + \alpha_2 m_2 u_{2x} - v(\alpha_1 m_1 + \alpha_2 m_2) = \alpha_3 m_3 u_{3x} + \alpha_4 m_4 u_{4x} - v(\alpha_3 m_3 + \alpha_4 m_4) \tag{6.11}$$

Subtracting equation (6.9), which expresses the law of conservation of momentum in Σ, from equation (6.11), which is derived from the law of conservation of momentum in Σ', we obtain

$$v\,(\alpha_1 m_1 + \alpha_2 m_2) = v(\alpha_3 m_3 + \alpha_4 m_4).$$

Multiplying both sides by c^2/v gives

$$\alpha_1 m_1 c^2 + \alpha_2 m_2 c^2 = \alpha_3 m_3 c^2 + \alpha_4 m_4 c^2. \tag{6.12}$$

Equation (6.12) must be satisfied if linear momentum, defined by equation (6.1), is to be conserved in *both* Σ and Σ'. Using equation (6.6), equation (6.12) can be rewritten in the form

$$E_1 + E_2 = E_3 + E_4, \tag{6.13}$$

where E_1 is the (total) energy of particle 1 in Σ etc. Equation (6.13) is the relativistic form of the *law of conservation of energy*. It must be satisfied if relativistic momentum, defined by equation (6.1), is to be conserved in both Σ

and Σ' when the relativistic velocity transformations are used to transform from Σ' to Σ. Using equation (6.5), equation (6.13) can be rewritten in the form

$$T_1 + m_1c^2 + T_2 + m_2c^2 = T_3 + m_3c^2 + T_4 + m_4c^2. \tag{6.14}$$

It should be stressed that, in addition to the kinetic energies T_1, T_2, T_3 and T_4 of the particles, the quantities m_1c^2, m_2c^2, m_3c^2 and m_4c^2 must be included in equations (6.13) and (6.14) when the law of conservation of energy is applied in relativistic mechanics. This is true even if the particle is at rest. For example, if particle 2 in Fig. 6.1(a) is at rest in Σ before the event, we must still include $E_2 = m_2c^2$ when the law of conservation of energy is applied in Σ. This suggests that, in addition to its kinetic energy, a particle of mass m makes a contribution mc^2 to the total energy of a system. This will be confirmed by applying equation (6.13) to interpret the dynamics of positron annihilation, electron pair production, nuclear reactions, particle production and particle decays. The experimental confirmation of equation (6.13) allows us to conclude that a stationary particle of mass m makes a contribution E^0, called the *rest energy*, to the relativistic form of the law of conservation of energy, where

$$E^0 = mc^2. \tag{6.15}$$

According to equation (6.15), the rest energy of a particle is a measure of its mass, and vice versa. The relationship between rest energy and mass will be developed further in Section 6.9. Equation (6.5) can now be interpreted as saying that the (total) energy E of a particle is equal to the sum of its kinetic energy T and its rest energy mc^2.

It is straightforward to show, starting from equation (6.9) and using equations (5.9) and (5.16), that, if momentum is to be conserved in both Σ and Σ', when the relativistic velocity transformations are used to transform from Σ to Σ', we must have in Σ'

$$E_1' + E_2' = E_3' + E_4', \tag{6.16}$$

where E_1' is the (total) energy of particle 1 in Σ' etc. It is straightforward to adapt the above analysis to the general case where there are arbitrary numbers of particles before and after the event.

For convenience, the laws of conservation of linear momentum and conservation of energy will now be summarized.

The law of conservation of linear momentum If the momenta of the particles present before the event are $\boldsymbol{p}_1, \boldsymbol{p}_2, \cdots, \boldsymbol{p}_i, \cdots$, and the momenta of the particles present after the event are $\bar{\boldsymbol{p}}_1, \bar{\boldsymbol{p}}_2, \cdots, \bar{\boldsymbol{p}}_j, \cdots$, where the momentum of each particle is defined by equation (6.1), then for point events

$$\sum_i \boldsymbol{p}_i = \sum_j \bar{\boldsymbol{p}}_j \,. \tag{6.17}$$

The law of conservation of energy If $E_1, E_2, ..., E_i, ...$ are the (total) energies of the particles present before the event and $\bar{E}_1, \bar{E}_2, ..., \bar{E}_j, ...$ are the (total) energies of the particles present after the event then for point events

$$\sum_i E_i = \sum_j \bar{E}_j \, . \tag{6.18}$$

Equations (6.17) and (6.18) can be applied in *any* inertial reference frame, such as the laboratory frame. The momentum $\boldsymbol{p}_i$ and the (total) energy E_i of the ith particle are related by equation (6.7):

$$E_i^2 = c^2 p_i^2 + m_i^2 c^4. \tag{6.19}$$

Equation (6.19) can be used to determine the momentum of a particle from its (total) energy, and vice versa. It will be shown in this chapter that predictions made on the basis of equations (6.17) and (6.18) are in excellent agreement with experimental results.

6.3 ELASTIC COLLISION OF TWO PARTICLES

To illustrate the application of the conservation laws of relativistic mechanics, consider the elastic collision of particles 1 and 2 of masses m_1 and m_2 respectively, as shown in Fig. 6.2. The problem will be solved in the laboratory frame Σ. The momentum of particle 1 in Σ before collision is $p_1 = \alpha_1 m_1 u_1$ directed along the x axis. The (total) energy of particle 1 before the collision is $E_1 = \alpha_1 m_1 c^2$, and its kinetic energy T_1 is $m_1 c^2 (\alpha_1 - 1)$. Particle 2 is at rest in Σ before the collision, so that $p_2 = 0$ and $E_2 = m_2 c^2$. Choose the direction of the y axis such that the paths

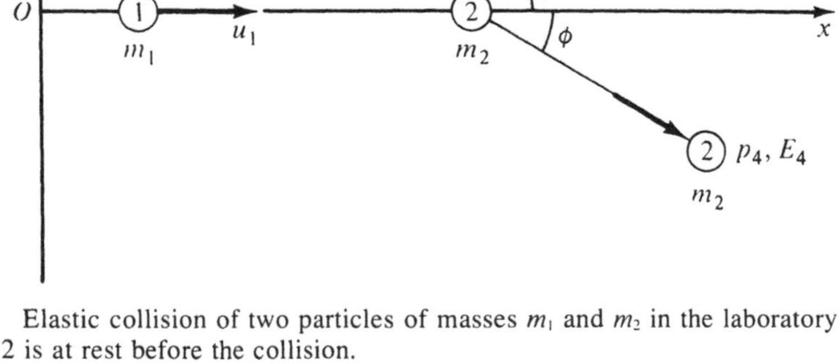

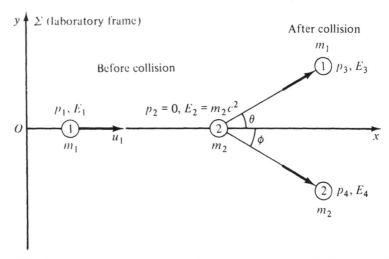

Fig. 6.2 Elastic collision of two particles of masses m_1 and m_2 in the laboratory frame. Particle 2 is at rest before the collision.

of particles 1 and 2 are in the (x, y) plane after the collision, as shown in Fig. 6.2. Let the paths of particles 1 and 2 be inclined at angles of θ and ϕ respectively to the x axis. Suppose that the momenta of particles 1 and 2 after the collision are $\mathbf{p}_3$ and $\mathbf{p}_4$ respectively and that their (total) energies after the collision are E_3 and E_4 respectively. Applying the law of conservation of momentum in the $+x$ and $+y$ directions of Σ, we have

$$p_1 = p_3 \cos\theta + p_4 \cos\phi. \tag{6.20}$$

$$0 = p_3 \sin\theta - p_4 \sin\phi. \tag{6.21}$$

Applying the law of conservation of energy, equation (6.18), remembering to include the rest energy m_2c^2 of particle 2, which is at rest in Σ before the collision, we have

$$E_1 + m_2c^2 = E_3 + E_4\,. \tag{6.22}$$

The (total) energies of the particles are related to their momenta by equation (6.7), which can be used to eliminate E_1, E_3 and E_4 in equation (6.22) to give

$$(c^2p_1^2 + m_1^2c^4)^{1/2} + m_2c^2 = (c^2p_3^2 + m_1^2c^4)^{1/2} + (c^2p_4^2 + m_2^2c^4)^{1/2}. \tag{6.23}$$

Assuming that m_1, m_2, p_1 and $p_2 = 0$ are given, there are four unknowns in the three equations (6.20), (6.21) and (6.23), namely p_3, p_4, θ and ϕ. To solve the problem, we must be given one of these unknowns or a relationship between them. As an example, we shall consider the special case when $m_1 = m_2 = m$ and $\theta = \phi$ after the collision in Fig. 6.2.

It follows from equation (6.21) that if $\theta = \phi$ then $p_3 = p_4$. Since the particles have the same mass m and the same momentum $p_3 = p_4$, it follows from equation (6.7) that $E_3 = E_4$. Hence equation (6.22) can be rewritten as

$$E_1 + mc^2 = 2E_3\,. \tag{6.24}$$

Putting $\phi = \theta$ and $p_4 = p_3$ in equation (6.20), and then multiplying by c, we obtain

$$cp_1 = 2cp_3 \cos\theta.$$

Squaring and using equation (6.7), we have

$$\cos^2\theta = \frac{c^2p_1^2}{4c^2p_3^2} = \frac{E_1^2 - m^2c^4}{4\,(E_3^2 - m^2c^4)}$$

$$= \frac{(E_1 + mc^2)\,(E_1 - mc^2)}{(2E_3 + 2mc^2)\,(2E_3 - 2mc^2)}\,. \tag{6.25}$$

Using equation (6.24) to substitute $E_1 + mc^2$ for $2E_3$ in equation (6.25) and taking the square root, we obtain

$$\cos\theta = \left(\frac{E_1 + mc^2}{E_1 + 3mc^2}\right)^{1/2} = \left(\frac{T_1 + 2mc^2}{T_1 + 4mc^2}\right)^{1/2}. \tag{6.26}$$

If the kinetic energy T_1 of the incident particle 1 before the collision is very much less than mc^2 then $\cos\theta$ tends to $(2mc^2/4mc^2)^{1/2} = 2^{-1/2}$, so that in the low-velocity limit $\theta = 45°$. In this case the angle between the paths of particles 1 and 2 after the collision is 90°. (According to Newtonian mechanics, the angle between the paths of particles 1 and 2 is always 90°, even when ϕ is not equal to θ.)

As a typical relativistic example, put $T_1 = mc^2$ so that $E_1 = 2mc^2$. In this case, according to equation (6.26), $\cos\theta = (3mc^2/5mc^2)^{1/2} = 0.7746$, giving $\theta = 39.2°$. Here the angle 2θ between the paths of particles 1 and 2 after the collision is 78.4°.

In the extreme relativistic limit, when $T_1 >> mc^2$, $\cos\theta$ tends to 1, so that θ and hence 2θ tend to zero. The above results show that according to relativistic mechanics, as the kinetic energy T_1 of the incident particle 1 increases, the total angle between the paths of particles 1 and 2 after the collision becomes smaller and smaller, whereas according to Newtonian mechanics, the total angle should remain equal to 90°. The experimental results on the elastic scattering of atomic particles, such as electrons, are in agreement with the predictions of relativistic mechanics. (For example, see Champion (1932).)

6.4 COMPTON SCATTERING OF X RAYS

It was shown in Section 2.9 that it is consistent with the theory of relativistic mechanics to assume that a photon of energy $h\nu$ has a momentum $h\nu/c$, where h is Planck's constant. Relativistic mechanics was applied successfully by Compton to interpret what is now known as the Compton effect. He showed experimentally that, when a monochromatic beam of X rays is scattered by a light element such as carbon, the scattered radiation consists of two components, one

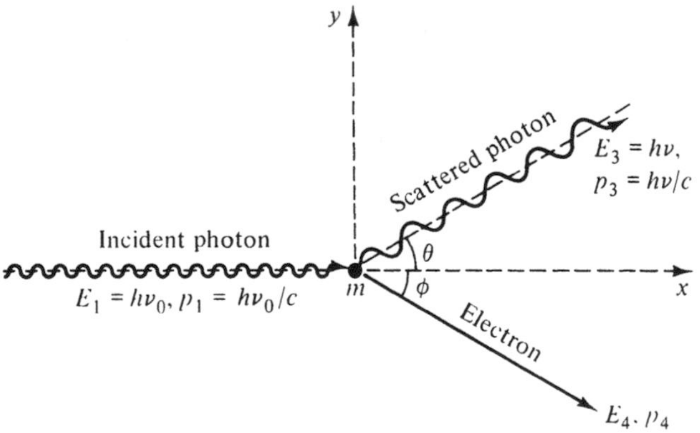

Fig. 6.3 The Compton effect. A photon of energy $h\nu_0$ and momentum $h\nu_0/c$ is incident on a stationary electron of mass m. After collision, a photon has energy $h\nu$ (which is less than $h\nu_0$) and momentum $h\nu/c$. The electron recoils with momentum p_4 and (total) energy E_4.

having the same wavelength as the incident radiation and the other having a slightly longer wavelength. In order to account for the presence of the longer-wavelength radiation, Compton suggested that the scattering process could be treated as an interaction between a single photon and a free electron. Let a photon of energy $E_1 = h\nu_0$ and momentum $p_1 = h\nu_0/c$ moving along the positive x direction be incident upon a stationary electron of mass m in an inertial frame Σ, as shown in Fig. 6.3. Choose the direction of the y axis such that the scattered photon and recoiling electron are in the (x, y) plane after the event. Suppose that the scattered photon has energy $E_3 = h\nu$ and moves off at an angle θ to the x axis after collision and that the electron moves off at an angle ϕ with momentum p_4 and (total) energy E_4 (Fig. 6.3). According to the law of conservation of energy,

$$E_1 + mc^2 = E_3 + E_4$$

or

$$E_1 + mc^2 - E_3 = E_4 .$$

On squaring both sides, we have

$$E_1^2 + m^2c^4 + E_3^2 + 2E_1mc^2 - 2E_1E_3 - 2E_3mc^2 = E_4^2 . \tag{6.27}$$

Application of the law of conservation of momentum gives

$$cp_1 = cp_3 \cos\theta + cp_4 \cos\phi, \tag{6.28}$$

$$0 = cp_3 \sin\theta - cp_4 \sin\phi. \tag{6.29}$$

For the incident photon $cp_1 = h\nu_0 = E_1$, and for the scattered photon $cp_3 = h\nu = E_3$. Hence equations (6.28) and (6.29) can be rewritten as

$$E_1 - E_3 \cos\theta = cp_4 \cos\phi, \tag{6.30}$$

$$E_3 \sin\theta = cp_4 \sin\phi. \tag{6.31}$$

On squaring both sides of equations (6.30) and (6.31), and then adding and using the relations $\sin^2\theta + \cos^2\theta = 1$, $\sin^2\phi + \cos^2\phi = 1$ and $(cp_4)^2 = E_4^2 - m^2c^4$, we have

$$E_1^2 - 2E_1E_3 \cos\theta + E_3^2 = E_4^2 - m^2c^4. \tag{6.32}$$

Subtracting equation (6.32) from equation (6.27) gives

$$2E_1mc^2 - 2E_1E_3 - 2E_3mc^2 + 2E_1E_3 \cos\theta = 0.$$

Dividing by $2E_1E_3$ and rearranging we obtain

$$1 - \cos\theta = mc^2\left(\frac{1}{E_3} - \frac{1}{E_1}\right). \tag{6.33}$$

Putting

$$E_1 = h\nu_0 = \frac{hc}{\lambda_0}, \qquad E_3 = h\nu = \frac{hc}{\lambda},$$

where λ_0 and λ are the wavelengths of the incident and scattered photons respectively, and then rearranging, we finally obtain:

$$\Delta\lambda = \lambda - \lambda_0 = \frac{h}{mc}(1 - \cos\theta)\,. \tag{6.34}$$

Substituting the numerical values for the constants h, m and c, we have

$$\Delta\lambda = 0.00242\,(1 - \cos\theta)\ \text{nm}. \tag{6.35}$$

According to equation (6.35), the wavelength of the scattered photon should be greater than the wavelength of the incident photon by $0.00242\,(1 - \cos\theta)$ nm. The predicted dependence of $\Delta\lambda$ on θ has been confirmed experimentally. The kinetic energies of the recoil electrons have also been measured and found to be in agreement with the calculated values. Thus, by attributing a momentum $h\nu/c$ to a photon of energy $h\nu$, and using the theory of relativistic mechanics, Compton was able to account for the Compton effect. The calculation of the cross-section for the scattering process had to await the application of the theory of relativistic quantum mechanics by Klein and Nishina.

6.5 NUCLEAR REACTIONS (INTRODUCTION)

In the two examples of the application of the laws of conservation of momentum and of energy that we have considered so far, namely the elastic collision of two particles of equal mass and the Compton scattering of X rays, the rest energies of the particles were the same before and after the events. We now go on to consider examples where the rest energies are changed in the events. The most familiar examples are nuclear reactions.

Consider the nuclear reaction

$$A + B \rightarrow C + D + Q,$$

where particles A and B react to produce particles C and D, and $Q = T_{\text{final}} - T_{\text{initial}}$ is the increase in kinetic energy. Let m_A, m_B, m_C and m_D be the masses of the reactants, which can be measured accurately using mass spectrometers, for example of the type discussed in Section 2.12. Let T_A and T_B be the kinetic energies of particles A and B before the reaction and T_C and T_D those of particles C and D after the reaction. Applying the law of conservation of energy, we obtain

$$E_A + E_B = E_C + E_D\,.$$

Putting $E_A = T_A + m_A c^2$ etc., we have

$$T_A + m_A c^2 + T_B + m_B c^2 = T_C + m_C c^2 + T_D + m_D c^2,$$

and, rearranging,

$$Q = (T_C + T_D) - (T_A + T_B) = (m_A + m_B - m_C - m_D)\,c^2. \tag{6.36}$$

This equation can be used to calculate the Q value of the reaction from the values of the masses of the particles. Equation (6.36), which was derived from the *law of conservation of energy*, can be rewritten in the simplified form

$$Q = \Delta T = -\Delta (mc^2) = -c^2 \Delta m. \quad (6.37)$$

where ΔT is the increase in the kinetic energy in the reaction and Δm is the increase in the masses of the particles. Equation (6.37) was derived using the law of conservation of energy, and states that Q, the increase in kinetic energy, is equal to the decrease in the rest energy. The decrease in the rest energy can be calculated from the experimentally determined masses of the particles. We shall go on in Section 6.10 to treat the atomic nucleus as a composite particle, and it will be shown that the changes in the rest energy are due mainly to changes in the binding energies of the protons and neutrons inside the nuclei. There is no need, however, to enquire about the origin of the changes in rest energy when applying equation (6.37), which was derived from the law of conservation of energy.

The first experimental verification of the relation $\Delta T = -c^2 \Delta m$ was obtained by Cockroft and Walton, who accelerated protons up to an energy of 0.25 MeV. They observed that when these protons were incident upon a lithium target, two α particles were sometimes emitted, in opposite directions. From the ranges of the α particles, Cockroft and Walton estimated that the kinetic energy of each particle was 8.6 MeV. The gain in kinetic energy in the reaction was

$$Q = (2 \times 8.6 - 0.25)\ \text{MeV} = 16.95\ \text{MeV},$$

where the 0.25 MeV represents the kinetic energy of the incident proton. Cockroft and Walton suggested that the reaction taking place was

$${}^{7}_{3}\text{Li} + {}^{1}_{1}\text{H} \rightarrow {}^{8}_{4}\text{Be} \rightarrow {}^{4}_{2}\text{He} + {}^{4}_{2}\text{He} + Q.$$

It will be shown in Section 6.11, using recent values of atomic masses, that the Q value calculated using $Q = \Delta T = -c^2 \Delta m$ is 17.35 MeV, which is in agreement with the results obtained by Cockroft and Walton. Many similar experiments have been performed, and in all cases it is possible to estimate the energy released or absorbed in a nuclear reaction from the difference between the measured atomic masses of the particles before and after the reaction, using the relation $\Delta T = -c^2 \Delta m$.

6.6 CREATION AND ANNIHILATION OF PARTICLES

According to relativistic mechanics, a particle of mass m makes a contribution of mc^2, called the rest energy, to the energy balance when the law of conservation of energy, equation (6.18), is applied. This suggests that an amount of energy equal to mc^2 must be expended in creating a particle of mass m, while if a particle of mass m is annihilated, an amount of energy equal to mc^2 is available for conversion into some other form of energy. Experimental evidence confirming the above suggestions will now be given.

6.6.1 Electron pair production

The positively charged electron, now called the *positron*, was discovered by Anderson in 1932. Soon afterwards, electron pair production was discovered. If a photon has energy $h\nu > 2mc^2$, where m is the mass of the electron, then the photon can produce an electron pair, i.e. a positive and negative electron. Electron pair production cannot take place spontaneously in a vacuum—the presence of a charged particle, such as an atomic nucleus or an electron, is necessary in order that *both* momentum and energy be conserved in the process of pair production (see Problem 6.9).

It will be assumed that electron pair production takes place in the electric field of a stationary charged particle of mass M. Let $\boldsymbol{p}_3$, $\boldsymbol{p}_4$ and $\boldsymbol{p}_r$ be the momenta, and T_3, T_4, and T_r the kinetic energies of the positron, electron and recoiling particle respectively. Let the energy of the incident photon be $E_1 = h\nu$, such that the incident momentum is $p_1 = h\nu/c$. According to the law of conservation of momentum, equation (6.17),

$$\boldsymbol{p}_1 = \boldsymbol{p}_3 + \boldsymbol{p}_4 + \boldsymbol{p}_r. \tag{6.38}$$

According to the law of conservation of energy, equation (6.18),

$$h\nu + Mc^2 = T_3 + mc^2 + T_4 + mc^2 + T_r + Mc^2$$

and therefore

$$h\nu = 2mc^2 + (T_3 + T_4 + T_r)\ . \tag{6.39}$$

Before the pair production takes place, we have a photon of energy $h\nu$ plus the rest energy Mc^2 of the stationary charged particle. After the event, in addition to Mc^2, we have an extra two electrons, each of rest energy mc^2, plus the kinetic energies of the positron, electron and the recoiling charged particle. In pair production some of the radiant or electromagnetic energy, in the form of a photon, is expended in creating two electrons, each of rest energy $mc^2 = 0.511$ MeV. It will be shown in Section 7.9 that in the limit when $M >> m$, which is true when the recoiling particle is an atomic nucleus, the minimum energy needed by the incident photon to produce an electron pair is $h\nu = 2mc^2 =$ 1.022 MeV. It will also be shown there that in the case when pair production takes place in the electric field of an electron, the minimum photon energy needed to produce an electron pair is $h\nu = 4mc^2 = 2.044$ MeV.

6.6.2 Positron annihilation

When a positron slows down it is attracted to a negative electron and forms positronium. The electron and positron move in orbits around their centre of mass in a way similar to the hydrogen atom, except that the masses of the two particles are equal in this case. In the lowest quantum state the quantum numbers n and l have the values 1 and 0 respectively. If the spin of the electron and the positron are opposite in direction the singlet ^{1}S state of total spin 0 is formed. If the spins are parallel then the triplet ^{3}S state of total spin 1 is formed. After an average lifetime of 8×10^{-9} s in the ^{1}S state, the positron and electron annihilate each other according to the reaction

$$e^+ + e^- = 2h\nu.$$

Since two photons are emitted, it is possible to conserve both energy and linear momentum. The energy of each photon should be equal to $mc^2 = 0.511$ MeV; this is confirmed by experiment. In the triplet ^{3}S state three photons are emitted, the lifetime being 7×10^{-6} s. Since three photons are emitted, spin and angular momentum can be conserved since the spin of the photon is one unit. If linear momentum is to be conserved, the three photons must be coplanar; this is confirmed by experiment. In this case the energy is distributed statistically between the three photons. Positron annihilation is an example of the conversion of rest energy into radiant energy. The fact that three rather than two photons, each of spin 1 are emitted when positronium is in the ^{3}S state of total spin 1 is an example of the law of conservation of angular momentum.

6.6.3 Particle production

When fast protons or neutrons, for example, collide with atomic nuclei, pions (π) can be produced, provided that the energy of the incident particle is high enough. The charged pions ($\pi^{\pm}$) produced have masses of 139.6 MeV/c^2, while the neutral pions (π^0) have masses of 135.0 MeV/c^2. In these collisions some of the kinetic energy of the incident particle is converted into the rest energies of the pions produced. At higher energies K mesons, nucleons, antiprotons and other baryons etc. can be produced. Experiments show that the laws of conservation of linear momentum and energy are valid in these collisions.

A full discussion of high-energy physics goes beyond the scope of this book. The interested reader is referred to the books by Dodd (1984) and Perkins (1987). It is found that extra conservation laws are applicable in certain cases, such as the laws of conservation of charge, parity and baryon number. The full theories must be used if one wants to calculate the probability that a particular process takes place. However, if the theories used are Lorentz-covariant, that is if they obey the principle of relativity when the Lorentz transformations are used, then the dynamics of the collisions must always be consistent with the relativistic laws of conservation of energy and conservation of momentum.

6.6.4 Example: the creation of a pion

To illustrate how the laws of conservation of momentum and energy can be applied to events in which new particles are created, consider the reaction:

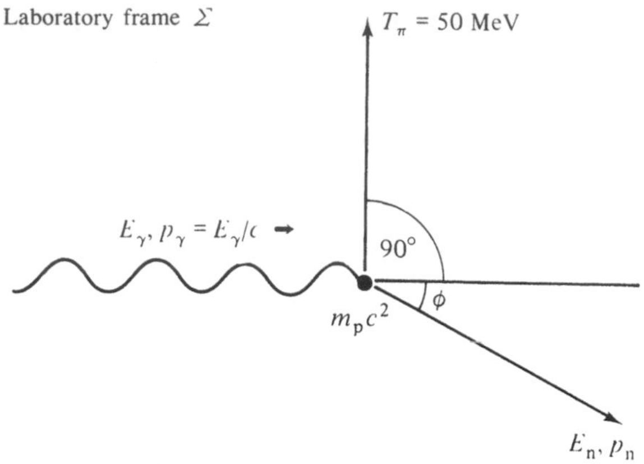

Fig. 6.4 An incident γ ray collides with a stationary proton to give a charged pion and a neutron.

$$\gamma + p \rightarrow n + \pi^+,$$

where an incident photon (γ ray) denoted γ, with energy E_γ, collides with a stationary proton (p), producing a neutron (n) and a positive pion (π^+). We shall determine E_γ, given that in the laboratory frame Σ the pion is emitted at an angle of 90° to the direction of the incident photon with a kinetic energy $T_\pi = 50$ MeV, as shown in Fig. 6.4. Let the neutron be emitted at an angle φ to the direction of the incident photon with energy E_n and momentum p_n.

Since $m_\pi c^2 = 139.6$ MeV and $T_\pi = 50$ MeV, the (total) energy of the pion is

$$E_\pi = 50 + 139.6 \text{ Mev} = 189.6 \text{ MeV}.$$

Using equation (6.7), we find that its momentum is given by

$$cp_\pi = (E_\pi^2 - m_\pi^2 c^4)^{1/2} = [(189.6 \text{ MeV})^2 - (139.6 \text{ MeV})^2]^{1/2} = 128.3 \text{ MeV}.$$

Applying the law of conservation of energy, we have

$$E_\gamma + m_p c^2 = E_\pi + E_n.$$

Putting $E_\pi = 189.6$ MeV and $m_p c^2 = 938.0$ MeV gives

$$E_\gamma + 938.0 \text{ MeV} = 189.6 \text{ MeV} + E_n$$

or

$$E_n = E_\gamma + 748.4 \text{ MeV}. \tag{6.40}$$

From conservation of momentum in the direction of motion of the incident photon, since $cp_\gamma = E_\gamma$

$$cp_\gamma = E_\gamma = cp_n \cos\phi\,. \tag{6.41}$$

From conservation of momentum in the direction perpendicular to the direction of motion of the incident photon, and the result $cp_\pi = 128.3$ MeV,

$$cp_{\pi} = 128.3 \text{ MeV} = cp_{n}\sin\phi\,. \tag{6.42}$$

Squaring and adding equations (6.41) and (6.42), using $\sin^2\phi + \cos^2\phi = 1$, and then applying equation (6.7), we obtain

$$E_{\gamma}^2 + (128.3 \text{ MeV})^2 = c^2 p_n^{\,2} = E_n^{\,2} - m_n^{\,2}c^4.$$

Putting $m_n c^2 = 939.5$ MeV and rearranging gives

$$E_n^{\,2} = E_{\gamma}^2 + (128.3 \text{ MeV})^2 + (939.5 \text{ MeV})^2. \tag{6.43}$$

Squaring both sides of equation (6.40) gives

$$E_n^{\,2} = E_{\gamma}^2 + 1496.8 \text{ MeV} \times E_{\gamma} + (748.4 \text{ MeV})^2. \tag{6.44}$$

From equations (6.43) and (6.44), we find

$$\begin{aligned} E_{\gamma} &= [(128.3 \text{ MeV})^2 + (939.5 \text{ MeV})^2 - (748.4 \text{ MeV})^2]/1496.8 \text{ MeV} \\ &= 226.5 \text{ MeV}. \end{aligned}$$

Hence the energy of the incident γ ray (photon) is 226.5 MeV. It is left as an exercise for the reader to extend the calculation to show that $E_n = 974.9$ MeV, $T_n = 35.4$ MeV and $\phi = 29.5°$.

6.6.5 Particle decays

Let a particle of mass M decay at *rest* into a particle of mass m_1 momentum p_1 and (total) energy E_1, and into a particle of mass m_2, momentum p_2 and (total) energy E_2 as shown in Fig. 6.5. If linear momentum is to be conserved, m_1 and m_2 must go off in opposite directions, such that

$$p_1 = p_2\,. \tag{6.45}$$

The law of conservation of energy gives

$$E_1 + E_2 = Mc^2. \tag{6.46}$$

Multiplying both sides of equation (6.45) by c and squaring, we obtain

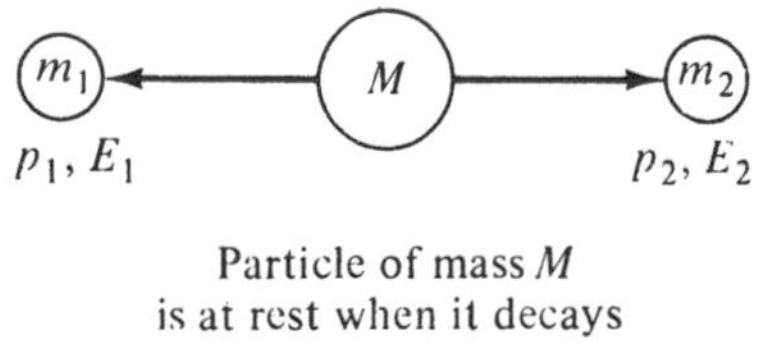

Fig. 6.5 A particle of mass M decays at rest into two particles of masses m_1 and m_2, which go in opposite directions, so as to conserve linear momentum.

$$c^2p_1^{\,2} = c^2p_2^{\,2}\,. \tag{6.47}$$

Using equation (6.7)

$$c^2p_1^{\,2} = E_1^{\,2} - m_1^{\,2}c^4, \qquad c^2p_2^{\,2} = E_2^{\,2} - m_2^{\,2}c^4.$$

Hence

$$E_1^{\,2} - m_1^{\,2}c^4 = E_2^{\,2} - m_2^{\,2}c^4,$$

or

$$E_1^{\,2} - E_2^{\,2} = (E_1 + E_2)\,(E_1 - E_2) = (m_1^{\,2} - m_2^{\,2})\,c^4. \tag{6.48}$$

Dividing equation (6.48) by equation (6.46), we find

$$E_1 - E_2 = \frac{(m_1^{\,2} - m_2^{\,2})\,c^2}{M}\,. \tag{6.49}$$

Adding equations (6.46) and (6.49) gives

$$E_1 = \frac{(M^2 + m_1^{\,2} - m_2^{\,2})\,c^2}{2M}\,. \tag{6.50}$$

While subtracting equation (6.49) from equation (6.46) gives

$$E_2 = \frac{(M^2 + m_2^{\,2} - m_1^{\,2})\,c^2}{2M}\,. \tag{6.51}$$

Using equations (6.7) and (6.47), we find that the momenta p_1 and p_2 of the two particles are given by

$$cp_1 = cp_2 = (E_1^{\,2} - m_1^{\,2}c^4)^{1/2} = (E_2^{\,2} - m_2^{\,2}c^4)^{1/2}. \tag{6.52}$$

The theory developed above can be applied to radioactive decay, photon emission, meson decay, etc. For example, a charged pion (π) decays into a muon (μ) and a neutrino. Experiments have shown that $m_\pi c^2 = 139.6$ MeV, $m_\mu c^2 = 105.7$ MeV and the neutrino has almost if not zero mass. If the pion decays at rest, then, substituting into equation (6.50),

$$E_\mu = \frac{(139.6\text{ MeV})^2 + (105.7\text{ MeV})^2 - (0\text{ MeV})^2}{2 \times 139.6\text{ MeV}} = 109.8\text{ MeV}.$$

Hence the kinetic energy of the muon is

$$T_\mu = E_\mu - m_\mu c^2 = 109.8\text{ MeV} - 105.7\text{ MeV} = 4.1\text{ MeV}.$$

This result is confirmed by experiment. It follows from equation (6.52) that

$$cp_\mu = 29.8\text{ MeV}.$$

Some particles decay into three particles, for example a muon decays into an electron and two neutrinos. Sometimes K mesons decay into three pions (π); for example,

$$K^+ \rightarrow \pi^+ + \pi^+ + \pi^-.$$

When the K^+ meson decays in a nuclear emulsion, the tracks of the three pions are visible, and if they end in the nuclear emulsion their energies can be estimated from their ranges in the emulsion. It is found that the three pions are coplanar, as they should be if linear momentum is conserved. It is also found that, although the momentum and energy available is distributed statistically between the three pions, both energy and linear momentum are conserved in the decays. It is possible to interpret the dynamics of all particle decays in terms of the laws of conservation of energy and of linear momentum. Other conservation laws, such as those of spin and angular momentum and of baryon and lepton numbers are also valid (for further details of these other conservation laws see Dodd (1984) Perkins (1987)).

6.7 THE TRANSFORMATIONS OF RELATIVISTIC MECHANICS

Relativistic mechanics was applied in Chapter 2 to the motions of electric charges in electric and magnetic fields in one particular inertial reference frame, generally the laboratory frame. Similarly in Sections 6.3–6.6 relativistic mechanics was applied in one inertial reference frame, again generally the laboratory frame. It is sometimes more convenient to carry out the calculations, or to present the experimental results, in a different inertial reference frame, such as the zero-momentum frame. To convert the results to or from the laboratory frame requires the transformations for momentum, energy and force. These will now be derived.

6.7.1 Transformation of momentum and energy

In the laboratory frame Σ, according to equations (6.1) and (6.6), the momentum and (total) energy of a particle of (invariant) mass m moving with velocity $\boldsymbol{u}$ are

$$p_x = \alpha m u_x, \qquad p_y = \alpha m u_y, \qquad p_z = \alpha m u_z, \qquad E = \alpha m c^2, \tag{6.53}$$

where

$$\alpha = \frac{1}{(1 - u^2/c^2)^{1/2}}, \tag{6.54}$$

and where u is the speed of the particle. The corresponding quantities measured in an inertial reference frame Σ' that is moving with uniform velocity v relative to Σ along their common x axis are

$$p_x' = \alpha' m u_x', \qquad p_y' = \alpha' m u_y', \qquad p_z' = \alpha' m u_z', \qquad E' = \alpha' m c^2, \tag{6.55}$$

where

$$\alpha' = \frac{1}{(1 - u'^2/c^2)^{1/2}}, \tag{6.56}$$

and where u' is the speed of the particle in Σ'. The mass m and c, the speed of light in empty space, are invariants, i.e. they have the same values in all inertial reference frames.

According to equations (5.6) and (5.14),

$$u_x' = \frac{u_x - v}{1 - vu_x/c^2},$$

$$\alpha' = \gamma\alpha(1 - vu_x/c^2) \tag{6.57}$$

where $\gamma = (1 - v^2/c^2)^{-1/2}$. Hence the product $\alpha' u_x'$ is given by

$$\alpha' u_x' = \gamma\alpha\,(u_x - v)\,. \tag{6.58}$$

Substituting for $\alpha' u_x'$ in the expression $p_x' = \alpha' m u_x'$, we obtain

$$\begin{aligned} p_x' &= \gamma\alpha\,(u_x - v)\,m \\ &= \gamma\,(\alpha m u_x - \alpha m v)\,. \end{aligned}$$

It follows from equations (6.53) that $\alpha m u_x = p_x$ and $\alpha m = E/c^2$. Hence

$$p_x' = \gamma\,(p_x - vE/c^2)\,. \tag{6.59}$$

Substituting for α' from equation (6.57) and for u_y' from equation (5.7) into the expression $p_y' = \alpha' m u_y'$, we obtain

$$p_y' = \gamma\alpha\,(1 - vu_x/c^2)\,m\,\frac{u_y}{\gamma\,(1 - vu_x/c^2)} = \alpha m u_y = p_y. \tag{6.60}$$

Similarly,

$$p_z' = p_z\,. \tag{6.61}$$

Substituting for α' from equation (6.57) into the equation $E' = \alpha' mc^2$, we obtain

$$\begin{aligned} E' &= \gamma\alpha\,(1 - vu_x/c^2)mc^2 = \gamma(\alpha mc^2 - v\alpha m u_x) \\ &= \gamma\,(E - vp_x)\,. \end{aligned} \tag{6.62}$$

Collecting the transformations, we have

$$\left.\begin{aligned} p_x' &= \gamma\,(p_x - vE/c^2), \\ p_y' &= p_y, \qquad p_z' = p_z\,, \\ E' &= \gamma\,(E - vp_x)\,. \end{aligned}\right\} \tag{6.63}$$

The inverse transformations are

$$\left.\begin{aligned} p_x &= \gamma\,(p_x' + vE'/c^2),\\ p_y = p_y', \qquad & p_z = p_z',\\ E &= \gamma\,(E' + vp_x'). \end{aligned}\right\} \tag{6.64}$$

It can be seen that p_x', p_y', p_z' and E'/c^2 transform in the same way as the space and time coordinates x', y', z' and t'. For example, corresponding to equation (6.59), $x' = \gamma\,(x - vt)$, while, corresponding to equation (6.62), $t' = \gamma\,(t - vx/c^2)$. Now

$$\begin{aligned} p'^2 - E'^2/c^2 &= p_x'^2 + p_y'^2 + p_z'^2 - E'^2/c^2\\ &= \gamma^2\,(p_x - vE/c^2)^2 + p_y^2 + p_z^2 - \gamma^2\,(E - vp_x)^2/c^2\\ &= p_x^2 + p_y^2 + p_z^2 - E^2/c^2\\ &= p^2 - E^2/c^2. \end{aligned} \tag{6.65}$$

Hence, the quantity $p^2 - E^2/c^2$ is an invariant; that is, it has the same value in all inertial reference frames. It follows from equation (6.7) that its numerical value is $-m^2c^2$.

6.7.2 Example of the energy and momentum transformations: the Doppler effect (photon model)

Consider again the experimental conditions shown previously in Fig. 5.3 and again in Fig. 6.6. A light source is moving with uniform velocity v along the x

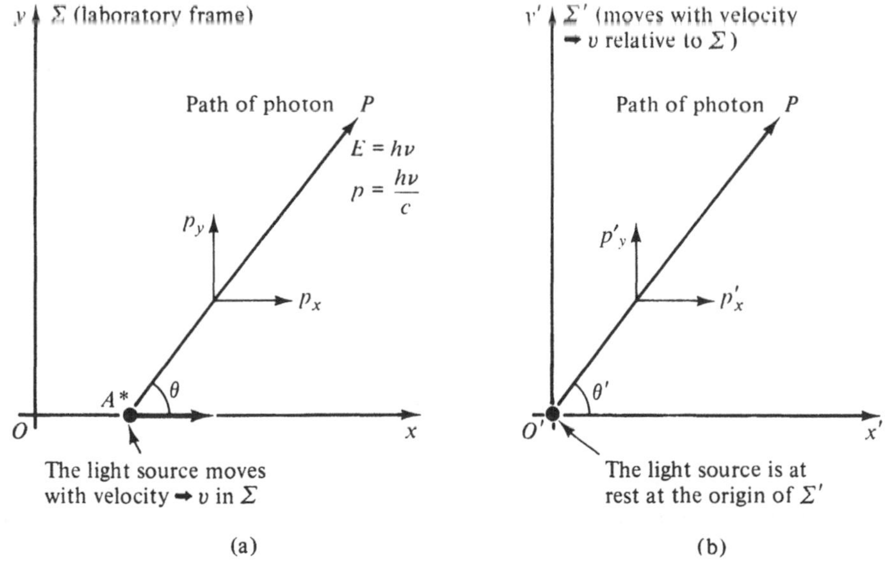

Fig. 6.6 The Doppler effect using the photon model of light.

axis of the laboratory frame Σ, as shown in Fig. 6.6(a). A photon of energy $h\nu$ and momentum $h\nu/c$ is emitted in the (x, y) plane at an angle θ to the x axis of Σ, when the light source is at the point A^*. In Σ

$$p_x = \frac{h\nu}{c}\cos\theta, \qquad p_y = \frac{h\nu}{c}\sin\theta, \qquad p_z = 0, \qquad E = h\nu . \tag{6.66}$$

In the inertial reference frame Σ' that is moving with uniform velocity v relative to Σ along their common x axis the light source is at rest at the origin O' of Σ'. Let the energy and momentum of the photon in Σ' be $h\nu_0$ and $h\nu_0/c$ respectively, and let the photon be emitted at an angle θ' to the x' axis in Σ'. In Σ'

$$p_x' = \frac{h\nu_0}{c}\cos\theta' ; \qquad p_y' = \frac{h\nu_0}{c}\sin\theta' ; \qquad p_z' = 0 ; \qquad E' = h\nu_0 . \tag{6.67}$$

From equation (6.62), $E' = \gamma(E - vp_x)$. Substituting for E', E and p_x we obtain,

$$h\nu_0 = \gamma\left(h\nu - \frac{vh\nu}{c}\cos\theta \right). \tag{6.68}$$

Cancelling h and rearranging, we find

$$\nu = \frac{\nu_0}{\gamma[1 - (v/c)\cos\theta]} . \tag{6.69}$$

Equation (6.69) is the same as equation (5.24), which was derived using the wave model of light. It gives the frequency of light measured by an observer at rest at P in the laboratory frame Σ in Fig. 6.6(a).

From equations (6.60) and (6.64),

$$p_y = p_y' , \tag{6.60}$$

$$p_x = \gamma(p_x' + vE'/c^2) , \tag{6.64}$$

and, on substituting from equations (6.66) and (6.67)

$$\frac{h\nu}{c}\sin\theta = \frac{h\nu_0}{c}\sin\theta' , \tag{6.70}$$

$$\frac{h\nu}{c}\cos\theta = \gamma\left(\frac{h\nu_0}{c}\cos\theta' + \frac{vh\nu_0}{c^2} \right). \tag{6.71}$$

Dividing equation (6.70) by equation (6.71), we obtain

$$\tan\theta = \frac{\sin\theta'}{\gamma(\cos\theta' + v/c)} . \tag{6.72}$$

Equation (6.72) is the same as equation (5.29). Thus both the Doppler effect and the aberration of light can be interpreted using the photon model of light in terms of the different measures of the energy and momentum of the photon in Σ and Σ'.

6.7.3 The force transformations

In the inertial frame Σ the force acting on a particle is defined as a vector

$$\boldsymbol{f} = \frac{\mathrm{d}\boldsymbol{p}}{\mathrm{d}t} \tag{6.73}$$

having components

$$f_x = \frac{\mathrm{d}p_x}{\mathrm{d}t} \quad \text{etc.} \tag{6.74}$$

Similarly, in the inertial frame Σ′ the force acting on the particle is defined as a vector having components

$$f_x' = \frac{\mathrm{d}p_x'}{\mathrm{d}t'} \quad \text{etc.} \tag{6.75}$$

Substituting for p_x' from equation (6.59),

$$f_x' = \gamma \frac{\mathrm{d}}{\mathrm{d}t'}\left(p_x - \frac{vE}{c^2}\right) = \gamma \frac{\mathrm{d}t}{\mathrm{d}t'}\frac{\mathrm{d}}{\mathrm{d}t}\left(p_x - \frac{vE}{c^2}\right). \tag{6.76}$$

Now,

$$\frac{\mathrm{d}t}{\mathrm{d}t'} = \left(\frac{\mathrm{d}t'}{\mathrm{d}t}\right)^{-1} = \left\{\frac{\mathrm{d}}{\mathrm{d}t}\gamma(t - vx/c^2)\right\}^{-1}$$

$$= \frac{1}{\gamma(1 - vu_x/c^2)}. \tag{6.77}$$

Substituting into equation (6.76), we have

$$f_x' = \frac{1}{1 - vu_x/c^2}\left(\frac{\mathrm{d}p_x}{\mathrm{d}t} - \frac{v}{c^2}\frac{\mathrm{d}E}{\mathrm{d}t}\right). \tag{6.78}$$

It follows from the definition of force that

$$\frac{\mathrm{d}p_x}{\mathrm{d}t} = f_x. \tag{6.79}$$

If all the work dW done by a force $\boldsymbol{f}$ in a displacement d$\boldsymbol{l}$ of the particle goes into increasing the kinetic energy T of the particle then

$$\mathrm{d}W = \mathrm{d}T = \boldsymbol{f} \bullet \mathrm{d}\boldsymbol{l}.$$

Dividing by dt, we find for a particle moving with velocity $\boldsymbol{u} = \mathrm{d}\boldsymbol{l}/\mathrm{d}t$

$$\frac{\mathrm{d}T}{\mathrm{d}t} = \frac{\mathrm{d}}{\mathrm{d}t}(E - mc^2)$$

$$= \frac{\mathrm{d}E}{\mathrm{d}t} = \boldsymbol{f} \bullet \frac{\mathrm{d}\boldsymbol{l}}{\mathrm{d}t} = \boldsymbol{f} \bullet \boldsymbol{u}$$

$$= f_x u_x + f_y u_y + f_z u_z \,. \tag{6.80}$$

Substituting from equations (6.79) and (6.80) for dp_x/dt and dE/dt into equation (6.78) and rearranging, we finally obtain

$$f_x' = f_x - \frac{v u_y}{c^2 (1 - v u_x/c^2)} f_y - \frac{v u_z}{c^2 (1 - v u_x/c^2)} f_z. \tag{6.81}$$

Since

$$p_y' = p_y \,,$$

we have

$$f_y' = \frac{dp_y'}{dt'} = \frac{dp_y}{dt'} = \frac{dt}{dt'} \frac{dp_y}{dt}$$

$$= \frac{dt}{dt'} f_y.$$

Using equation (6.77), we obtain

$$f_y' = \frac{f_y}{\gamma (1 - v u_x/c^2)} \,. \tag{6.82}$$

Similarly,

$$f_z' = \frac{f_z}{\gamma (1 - v u_x/c^2)} \,. \tag{6.83}$$

★ 6.8 THE ZERO-MOMENTUM FRAME

★ 6.8.1 Introduction

As an introduction to the zero-momentum frame, consider the collision of two particles of equal masses m, as shown in Figs. 6.7(a,b). Let particle 1 have velocity u_1, momentum p_1, kinetic energy T_1 and (total) energy $E_1 = T_1 + mc^2$ relative to the laboratory frame Σ before the collision, as shown in Fig. 6.7(a). Let particle 1 move parallel to the x axis of Σ and collide with particle 2, which is stationary in the laboratory frame Σ before the collision. Hence for particle 2 before the collision, we have, relative to Σ, $u_2 = 0$, $p_2 = 0$, $T_2 = 0$ and $E_2 = mc^2$. The total linear momentum in the laboratory frame Σ before the collision is $p_1 + p_2 = p_1 + 0 = p_1$.

The zero-momentum frame Σ^0 is an inertial reference frame that is moving at such a velocity v relative to Σ that the total linear momentum of the colliding particles, measured in Σ^0, is zero before the collision. (The zero-momentum frame is often called the centre-of-mass system, particularly in Newtonian

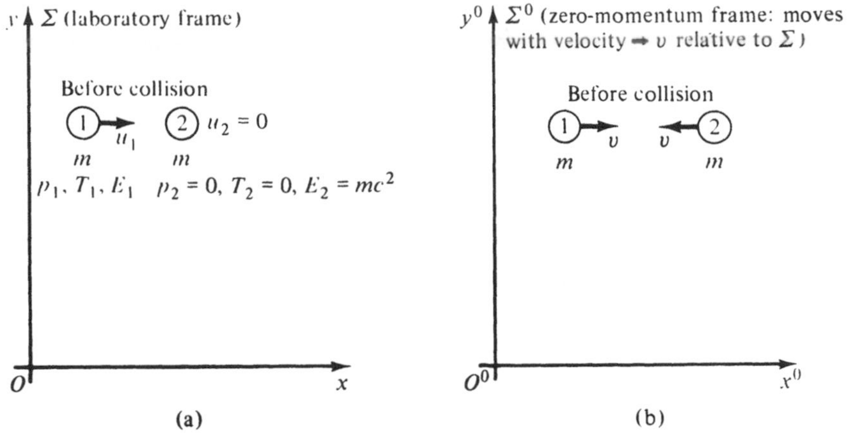

Fig. 6.7 Collision of two particles having equal masses m. (a) Particle 2 is at rest in Σ before the collision, while particle 1 moves with velocity u_1 and momentum p_1 relative to Σ before the collision. (b) In the zero-momentum frame Σ^0 the total momentum is zero before the collision, and particles 1 and 2 approach each other with equal speeds v relative to Σ^0.

mechanics.) If the total momentum is zero in Σ^0 before the collision, particles 1 and 2 must approach each other with equal but opposite momenta in Σ^0, so that, since they both have the same mass m, they must approach each other with equal and opposite speeds in Σ^0, as shown in Fig. 6.7(b).

The speed v of the zero-momentum frame Σ^0 relative to the laboratory frame Σ will now be determined. Since before the collision the momentum and (total) energy of particle 1 relative to Σ are p_1 and E_1 respectively, according to equation (6.59) the momentum of particle 1 in the zero-momentum frame Σ^0 is

$$p_1^0 = \gamma\,(p_1 - vE_1/c^2)\,. \tag{6.84}$$

Similarly, since $p_2 = 0$ and $E_2 = mc^2$ relative to Σ before the collision, in the zero-momentum frame Σ^0 we have

$$p_2^0 = \gamma\,(p_2 - vE_2/c^2) = -\,\gamma v mc^2/c^2. \tag{6.85}$$

Adding equations (6.84) and (6.85),

$$p_1^0 + p_2^0 = \gamma\,[p_1 - v\,(E_1 + mc^2)/c^2].$$

If the total momentum in the zero-momentum frame Σ^0 is to be zero before the collision, $p_1^0 + p_2^0$ must be zero, and so

$$\gamma\,[p_1 - v\,(E_1 + mc^2)/c^2] = 0.$$

Hence the speed of the zero-momentum frame Σ^0 relative to the laboratory frame Σ is

$$v = \frac{c^2 p_1}{E_1 + mc^2} = \frac{c^2 p_1}{T_1 + 2mc^2} . \tag{6.86}$$

Since particle 2 is at rest in the laboratory frame Σ before the collision, it follows from equation (5.6) that its velocity in the zero-momentum frame Σ^0 before the collision is

$$u_{2x}^0 = \frac{u_{2x} - v}{1 - vu_{2x}/c^2} = -v , \tag{6.87}$$

where v is given by equation (6.86). Since particles 1 and 2 have the same mass m and equal and opposite momenta in Σ^0 before the collision, it follows from equation (6.1) that the velocity of particle 1 in Σ^0 before the collision is

$$u_{1x}^0 = +v . \tag{6.88}$$

Hence particles 1 and 2 approach each other with equal, but oppositely directed, speeds v, given by equation (6.86), in the zero-momentum frame Σ^0 before the collision, as shown in Fig. 6.7(b).

It is straightforward for the reader to repeat the above argument to show that if the masses of particles 1 and 2 were not equal but were m_1 and m_2 respectively then the velocity of the zero-momentum frame Σ^0 relative to the laboratory frame Σ would be

$$v = \frac{c^2 p_1}{E_1 + m_2 c^2} = \frac{c^2 p_1}{T_1 + m_1 c^2 + m_2 c^2} . \tag{6.89}$$

In this latter case the particles still approach each other with equal and oppositely directed momenta in the zero-momentum frame Σ^0, but they do not approach each other with equal but oppositely directed speeds.

★ 6.8.2 An example of an elastic collision considered in the zero-momentum frame

Consider again the example of the elastic collision of two particles, of equal masses m, discussed previously in Section 6.3, and shown again in Figs. 6.8(a,b). As a typical example, it will be assumed that the kinetic energy T_1 of particle 1 in the laboratory frame Σ, shown in Fig. 6.8(a), is numerically equal to mc^2, so that the (total) energy of particle 1 in Σ before the collision is

$$E_1 = T_1 + mc^2 = 2mc^2.$$

Using equation (6.7), we find that the momentum p_1 of particle 1 in the laboratory frame Σ before the collision is

$$cp_1 = (E_1^2 - m^2c^4)^{1/2} = 1.732\, mc^2.$$

Particle 2 is at rest in Σ before the collision, so that

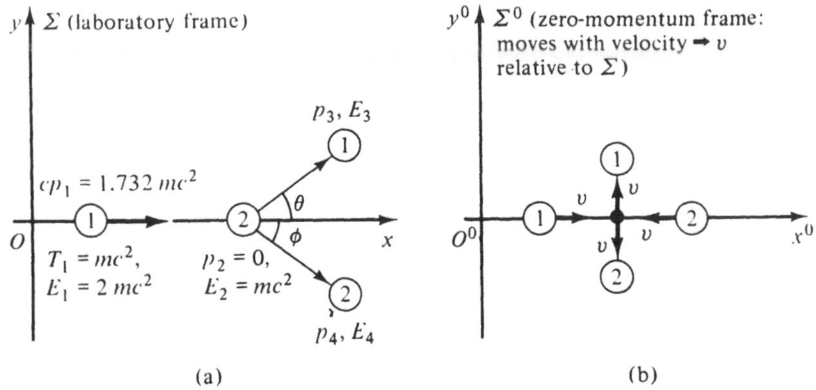

Fig. 6.8 (a) Elastic collision of a particle of mass m and kinetic energy mc^2 with a stationary particle, also of mass m. (b) The zero-momentum frame.

$$E_2 = mc^2, \qquad p_2 = 0.$$

It follows from equation (6.86) that the velocity of the zero-momentum frame (Fig. 6.8(b)) relative to the laboratory frame Σ (Fig. 6.8(a)) is

$$v = \frac{c^2 p_1}{E_1 + mc^2} = \frac{c^2 \times 1.732\,mc}{3mc^2} = 0.577c\,. \tag{6.90}$$

Therefore

$$\gamma = (1 - v^2/c^2)^{-1/2} = 1.225\,. \tag{6.91}$$

Both particles 1 and 2 are moving with speed $v = 0.577c$ along the x^0 axis of the zero-momentum frame Σ^0 before the collision, as shown in Fig. 6.8(b). The magnitudes of the momenta p_1^0 and p_2^0 of the particles 1 and 2 in Σ^0 before the collision are

$$p_1^0 = p_2^0 = \alpha_1^0 mv = \gamma mv = 1.225\,m \times 0.577c = 0.707mc\,.$$

The (total) energies E_1^0 and E_2^0 of particles 1 and 2 in Σ^0 before the collision are

$$E_1^0 = E_2^0 = \gamma mc^2 = 1.225mc^2.$$

It will be assumed that after the collision particles 1 and 2 rebound with equal speeds $v = 0.577c$, and hence equal momenta $p_3^0 = p_4^0 = 0.707mc$ and equal (total) energies $E_3 = E_4 = 1.225mc^2$ in the zero-momentum frame Σ^0 at angles of 90° to the x^0 axis, as shown in Fig. 6.8(b). The components of the momentum $\mathbf{p}_3^0$ of particle 1 in Σ^0 after the collision are

$$p_{3x}^0 = 0, \qquad p_{3y}^0 = 0.707mc, \qquad p_{3z}^0 = 0.$$

Applying the inverse momentum transformations, equations (6.64), we find that the components of the momentum $\mathbf{p}_3$ of particle 1 in the laboratory frame Σ after the collision are

$$p_{3x} = \gamma(p_{3x}^0 + vE_3^0/c^2) = \gamma v E_3^0/c^2$$
$$= 1.225 \times 0.577c \times 1.225mc^2/c^2 = 0.866mc,$$
$$p_{3y} = p_{3y}^0 = 0.707mc.$$

It can be seen from Fig. 6.8(a) that

$$\tan\theta = \frac{p_{3y}}{p_{3x}} = \frac{0.707}{0.866} = 0.817,$$

and therefore

$$\theta = 39.2°.$$

The magnitude of the momentum p_3 of particle 1 in the laboratory frame Σ after the collision is

$$p_3 = (p_{3x}^2 + p_{3y}^2)^{1/2} = 1.118mc.$$

Using equation (6.7), we have

$$E_3 = (c^2p_3^2 + m^2c^4)^{1/2} = 1.5mc^2.$$

The reader can check this value of E_3 using the inverse relativistic energy transformation, equation (6.64), and show that, in this symmetrical case, ϕ is also equal to 39.2° so that $\theta + \phi = 78.4°$, in agreement with the results derived in Section 6.3.

As an exercise, the reader should assume that particles 1 and 2 rebound at angles of 45° and 135° to the x^0 axis of Σ^0 in Fig. 6.8 after the collision and show that in this case $\theta = 18.7°$ and $\phi = 63.1°$, so that $\theta + \phi = 81.8°$ and $E_3 = 1.85mc^2$ and $E_4 = 1.15mc^2$.

★ 6.8.3 Decay of a pion in flight

A positively charged pion (π^+) of mass 139.6 MeV/c^2 is moving with a velocity of $0.8c$ along the x axis of the laboratory frame Σ when it decays into a muon (μ^+) of mass 105.7 MeV/c^2 and a neutrino of zero mass. The maximum and minimum values of the kinetic energy of the muon in the laboratory frame will be calculated.

Consider the zero-momentum frame Σ^0, which is moving with velocity $v = 0.8c$ along the x axis of the laboratory frame Σ in the direction of motion of the pion. The pion is at rest in Σ^0 when it decays. Hence the analysis of the decay of a stationary pion given in Section 6.6.5 is applicable in the zero-momentum frame Σ^0. It was shown there that the (total) energy of the muon in Σ^0 is $E_\mu^0 = 109.8$ MeV, and its momentum in Σ^0 is given by $cp_\mu^0 = 29.8$ MeV. Let the muon be emitted at an angle θ' to the x^0 axis of Σ^0. According to the inverse energy transformation, the (total) energy E_μ of the muon in the laboratory frame Σ is

$$E_\mu = \gamma(E_\mu^0 + vp_{\mu x}^0) = \gamma(E_\mu^0 + vp_\mu^0 \cos\theta') \tag{6.92}$$

It follows from equation (6.92) that E_μ is at a maximum when $\cos\theta' = 1$, that is when $\theta' = 0°$, and E_μ is a minimum when $\theta' = 180°$. Since $v = 0.8c$ and $\gamma = \frac{5}{3}$, it follows from equation (6.92) that, when $\theta' = 0°$

$$(E_\mu)_{\max} = \gamma\left(E_\mu^0 + \frac{v}{c}cp_\mu^0\right) = \tfrac{5}{3}(109.8\text{ MeV} + 0.8 \times 29.8\text{ MeV})$$

$$= 222.7\text{ MeV}.$$

Therefore

$$(T_\mu)_{\max} = (E_\mu)_{\max} - m_\mu c^2 = 117\text{ MeV}.$$

When $\theta' = 180°$, equation (6.92) gives

$$(E_\mu)_{\min} = 143.3\text{ MeV}, \qquad (T_\mu)_{\min} = 37.6\text{ MeV}.$$

When $\theta' = 60°$, equation (6.92) gives

$$E_\mu = 202.9\text{ MeV}, \qquad T_\mu = 97.2\text{ MeV}.$$

It is left as an exercise for the reader to use the momentum transformations to show that in the last case the muon is emitted at an angle of 8.6° to the x axis in Σ.

★ 6.8.4 Energy available in the zero-momentum frame

Consider the collision of two particles of equal masses m, as shown in Figs. 6.9(a,b). Let particle 1 have momentum $\boldsymbol{p}_1$, kinetic energy T_1 and (total) energy $E_1 = T_1 + mc^2$ relative to the laboratory frame Σ before the collision, as shown in Fig. 6.9(a). Let particle 1 move parallel to the x axis of Σ and collide with particle 2, which is stationary in the laboratory frame Σ before the collision. For particle 2 before the collision $p_2 = 0$, $T_2 = 0$ and $E_2 = mc^2$ in Σ. The total momentum in the laboratory frame Σ before the collision is $p_1 + p_2 = p_1 + 0 = p_1$. If momentum is to be conserved, it is impossible for the particles to be at rest in Σ after the collision, since they must have a total momentum p_1 in Σ after the collision.

In the zero-momentum frame Σ^0, shown in Fig. 6.9(b), the particles approach each other with equal speeds v, where v is given by equation (6.86). Since the total momentum is zero in Σ^0 before the collision, it is consistent with the law of conservation of momentum in Σ^0 for all the particles present after the collision to be at rest in Σ^0, as shown in Fig. 6.9(b).

According to equation (6.6), the (total) energy of a particle of mass m moving with speed v is $mc^2(1 - v^2/c^2)^{-1/2}$, so that the (total) energy of each of the colliding particles in the zero-momentum frame Σ^0 before the collision is $mc^2(1 - v^2/c^2)^{-1/2}$. Hence the total energy E^0 available in the zero-momentum frame Σ^0 before the collision is given by

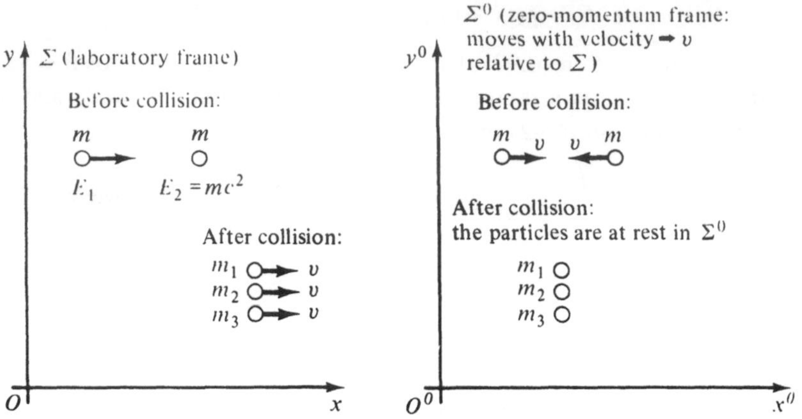

Fig. 6.9 (a) A particle of mass m, kinetic energy T_1 and (total) energy E_1 collides with a stationary particle, also of mass m, creating particles of masses m_1, m_2 and m_3. (b) The zero-momentum frame in which the particles approach each other with equal speeds v before the collision, creating particles of masses m_1, m_2 and m_3, all of which are at rest in Σ^0. These particles all have a velocity v parallel to the x axis in Σ, as shown in (a).

$$E^0 = \frac{2mc^2}{(1 - v^2/c^2)^{1/2}}. \tag{6.93}$$

From equation (6.86),

$$v = \frac{c^2 p_1}{E_1 + mc^2},$$

and so

$$1 - \frac{v^2}{c^2} = 1 - \frac{c^4 p_1^2}{c^2 (E_1 + mc^2)^2}$$

$$= \frac{E_1^2 + 2E_1 mc^2 + m^2c^4 - c^2 p_1^2}{(E_1 + mc^2)^2}.$$

From equation (6.7),

$$E_1^2 - c^2 p_1^2 = m^2 c^4.$$

Hence

$$1 - \frac{v^2}{c^2} = \frac{2E_1 mc^2 + 2m^2c^4}{(E_1 + mc^2)^2} = \frac{2mc^2}{E_1 + mc^2},$$

and so

$$\frac{1}{(1 - v^2/c^2)^{1/2}} = \frac{(E_1 + mc^2)^{1/2}}{(2mc^2)^{1/2}}. \tag{6.94}$$

Substituting in equation (6.93), we obtain

$$E^0 = 2mc^2 \frac{(E_1 + mc^2)^{1/2}}{(2mc^2)^{1/2}}$$

$$= (2mc^2E_1 + 2m^2c^4)^{1/2}. \qquad (6.95)$$

Since $E_1 = T_1 + mc^2$ in Σ, equation (6.95) can be rewritten as

$$E^0 = (2mc^2T_1 + 4m^2c^4)^{1/2}, \qquad (6.96)$$

where T_1 and E_1 are the kinetic energy and the (total) energy of particle 1 in the *laboratory frame* Σ before the collision. The *total* energy E^0 in the zero-momentum frame Σ^0 includes the total rest energy $2mc^2$ of the colliding particles.

Since the total momentum of the particles is zero in Σ^0 before the collision, it is possible to create particles of masses m_1, m_2, m_3, ..., that are all at rest in Σ^0, as shown in Fig. 6.9(b), and still conserve momentum in Σ^0. The sum of the (total) energies of the stationary particles in Σ^0 is then

$$E^0 = m_1c^2 + m_2c^2 + m_3c^2 + \dots . \qquad (6.97)$$

The threshold (minimum) kinetic energy T_1 required in the laboratory frame Σ to produce the particles of masses m_1, m_2, m_3, ..., can be obtained by equating the right-hand sides of equations (6.96) and (6.97). Squaring then gives

$$(E^0)^2 = 2mc^2T_1 + 4m^2c^4 = (m_1c^2 + m_2c^2 + m_3c^2 + \dots)^2. \qquad (6.98)$$

If the particles had finite velocities in Σ^0, one would have to add their kinetic energies to the right-hand side of equation (6.97), thereby increasing the value of E^0 required in the zero-momentum frame Σ^0, and, according to equation (6.96), increasing the value of T_1, the kinetic energy that the incident particle must have in the laboratory frame Σ.

As an example, consider the case of a proton (p^+) of kinetic energy T_1 in the laboratory frame that collides with a proton that is at rest in the laboratory frame, producing a neutral pion (π^0) according to the reaction

$$p^+ + p^+ \rightarrow p^+ + p^+ + \pi^0. \qquad (6.99)$$

The minimum value of T_1 to produce a neutral pion of mass given by $m_\pi c^2 =$ 135.0 MeV will be calculated. The minimum value of T_1 is obtained when the two protons and the π^0 are at *rest* in the zero-momentum frame Σ^0 after the collision. In this case the only energy in the zero-momentum frame Σ^0 after the collision is equal to the sum of the rest energies of the two protons and the π^0. Hence the total energy in the zero-momentum frame Σ^0 after the collision is

$$E^0 = 2m_pc^2 + m_\pi c^2. \qquad (6.100)$$

The rest energy of a proton is $m_pc^2 = 938$ MeV. Substituting in equation (6.98), we have

$$2m_pc^2T_1 + 4m_p^2c^4 = 4m_p^2c^4 + 4m_pm_\pi c^4 + m_\pi^2c^4,$$

and therefore

$$T_1 = 2m_\pi c^2 + \frac{m_\pi^2c^2}{2m_p} = m_\pi c^2\left(2 + \frac{m_\pi}{2m_p}\right).$$

Substituting for $m_\pi c^2$ and m_pc^2,

$$T_1 = 135\ \text{MeV} \times \left(2 + \frac{135}{2 \times 938}\right) = 280\ \text{MeV}.$$

In order to produce a π^0 in a proton–proton collision, the proton accelerator must accelerate protons to kinetic energies above 280 MeV. The Berkeley frequency-modulated cyclotron, discussed in Section 2.7, was able to do this, and pions were first produced in the laboratory using this machine.

The next generation of proton accelerators was designed to produce proton pairs according to the reaction

$$p^+ + p^+ \rightarrow p^+ + p^+ + p^+ + p^-.$$

The antiproton (p^-) is a particle of the same mass as a proton, but with the opposite charge, i.e. a negative charge of -1.602×10^{-19} C. The threshold kinetic energy for the above reaction will now be calculated. If all four product particles are at rest in the zero-momentum frame after the collision then the post-collision total energy in that frame is

$$E^0 = 4m_pc^2. \tag{6.101}$$

Substituting in equation (6.98), we have

$$2T_1m_pc^2 + 4m_p^2c^4 = 16m_p^2c^4,$$

and therefore

$$T_1 = 6m_pc^2.$$

Since for a proton $m_pc^2 = 938$ MeV, the threshold kinetic energy (i.e. the minimum kinetic energy that the incident proton must have in the laboratory frame) is 5.63 GeV. The Berkeley Bevatron accelerated protons to kinetic energies of 6 GeV. The antiproton was discovered using this machine by Chamberlain, Segrè, Wiegand and Ypsilantis in 1955.

The original CERN proton synchrotron accelerated protons up to kinetic energies of 30 GeV. At these energies, T_1 is much greater, than $m_pc^2 = 0.938$ GeV, and equation (6.96) can be rewritten in the approximate form

$$E^0 \approx (2m_pc^2T_1)^{1/2} \tag{6.102}$$

for proton–proton collisions. At high energies the energy available in the zero-momentum frame goes up as the square root of the kinetic energy the indicent particle has in the laboratory frame. For $T_1 = 30$ GeV and $m_pc^2 = 0.938$ GeV,

$$E^0 \approx (2 \times 0.938\ \text{GeV} \times 30\ \text{GeV})^{1/2} \approx 7.5\ \text{GeV}.$$

For $T_1 = 270$ GeV the energy available in the zero-momentum frame is

$$E^0 \approx (2 \times 0.938\ \text{GeV} \times 270\ \text{GeV})^{1/2} = 22.5\ \text{GeV}.$$

For the tevatron, for which $T_1 = 1000$ GeV, the energy available in the zero-momentum frame is

$$E^0 \approx (2 \times 0.938\ \text{GeV} \times 1000\ \text{GeV})^{1/2} = 43.3\ \text{GeV}.$$

In order to obtain higher available energies in the zero-momentum frame, colliding beams of particles have been used. The $W^{\pm}$ and Z^0 particles were discovered at CERN using beams of protons and antiprotons that collided head-on. The (total) energy of both the protons and the negative protons in the laboratory frame was 270 GeV. Since two particles of equal masses approached each other with equal speeds in this experiment, the laboratory frame was also the zero-momentum frame, so that the total energy E^0 available in the zero-momentum frame was $2 \times 270\ \text{GeV} = 540\ \text{GeV}$. This greatly exceeded the $E^0 = 22.5$ GeV available in the zero-momentum frame when a proton of kinetic energy 270 GeV strikes a stationary proton.

According to equation (6.96), a moving proton needs a kinetic energy of $T_1 = 155$ TeV to give an energy of $E^0 = 540$ GeV in the zero-momentum frame when it collides with a stationary proton.

The calculation of threshold energies is much quicker using 4-vector methods (see Section 7.9).

★ 6.8.5 Discussion

The relativistic transformations are not generally used to relate the measurements made by *different* inertial observers in *different* inertial reference frames on the same event. Experiments on high-energy physics are normally carried out in terrestrial laboratories. If a problem is solved in the zero-momentum frame Σ^0 then, using different values for the velocity v of the zero-momentum frame relative to the laboratory, we can work out what should happen under different experimental conditions in the laboratory. High-energy physicists often convert the results of experiments performed in the laboratory into the corresponding values in the zero-momentum frame, as a convenient way of presenting their results.

★ 6.9 INELASTIC COLLISIONS AND THE INERTIA OF ALL FORMS OF ENERGY

Consider the completely inelastic collision shown in Figs. 6.10(a,b). So far in our discussions of relativistic mechanics we have not discussed the structure and composition of the particles, and the theory has only been applied to single atomic particles, such as protons and electrons. For purposes of discussion, it will now be assumed in this section that the colliding particles in Figs. 6.10(a,b) are small particles of matter composed of many atoms. It will be assumed that the particles are still small enough for the collision to be considered as taking place at one point of space at one instant of time, so that the Lorentz transformations can be applied. (A brief review of bodies of finite extent is given in Section 6.13.)

In the laboratory frame Σ, particles 1 and 2 of the same mass m are moving with momenta $\boldsymbol{p}_1$ and $\boldsymbol{p}_2$ respectively parallel to the x axis of Σ before the collision, as shown in Fig. 6.10(a). After the collision, the particles stick together to form a composite particle.

In the zero-momentum frame Σ^0, which is moving with uniform velocity v relative to the laboratory frame Σ, particles 1 and 2 of the same mass m move with velocities $+v$ and $-v$ parallel to the x^0 axis of Σ^0 before the collision, as shown in Fig. 6.10(b). Particles 1 and 2 stick together to form one composite particle. Since the total momentum in Σ^0 before the collision is zero, if momentum is conserved in the collision in Σ^0 then the composite particle must be at rest in Σ^0 after the collision. Let the (total) energies of particles 1 and 2 in the zero-momentum frame Σ^0 before the collision be E_1^0 and E_2^0 respectively, and let the corresponding kinetic energies be $T_1{}^0$ and $T_2{}^0$ respectively. Using equation (6.5), we have

$$E_1^0 + E_2^0 = T_1{}^0 + mc^2 + T_2{}^0 + mc^2$$

$$= 2mc^2 + T_1{}^0 + T_2{}^0. \tag{6.103}$$

Let the momenta of particles 1 and 2 in the zero-momentum frame Σ^0 before the collision be $\boldsymbol{p}_1^0$ and $\boldsymbol{p}_2^0$ respectively. Since the total momentum is zero in the

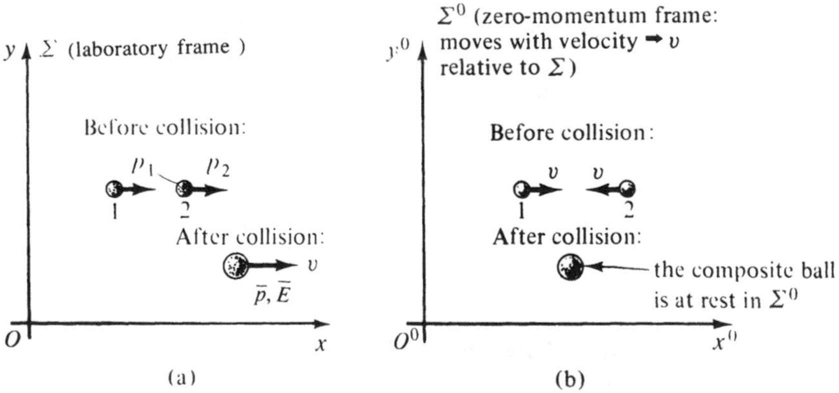

Fig. 6.10 Two spheres approach each other with equal speeds v in Σ^0, and are brought to rest relative to Σ^0 by the collision as shown in (b). The composite ball moves with uniform velocity v relative to Σ, as shown in (a).

zero-momentum frame before the collision, and particles 1 and 2 move parallel to the x^0 axis of Σ^0, we have

$$p_{1x}^0 + p_{2x}^0 = 0. \tag{6.104}$$

If it is assumed that all the kinetic energy in the zero-momentum frame Σ^0 is converted into heat then the heat Q^0 produced in the collision, measured in the zero-momentum frame Σ^0, is

$$Q^0 = T_1^0 + T_2^0. \tag{6.105}$$

Now consider the collision in the laboratory frame Σ, shown in Fig. 6.10(a). The momenta $\boldsymbol{p}_1$ and $\boldsymbol{p}_2$ of particles 1 and 2 in Σ before the collision can be related to their momenta and (total) energies in Σ^0 before the collision using the inverse of equation (6.59). We have

$$p_{1x} = \gamma\,(p_{1x}^0 + vE_1^0/c^2), \qquad p_{2x} = \gamma\,(p_{2x}^0 + vE_2^0/c^2)\,.$$

Since according to equation (6.104), $p_{1x}^0 + p_{2x}^0 = 0$, adding p_{1x} and p_{2x} gives

$$p_{1x} + p_{2x} = \gamma v\,(E_1^0 + E_2^0)/c^2.$$

Using equation (6.103), we find that the total momentum in the laboratory frame Σ before the collision is

$$p_{1x} + p_{2x} = \gamma v\,[2m + (T_1^0 + T_2^0)/c^2]. \tag{6.106}$$

Let the *composite* particle have momentum $\bar{\boldsymbol{p}}$ and (total) energy $\bar{E}$ in the laboratory frame Σ after the collision. If momentum is to be conserved in the $+x$ direction in the laboratory frame Σ, it follows from equation (6.106) that

$$\bar{p}_x = p_{1x} + p_{2x} = \gamma v\,[2m + (T_1^0 + T_2^0)/c^2]. \tag{6.107}$$

Since, according to equation (6.105), $T_1^0 + T_2^0$ is equal to the heat Q^0 generated in the collision, measured in Σ^0, the condition that must be satisfied if momentum is to be conserved in the laboratory frame Σ, namely equation (6.107), can be written in the form

$$\bar{p}_x = \gamma v\,(2m + Q^0/c^2)\,. \tag{6.108}$$

After the collision, the composite particle, which is at rest in Σ, moves with a velocity $u = v$ in the laboratory frame Σ. At first sight, one might expect that the mass of the composite particle *might* be equal to $2m$, the sum of the masses of the colliding particles. In this case the momentum of the composite particle, measured in the laboratory frame Σ in which the composite particle moves with velocity $u = v$ (such that $\alpha = \gamma$), would be

$$\bar{p}_x = \alpha\,(2m)\,u = 2\gamma m v. \tag{6.109}$$

Comparing equations (6.108) and (6.109), it can be seen that if the mass of the composite particle were $2m$ then momentum would *not* be conserved in the laboratory frame Σ. In order to conserve momentum in Σ, the quantity $2m$ in equation (6.109) must be replaced by $2m + Q^0/c^2$, such that the momentum of the composite particle in Σ after the collision becomes, with $u = v$ and $\alpha = \gamma$,

$$\begin{aligned}\bar{p}_x &= \alpha\,(2m + Q^0/c^2)\,u \\ &= \gamma\,(2m + Q^0/c^2)\,v.\end{aligned} \tag{6.110}$$

According to equation (6.110), if momentum is to be conserved in the laboratory frame Σ as well as in Σ^0, the heat generated in the collision must contribute to the momentum of the composite particle, measured in the laboratory frame Σ.

It follows from the inverse of equation (6.62) that the (total) energies E_1 and E_2 of particles 1 and 2 in the laboratory frame Σ before the collision are

$$E_1 = \gamma\,(E_1^0 + v p_{1x}^0)\,, \qquad E_2 = \gamma\,(E_2^0 + v p_{2x}^0)\,.$$

Since, according to equation (6.104), $p_{1x}^0 + p_{2x}^0 = 0$ we have

$$E_1 + E_2 = \gamma\,(E_1^0 + E_2^0)\,.$$

Using equation (6.103), we obtain for the sum of the (total) energies of particles 1 and 2 in Σ before the collision

$$E_1 + E_2 = \gamma\,(2mc^2 + T_1^0 + T_2^0)\,. \tag{6.111}$$

If the energy is to be conserved in the laboratory frame Σ, the total energy of the composite particle in Σ after the collision must be $E_1 + E_2$. Since, from equation (6.105), $T_1^0 + T_2^0 = Q^0$, in order to conserve energy in the laboratory frame Σ the (total) energy $\bar{E}$ of the composite particle measured in Σ after the collision must, with $u = v$ and $\alpha = \gamma$, be given by

$$\begin{aligned}\bar{E} = E_1 + E_2 &= \gamma\,(2m + Q^0/c^2)\,c^2 \\ &= \alpha\,(2m + Q^0/c^2)\,c^2.\end{aligned} \tag{6.112}$$

If we put

$$2m + Q^0/c^2 = M, \tag{6.113}$$

then equations (6.110) and (6.112) for the momentum $\bar{\boldsymbol{p}}$ and (total) energy $\bar{E}$ that the composite particle must have in the laboratory frame Σ after the collision, if momentum and energy are to be conserved in Σ, can be written as

$$\bar{\boldsymbol{p}} = \alpha M \boldsymbol{u}\,, \tag{6.114}$$

$$\bar{E} = \alpha M c^2. \tag{6.115}$$

These are the same as equations (6.1) and (6.6), which suggests that the composite particle has the momentum and (total) energy of a particle of mass M, given by equation (6.113). Multiplying both sides of equation (6.113) by c^2, we obtain

$$Mc^2 = 2mc^2 + Q^0. \tag{6.116}$$

The quantity Mc^2 is equal to the *sum* of the initial rest energies of the colliding particles, which is $2mc^2$, and the heat Q^0 generated in the inelastic collision, *measured in the zero-momentum frame in which the composite particle is at rest* after the collision. When heat is generated in the composite particle, not only are the kinetic energies of the atomic constituents increased, but the addition of heat can lead, for example, to thermal expansion thereby affecting the potential energies of the atomic constituents. This suggests that changes in the kinetic energies and potential energies of the particles making up the composite particle affect its mass M via the Q^0 term in equation (6.116). On the basis of these results, it will be assumed that a composite particle behaves like a particle of mass M given by

$$Mc^2 = E^0, \tag{6.117}$$

where E^0 is the total energy of the composite particle measured in the zero-momentum (centre-of-mass) frame Σ^0 in which the composite particle is at rest. The quantity E^0 is sometimes called the *rest energy* of the composite particle. The energy E^0 in Σ^0 includes the rest energies, kinetic energies and potential energies of all the particles making up the composite particle; that is

$$\begin{aligned} Mc^2 = E^0 &= \sum_i m_i c^2 + \sum_i T_i^0 + \sum_i U_i^0 \\ &= \sum_i m_i c^2 + \sum_i (T_i^0 + U_i^0)\,, \end{aligned} \tag{6.118}$$

where $m_i c^2$, T_i^0 and U_i^0 are the rest energy, kinetic energy and potential energy of the ith constituent particle, measured in the zero-momentum frame Σ^0, in which the composite particle is at rest. If $\Sigma_i\,(T_i^0 + U_i^0)$ changes then, according to equation (6.118), the mass M of the composite particle changes.

Since c^2 is a universal constant, it can be seen from equation (6.117) that M is proportional to E^0; that is, the mass of the composite particle is a measure of its total energy E^0, measured in the zero-momentum frame Σ^0 in which the composite particle is at rest. All forms of energy contribute to E^0, and hence contribute to the mass M of the particle. Since $M = E^0/c^2$, where c^2 is a universal constant, using M is equivalent to using E^0. Instead of using the mass M to express the momentum of the composite particle in the form $\boldsymbol{p} = \alpha M \boldsymbol{u}$, where $\boldsymbol{u}$ is its velocity, the momentum can be expressed in terms of the total energy E^0 of the composite particle measured in the zero-momentum frame Σ^0, or the total energy $E = \alpha E^0$, measured in the laboratory frame Σ, as follows:

$$\boldsymbol{p} = \alpha \frac{E^0}{c^2} \boldsymbol{u} = \frac{E}{c^2} \boldsymbol{u}\,. \tag{6.119}$$

Since the expression $\alpha M \boldsymbol{u}$ is almost invariably used for the momentum of a composite particle, we shall continue to follow this practice.

When the concepts of energy and inertial mass are introduced into Newtonian mechanics, they seem to be very different. For example, the energy of a particle is often defined in Newtonian mechanics as the capacity of the particle to do work, whereas its inertial mass is defined as a measure of the resistance of the particle to changes in its velocity. In relativistic mechanics the mass of a composite particle is a measure of its rest energy, that is its total energy E^0 measured in the inertial reference frame in which it is at rest. According to equation (6.119), all forms of energy contribute to the momentum of a moving composite particle.

When a composite particle is moving—that is, when its centre-of-mass is moving relative to an inertial reference frame—in addition to its rest-energy $E^0 = Mc^2$, it has a kinetic energy T due to the motion of its centre-of-mass relative to that inertial reference frame. The (total) energy of the composite particle in that frame is then $T + E^0 = T + Mc^2$.

In the case of a photon the (total) energy is $E = h\nu$. Putting $u = c$ and $E = h\nu$ in equation (6.119), we then have

$$p = \frac{E}{c^2}c = \frac{h\nu}{c}.$$

This is in agreement with equation (2.62).

Since c^2 is a universal constant, it is reasonable to use the energy units based on the electron volt, introduced in Section 2.10, to measure the mass of a particle. For example, high-energy physicists often say that the mass of an electron is 0.511 MeV. To avoid confusion, we have always expressed the mass of an electron in the form $m_e c^2 = 0.511$ MeV, or $m_e = 0.511$ MeV/c^2. Theoretical physicists sometimes use natural units, where c is put equal to unity, so that in these units $M = E^0$.

★ 6.10 THE ATOMIC NUCLEUS AS AN EXAMPLE OF A COMPOSITE PARTICLE

A simplified model will be used for the atomic nucleus. It will be assumed that inside a nucleus of mass number (atomic weight) A and atomic number Z there are Z protons and $N = A - Z$ neutrons. The term "nucleon" is applied collectively to both protons and neutrons when we do not wish to distinguish between them, so that one would say that there are A nucleons inside the nucleus.

Consider the atomic nucleus in the zero-momentum frame, in which it is at rest. There are very strong attractive forces between the nucleons inside the nucleus, and work would have to be done to separate them. The attractive force on a nucleon inside the nucleus can be represented approximately by a spherical potential-energy well of the type shown in Fig. 6.11(a) for the case of a neutron. Here the potential energy of the neutron is plotted against the distance of the neutron from the centre of the nucleus. The radius of the spherical potential well is typically $1.4 \times 10^{-15} A^{1/3}$ m. The zero of the energy scale is generally chosen to be the energy that the neutron would have if it were at rest at infinity.

For a typical atomic nucleus of mass number 100 the depth of the potential-energy well is approximately 36 MeV. Work equal to 36 MeV would have to be

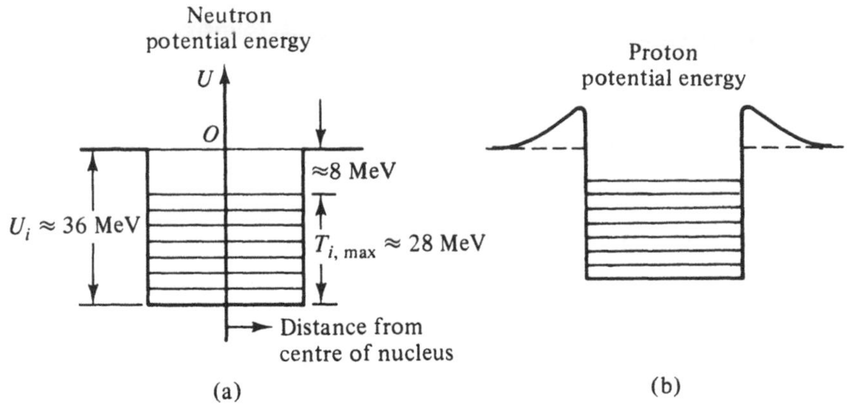

Fig. 6.11 (a) Potential energy well for the neutrons in a typical atomic nucleus. (b) The potential-energy well for the protons in a typical heavy nucleus. The potential-energy well for the protons is not as deep as for the neutrons in a nucleus. In the case of protons there is also a repulsive Coulomb potential barrier outside the nucleus due to the electric charges on the protons in the nucleus.

done to move a stationary neutron from the bottom of such a potential well and then take it to infinity at zero speed. Assuming that the potential energy of such a neutron is zero at infinity, the potential energy U_i^0 of the ith neutron, if it were *at rest* at the bottom of the potential-energy well, would be $U_i^0 = -36$ MeV. According to the shell model of the nucleus, the neutrons are not at rest, but occupy single-particle quantum states inside the potential well having kinetic energies T_i^0 up to about 28 MeV, as shown in Fig. 6.11(a). For a neutron in the highest occupied state $T_i^0 + U_i^0$ is negative and equal to about -8 MeV inside a nucleus of $A \approx 100$. One would have to do about 8 MeV of work to remove such a neutron from the nucleus. In the case of protons there is also a Coulomb electrostatic repulsion between the protons due to their electric charges, which modifies the potential-energy well, as sketched in Fig. 6.11(b).

According to equation (6.118), the total energy E^0 of all the nucleons inside the nucleus, measured in the zero-momentum frame in which the nucleus is at rest, is given by

$$E^0 = \sum_i m_i c^2 + \sum_i (T_i^0 + U_i^0)\,, \tag{6.120}$$

where m_i is the mass of the ith nucleon and T_i^0 and U_i^0 are respectively, the kinetic and potential energies of the ith nucleon, measured in the zero-momentum frame, in which the nucleus is at rest. If $M_{A,Z}$ is the mass of a nucleus consisting of Z protons and $N = A - Z$ neutrons then, since, from equation (6.117), $M_{A,Z}\,c^2 = E^0$, it follows from equation (6.120) that

$$M_{A,Z}c^2 = E^0 = Zm_p c^2 + Nm_n c^2 + \sum_i (T_i^0 + U_i^0), \tag{6.121}$$

where m_p and m_n are the masses of the proton and neutron respectively and the summation is over all the nucleons inside the nucleus. *The sum* $\Sigma_i (T_i^0 + U_i^0)$ *is negative*. In practice, instead of using $\Sigma_i (T_i^0 + U_i^0)$, it is more conventional to use the total binding energy of the nucleus, denoted by $B_{A,Z}$, which is *defined* as the total work that must be done to remove all the nucleons from the nucleus, such that they are at rest at infinite separations at infinity, when the total energy of the separated nucleons is equal to the sum of their individual rest energies, which is $Zm_pc^2 + Nm_nc^2$. Since the initial total energy of the nucleus in the zero-momentum frame is $E^0 = M_{A,Z}c^2$, and the work that must be done to separate the nucleons is $B_{A,Z}$, it follows from the law of conservation of energy that

$$M_{A,Z}c^2 + B_{A,Z} = Zm_pc^2 + Nm_nc^2. \tag{6.122}$$

Rearranging equation (6.122), we have

$$M_{A,Z}c^2 = Zm_pc^2 + Nm_nc^2 - B_{A,Z}. \tag{6.123}$$

Comparing equations (6.123) and (6.121), it can be seen that

$$B_{A,Z} = -\sum_i (T_i^0 + U_i^0). \tag{6.124}$$

If the nucleons inside the nucleus were more tightly bound, for example if the energy levels were lower down in Fig. 6.11(a), then $\Sigma_i (T_i^0 + U_i^0)$ would be more negative, the total binding energy $B_{A,Z}$ would be more positive, and, according to equations (6.121) and (6.123), $M_{A,Z}c^2$ would decrease.

In practice, mass spectrometers determine the masses of positive ions, in the way described in Section 2.12, using the equation:

$$\frac{d}{dt}(\alpha M\boldsymbol{u}) = q\boldsymbol{E} + q\boldsymbol{u}\times\boldsymbol{B}, \tag{6.125}$$

where M is now the mass of the positive ion. In the case of a positive ion, in addition to the A nucleons in the nucleus, there are $Z-1$ orbital electrons whose rest, kinetic and potential energies also contribute to the total energy E^0 in the zero-momentum frame, in which the positive ion is at rest, and hence to the measured mass $M = E^0/c^2$ of the positive ion. The rest energy of an electron is $m_ec^2 = 0.511$ MeV, whereas for a positive ion of $A \approx 100$, $Mc^2 \approx 100\,\text{GeV} = 10^5$ MeV. Thus the ratio m_ec^2/Mc^2 is of order 5×10^{-6}. A typical mass spectrometer measures the mass M of the ion to an accuracy better than 1 part in 10^6. Hence the rest energy m_ec^2 of the missing electron must be added to the measured value of Mc^2 for the positive ion in order to determine the rest energy $M_{A,Z}c^2$ of the *neutral* atom of mass number A and atomic number Z. The binding energy of the electron missing from the positive ion is typically of order 10 eV, which is only one part in 10^{10} of Mc^2, and so the contribution of the binding energy of the missing electron can generally be neglected when working to an accuracy of one part in 10^6. Hence the mass of the neutral atom is given, to a very good approximation, by $M_{A,Z} = M + m_e$, where M is the measured mass of the positive ion and m_e the mass of an electron.

★ 6.11 NUCLEAR REACTIONS (FULL THEORY)

It is conventional to work with the masses of the neutral atoms when calculating the Q of a nuclear reaction. For example, consider the reaction

$$^{7}_{3}\text{Li} + ^{1}_{1}\text{H} \rightarrow 2\ ^{4}_{2}\text{He} + Q \qquad (6.126)$$

where the symbols $^{7}_{3}$Li, $^{1}_{1}$H and $^{4}_{2}$He here denote the masses of the *neutral* lithium, hydrogen and helium *atoms* respectively, and therefore include the masses and binding energies of the orbital electrons. Note that there are four electrons on both sides of equation (6.126), so that the total rest energies of the electrons cancel. The difference in the total binding energy of the orbital electrons on the two sides of equation (6.126) can generally be neglected when working to an accuracy of 1 part in 10^6. The significant changes in the rest energies in equation (6.126) are due to the changes in the atomic nuclei. This shows that using the masses of the neutral atoms gives virtually the same Q values as if the nuclear masses were used.

It is also usual to use atomic mass units (symbol u). The atomic mass scale is defined by assuming that the mass of the neutral $^{12}_{6}$C atom is exactly 12 atomic mass units, that is 12 u. The conversion factor is

$$1\ \text{u} \equiv 931.50\ \text{MeV}/c^2. \qquad (6.127)$$

Recent values of the relevant atomic masses of the reactants in equation (6.126) are

$$^{1}_{1}\text{H} = 1.007\ \ 825\ \text{u},$$

$$^{4}_{2}\text{He} = 4.002\ 603\ \text{u},$$

$$^{7}_{3}\text{Li} = 7.016\ 003\ \text{u}.$$

Hence Δm is given by

$$\Delta m = 2 \times 4.002\ 603\ \text{u} - (7.016\ 003\ \text{u} + 1.007\ 825\ \text{u})$$

$$= -\ 0.018\ 622\ \text{u}.$$

Using equation (6.127), we find

$$\Delta\ (mc^2) = -\ 17.35\ \text{MeV}.$$

Using equation (6.37), we have

$$Q = \Delta T = -\ \Delta\ (mc^2) = +\ 17.35\ \text{MeV}. \qquad (6.128)$$

It was shown in Section 6.5 that this prediction is in agreement with the value obtained by Cockroft and Walton. Equation (6.37) was derived in Section 6.5 by applying the law of conservation of energy in the laboratory frame, and was interpreted there by saying that the gain in kinetic energy Q was accompanied by a decrease in the total rest energies of the particles. The origin of the change $\Delta\ (mc^2)$ in the total rest energy will now be considered.

The total number of protons, neutrons and electrons is generally the same before and after a low-energy nuclear reaction. For example, there are four protons, four neutrons and four electrons on both sides of equation (6.126). Consider again the general nuclear reaction given by

$$\mathrm{A} + \mathrm{B} \rightarrow \mathrm{C} + \mathrm{D} + Q\,. \tag{6.129}$$

Application of the law of conservation of energy in the laboratory frame gives

$$E_{\mathrm{A}} + E_{\mathrm{B}} = E_{\mathrm{C}} + E_{\mathrm{D}}\,. \tag{6.130}$$

According to equation (6.5), $E_{\mathrm{A}} = M_{\mathrm{A}}c^2 + T_{\mathrm{A}}$ where M_{A} is the mass and T_{A} the kinetic energy of composite particle A due to the motion of that particle *as a whole*—that is, due to the motion of its centre of mass in the laboratory frame. Hence equation (6.130) can be rewritten as

$$M_{\mathrm{A}}c^2 + T_{\mathrm{A}} + M_{\mathrm{B}}c^2 + T_{\mathrm{B}} = M_{\mathrm{C}}c^2 + T_{\mathrm{C}} + M_{\mathrm{D}}c^2 + T_{\mathrm{D}}\,. \tag{6.131}$$

From equation (6.123), we have

$$M_{\mathrm{A}}c^2 = \sum_i m_{\mathrm{A}_i}c^2 - B_{\mathrm{A}}\,, \tag{6.132}$$

where m_{A_i} is the mass of the ith constituent particle of the composite particle A and B_{A} is the total binding energy of A. Similar expressions hold for $M_{\mathrm{B}}c^2$, $M_{\mathrm{C}}c^2$ and $M_{\mathrm{D}}c^2$. Substituting for $M_{\mathrm{A}}c^2$ etc. into equation (6.131), we have

$$\sum_i m_{\mathrm{A}_i}c^2 - B_{\mathrm{A}} + T_{\mathrm{A}} + \sum_i m_{\mathrm{B}_i}c^2 - B_{\mathrm{B}} + T_{\mathrm{B}}$$
$$= \sum_i m_{\mathrm{C}_i}c^2 - B_{\mathrm{C}} + T_{\mathrm{C}} + \sum_i m_{\mathrm{D}_i}c^2 - B_{\mathrm{D}} + T_{\mathrm{D}}\,. \tag{6.133}$$

If the total number of constituent particles, namely protons, neutrons and electrons, is conserved in the reaction then

$$\sum_i m_{\mathrm{A}_i}c^2 + \sum_i m_{\mathrm{B}_i}c^2 = \sum_i m_{\mathrm{C}_i}c^2 + \sum_i m_{\mathrm{D}_i}c^2. \tag{6.134}$$

Substracting equation (6.134) from equation (6.133) and rearranging gives

$$T_{\mathrm{C}} + T_{\mathrm{D}} - (T_{\mathrm{A}} + T_{\mathrm{B}}) = B_{\mathrm{C}} + B_{\mathrm{D}} - (B_{\mathrm{A}} + B_{\mathrm{B}})\,. \tag{6.135}$$

According to this equation, the increase in the kinetic energies of the composite particles (atoms) measured in the laboratory frame is equal to the increase in the total binding energy of the composite particles measured in the individual zero-momentum frames in which the composite particles A, B, C and D are individually at rest. If we put $T_{\mathrm{C}} + T_{\mathrm{D}} - T_{\mathrm{A}} - T_{\mathrm{B}} = \Delta T$ and $B_{\mathrm{C}} + B_{\mathrm{D}} - B_{\mathrm{A}} - B_{\mathrm{B}} = \Delta B$, equation (6.135) can be written in the concise form

$$Q = \Delta T = \Delta B. \tag{6.136}$$

According to equation (6.124), the total binding energy of the composite particle A is

$$B_A = -\sum_i (T_{A_i}^0 + U_{A_i}^0),$$

where $T_{A_i}^0$ and $U_{A_i}^0$ are measured in the reference frame in which A is at rest. Hence equation (6.136) can be rewritten as

$$Q = \Delta T = \Delta B$$

$$= -\left[\sum_i (T_{C_i}^0 + U_{C_i}^0) + \sum_i (T_{D_i}^0 + U_{D_i}^0) - \sum_i (T_{A_i}^0 + U_{A_i}^0) - \sum_i (T_{B_i}^0 + U_{B_i}^0)\right]. \quad (6.137)$$

According to this equation, Q, the gain in the kinetic energies of the composite atomic particles, that is $T_C + T_D - T_A - T_B$ measured in the laboratory frame, is associated with an increase $\Delta B = B_C + B_D - B_A - B_B$ in the total binding energy, which is related by equation (6.137) to differences in the kinetic energies and potential energies of the constituent particles making up the composite particles C and D compared with A and B. According to equations (6.132) and (6.121), an increase in the total binding energy of a composite particle, which is due to a more negative value for the sum $\Sigma_i (T_i^0 + U_i^0)$ leads to a proportionate decrease in the mass of the composite particle. Hence the change $\Delta (mc^2) = (M_C + M_D - M_A - M_B) c^2$ can be used to determine the change in the total binding energy $\Delta B = -\Delta (mc^2)$, which can then be used to determine $Q = \Delta T$ using equation (6.136), which was derived using the law of conservation of energy. In practice, when determining Q values, there is no need to interpret the origins of the changes leading to $\Delta (mc^2)$. Application of the law of conservation of energy, in the simple way described in Section 6.5, leads directly to equation (6.37).

★ 6.12 THE BINDING ENERGY OF THE NUCLEUS

If the total binding energy of the orbital electrons in the atom is negligible compared with the total binding energy of the nucleons in the nucleus then the total binding energy $B_{A,Z}$ of a nucleus consisting of Z protons and $N = A - Z$ neutrons can be written in the approximate form

$$B_{A,Z} = Zm_H c^2 + Nm_n c^2 - M_{A,Z} c^2, \quad (6.138)$$

where m_H is now the mass of the neutral hydrogen atom, m_n is the mass of the neutron and $M_{A,Z}$ is the mass of the neutral atom. Note that the rest energies of the Z electrons in the term $Zm_H c^2$ are compensated by the rest energies of the Z electrons in the neutral atom, which are included in $M_{A,Z} c^2$. Using $m_H = 1.007825$ u, $m_n = 1.008665$ u and the conversion factor 1 u = 931.50 MeV, equation (6.138) becomes

$$B_{A,Z} = 931.50\,(1.007\,825\,Z + 1.008\,665\,N - M_{A,Z})\ \text{MeV}, \quad (6.139)$$

where $M_{A,Z}$ is the mass of the neutral atom, expressed in atomic mass units u. Dividing both sides of this equation by the mass number A, we obtain the *mean* binding energy per particle (nucleon) in the nucleus. The results for various

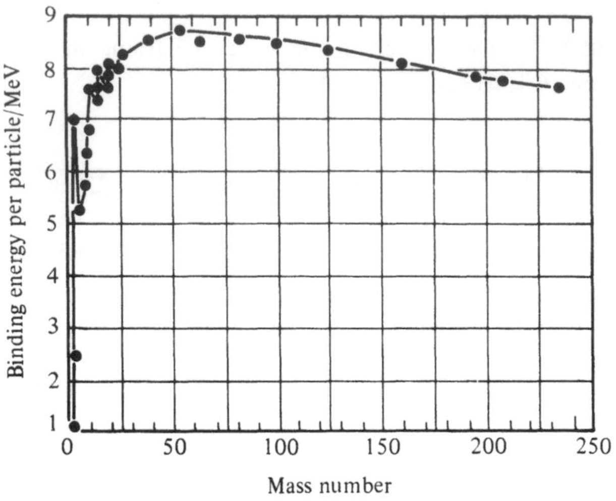

Fig. 6.12 The mean binding energy per nucleon for the most stable nucleus at each mass number.

values of Z are shown in Fig. 6.12. The maximum of the binding energy per particle curve is at $Z = 56$; that is, the individual nucleons are more tightly bound, on average, for $Z = 56$ than for $Z < 56$ or $Z > 56$. If two or more light nuclei ($Z << 56$) are combined to form a single nucleus of $Z < 56$, it follows from Fig. 6.12 that the mean binding energy per nucleon is generally greater in the combined nucleus than in the original separate nuclei, so that the total binding energy is increased and, according to equation (6.136), energy is released. This type of nuclear reaction is called a fusion or thermonuclear reaction. Thermo-nuclear reactions are important in energy production in stars. It is believed that there are two main cycles of nuclear reactions, namely the proton–proton chain, which predominates in small stars, of masses close to that of the Sun, and the CNO cycle, which predominates in more massive stars. The net result, in each case, is to produce a helium nucleus from four protons with a release of about 26.7 MeV of energy per helium nucleus formed.

Another important type of nuclear reaction is fission. It can be seen from Fig. 6.12 that the average binding energy per nucleon (proton or neutron) in the nucleus decreases at very high values of mass number. The protons and neutrons are not as tightly bound in very large nuclei such as uranium as in medium-sized nuclei of mass number about 100. If a nucleus, such as uranium, is split in two then, since the protons and neutrons will be more tightly bound in the resulting nuclei, energy will be released. A typical example of the fission of a ^{235}U nucleus is

$$^{235}_{92}U + ^{1}_{0}n \rightarrow ^{236}_{92}U \rightarrow ^{140}_{54}Xe + ^{94}_{38}Sr + 2^{1}_{0}n + \gamma + 200\ \text{MeV}.$$

6.13 "RIGID" BODIES OF FINITE SIZE

Experiments have shown that energy and linear momentum are conserved in all the phenomena described so far in this chapter. It appears that processes such as pair production do not take place unless there is a charged particle present so that energy and momentum can be conserved. It should be emphasized that all the phenomena considered were events taking place at approximately one point of space at one instant of time. When two particles such as two electric charges, separated in space at points x_1 and x_2 on the x axis of an inertial frame Σ, give rise to electromagnetic forces on each other that lead to changes in their momenta, if momentum is conserved in one inertial frame Σ at time t then, if the momenta of the individual charges are transformed to Σ', the transformed values of momenta refer to times $t'_1 = \gamma(t - vx_1/c^2)$ and $t'_2 = \gamma(t - vx_2/c^2)$ respectively, and so do not refer to the same time in Σ'. Since the momenta are changing continuously, it is not possible to formulate a law of conservation of momentum for spatially separated interacting charged particles that is valid in all inertial frames, unless one attributes momentum to the electromagnetic field (see Jackson (1975)). Similar arguments can be applied to colliding particles of *finite* extension. There are, however, a very wide range of phenomena that can be considered, to a good approximation, as events taking place at one point of space at one instant of time, and experiments have shown that in these events the laws of conservation of linear momentum and of energy are valid. It must be remembered that the last statement must be qualified to the extent of the uncertainty principle.

After discussing the mechanics of "point" particles, it is normal in Newtonian mechanics to extend the theory to rigid bodies of finite dimensions, the mechanics of continuous media, hydrodynamics etc. As fluids and large bodies in terrestrial conditions generally move with velocities very much less than the velocity of light, the deviations from Newtonian mechanics are usually not important. (However, in astrophysical contexts relativistic effects can be important in continuous mechanics—for example in the fluid dynamics of gas accretion onto neutron stars and black holes, as well as in the "jets" of highly relativistic gas emerging from objects such as quasars.) It is more convenient to develop these topics using tensor methods. For these reasons these topics will not be considered here but the interested reader is referred for example to Møller (1952) and Pauli (1958). The present discussion will be confined to a few comments about "rigid" bodies.

In Newtonian mechanics a rigid body is defined to be one in which the distance between any two particles making up the body is an invariant. If one of these two particles is acted on by a force that has a component along the line joining the particles, such that one of the particles is moved, then, according to Newtonian mechanics, this force must produce instantaneously a similar change in the position of the second particle, so that the separation of the two particles remains the same; otherwise the separation of the two particles would change and the body would not be rigid according to the classical definition. Therefore a rigid body, in the classical sense, should be able to transmit forces between its constituent parts with infinite speed. According to the theory of special relativity, energy and momentum (and hence forces) cannot be transmitted with a velocity exceeding that of light, so that there are no rigid bodies in the strict classical sense, and the concept of a rigid body is rarely introduced into the theory of special relativity. Sometimes such a concept is introduced for the purposes of discussion; for example, to quote McCrea (1952),

> We shall therefore now define *a rigid rod as one along which impulses are transmitted with speed c*. Since our rod is being treated as a one-dimensional body and since we are dealing only with longitudinal motion, no other condition is required. Moreover, since the theory permits the existence of no "more rigid" body of this sort, there is no objection to adopting the term *rigid* in this sense.

At first sight it appears that this new definition does away with the normal ideas of everyday solid bodies such as a steel rod. Consider the propagation of longitudinal waves in such a rod. Let the rod be uniform, and let it be composed of a material of density ρ and Young's modulus E. The velocity u of longitudinal waves along the rod is

$$u = \left(\frac{E}{\rho}\right)^{1/2} .$$

Typical values for steel are $\rho = 7.8 \times 10^3$ kg m^{-3} and $E = 2.0 \times 10^{11}$ N m^{-2}, so that $u \approx 5 \times 10^3$ m s^{-1} and $u/c \approx 1.7 \times 10^{-5}$. Hence the relativistic upper limit is not of any practical significance, even for the most rigid substances known.

A classical rigid body is often used in classical physics as the ideal reference frame for measuring coordinates. Such a reference frame cannot be used in relativity theory since lengths are not invariant. On the other hand, according to the theory of special relativity, light moves in straight lines with the same speed in all inertial frames, so that straight lines can be defined in terms of light paths, and distances measured using the radar methods described in Chapter 4.

6.14 HISTORICAL NOTE: THE CONCEPT OF MASS IN RELATIVISTIC MECHANICS

6.14.1 Introduction

The only mass that we introduced into our discussions of relativistic mechanics in Chapter 2 and this chapter was the mass m appearing in the definition of relativistic momentum by equations (2.1) and (6.1), namely

$$\mathbf{p} = \alpha m \mathbf{u} .$$

The mass m also appeared in equation (2.24) for the relativistic form of the Lorentz force, namely

$$f = \frac{\mathrm{d}}{\mathrm{d}t}(\alpha m\boldsymbol{u}) = q\boldsymbol{E} + q\boldsymbol{u} \times \boldsymbol{B}\,. \tag{6.140}$$

The mass m can be determined, whatever the speed of the particle, provided that the equations of relativistic mechanics are used. For example, the mass of a positive ion of any speed can be measured using a mass spectrometer in the way described in Section 2.12, provided that equation (6.140) is used to interpret the results. If relativistic mechanics is applied to the collision discussed in Section 1.2.3 then, instead of equation (1.8), we have

$$\frac{m_1}{m_2} = \frac{\alpha_4 u_4}{\alpha_1 u_1 - \alpha_3 u_3}\,. \tag{6.141}$$

If u_1, u_3 and u_4 can be measured, this equation can be used to determine m_1/m_2 whatever the speeds of the particles. In the extreme low-velocity limit when u_1, u_3 and u_4 are all very much less than c the values of α_1, α_3 and α_4 are all close to unity and equation (6.141) reduces to equation (1.8), showing that Newtonian mechanics can be used to determine m in the zero-velocity limit. However, it is not necessary to use the zero-velocity limit to determine m, provided that the equations of relativistic mechanics are used.

6.14.2 Longitudinal and transverse mass

In Newtonian mechanics the inertial mass of a particle can be defined, using equation (1.5), as the ration f/a of the magnitude of the force $\boldsymbol{f}$ acting on the particle to the magnitude of the acceleration $\boldsymbol{a}$ produced by that force. This definition of mass cannot be taken over into relativistic mechanics, since from equation (6.3)

$$\boldsymbol{f} = \frac{\mathrm{d}}{\mathrm{d}t}(\alpha m\boldsymbol{u}) = \alpha m\boldsymbol{a} + m\boldsymbol{u}\frac{\mathrm{d}\alpha}{\mathrm{d}t}\,. \tag{6.142}$$

In general $\boldsymbol{a}$ is not parallel to $\boldsymbol{f}$, as it is in Newtonian mechanics.

If the force $\boldsymbol{f}$ is always perpendicular to the velocity $\boldsymbol{u}$ of the particle, which is true for the case of a charged particle moving in a time-independent magnetic field, discussed in Section 2.6, then the speed of the charge is constant, so that $\alpha = (1 - u^2/c^2)^{-1/2}$ is constant and $\mathrm{d}\alpha/\mathrm{d}t$ is zero. In this case equation (6.142) reduces to

$$f/a = \alpha m = m_t\,. \tag{6.143}$$

In this context the quantity $m_t = \alpha m$ used to be called the *transverse* mass of the particle.

If the force $\boldsymbol{f}$ is always parallel to $\boldsymbol{u}$, which is true for the case of a charged particle accelerated from rest in a uniform electric field, discussed in Section 2.5, then the speed of the charged particle increases, and

$$\frac{d\alpha}{dt} = \frac{du}{dt}\frac{d\alpha}{du} = a\frac{d}{du}\left[\frac{1}{(1-u^2/c^2)^{1/2}}\right]$$
$$= \frac{au/c^2}{(1-u^2/c^2)^{3/2}} .$$

Substituting in equation (6.142), we find that in this case

$$f/a = \alpha^3 m = m_l . \tag{6.144}$$

The quantity $m_l = \alpha^3 m$ used to be called the *longitudinal* mass of the particle.

If f and $\boldsymbol{u}$ are neither parallel nor perpendicular to each other, the ratio f/a is given by a more complicated expression. The value of f/a varies from αm to $\alpha^3 m$, depending on the angle between f and $\boldsymbol{u}$. These results show that the ratio f/a cannot be used to define a unique mass in relativistic mechanics. For the want of a generally accepted term, we shall define the ratio f/a to be the *inertia* I of the particle. The value of I depends on the experimental conditions, but is always in the range $\alpha m \leqslant I \leqslant \alpha^3 m$. It can be seen that two factors, namely the mass m and $\alpha = (1-u^2/c^2)^{-1/2}$ contribute to the inertia I. The mass m is related to the total energy E^0 of the composite particle, measured in the zero-momentum frame in which the particle is at rest, by the equation $mc^2 = E^0$. All forms of energy contribute to E^0 and hence to m. Since α increases as the speed of the particle increases, so does the inertia I. This means that the acceleration produced by a given force decreases as the speed of the particle increases. This leads to the existence of a limiting speed for accelerated particles, which is the speed of light in empty space.

6.14.3 Relativistic (or momental) mass

Another mass used extensively in earlier treatments of relativistic mechanics is the relativistic mass m_r, given by

$$m_r = \frac{m}{(1-u^2/c^2)^{1/2}} . \tag{6.145}$$

When u is zero, the relativistic mass m_r is equal to m. For this reason m was often called the rest mass and equation (6.145) was often interpreted as the variation of the relativistic mass of a particle with its velocity. (Some authors use the symbol m_0 for rest mass and m for relativistic mass.) Using the relativistic mass m_r, given by equation (6.145), the expressions for the relativistic momentum and (total) energy of a single particle become

$$\boldsymbol{p} = m_r\boldsymbol{u} , \tag{6.146}$$

$$E = m_r c^2. \tag{6.147}$$

There is no intrinsic change in a particle when we change our standard of rest. For example, the mass of a proton should be equal to 1.6724×10^{-27} kg in all inertial reference frames, whatever their velocities relative to the laboratory frame. This was one of the reasons why we have avoided the use of the term "relativistic mass" in this book.

It is difficult to give an operational definition of the relativistic mass m_r. It is only when $\boldsymbol{f}$ and $\boldsymbol{u}$ are perpendicular that m_r is equal to the inertia $I = f/a$. One possible definition would be to define m_r as that quantity which when multiplied by the velocity gives the relativistic momentum of the particle. In this context the term *momental* mass would be more appropriate than relativistic mass. If the reader wishes to use the term "relativistic mass" then the approach that fits in best with our development of relativistic mechanics is to treat the relativistic mass m_r as an abbreviation for αm.

REFERENCES

Champion, F.C. (1932) *Proc. R. Soc. Lond.* **A136**, 630.
Dodd, J.E. (1984) *The Ideas of Particle Physics*. Cambridge University Press.
Jackson, J.D. (1975) *Classical Electrodynamics*, 2nd edn. Wiley, New York.
McCrea, W.H. (1952) *Sci. Proc. R. Dublin Soc.* **26**, 27.
Møller, C. (1952) *The Theory of Relativity*. Oxford University Press.
Pauli, W. (1958) *Theory of Relativity*. Pergamon Press, Oxford.
Perkins, D.H. (1987) *Introduction to High Energy Physics*, 3rd edn. Addison-Wesley, Reading, Massachusetts.
Rosser, W.G.V. (1964) *An Introduction to the Theory of Relativity*, p. 175. Butterworth, London.

PROBLEMS

6.1 A proton of kinetic energy 500 MeV collides elastically with a stationary proton. The protons rebound with equal energies. Apply the laws of conservation of momentum and energy to determine the angle between the paths of the protons after the collision. (The rest mass of a proton is 938 MeV/c^2.)

6.2 A particle of mass m_1 having kinetic energy T_1 collides with a stationary particle of mass m_2. If m_1 is scattered through an angle of 90° and has momentum $\bar{p}_1$ after the collision, show that after the collision m_2 has momentum

$$(\bar{p}_1^2 + 2T_1m_1 + T_1^2/c^2)^{1/2}$$

and makes an angle

$$\arctan\left[\bar{p}_1\,(2T_1m_1 + T_1^2/c^2)^{-1/2}\right]$$

with the direction of the incident particle.

6.3 A particle of mass M moving with velocity u collides with a stationary particle of mass m. If the particles stick together, show that the speed of the composite particle is

$$\frac{\alpha M}{\alpha M + m}u,$$

where $\alpha = (1 - u^2/c^2)^{-1/2}$. [*Hint*: Put the mass of the composite particle equal to M^*, then eliminate M^*.]

6.4 An incident photon of energy $+h\nu$ is scattered through an angle of 180° in a Compton-type collision with a stationary electron. Calculate the total energies after the collision of (a) the electron; (b) the photon. [*Hint*: Apply the laws of conservation of (total) energy and of linear momentum in the laboratory frame.]

6.5 A photon of energy 2.60 MeV undergoes a Compton-type collision with a stationary electron of mass 0.511 MeV/c^2, of the type shown in Fig. 6.3. The path of the scattered photon makes an angle $\theta = 30°$ with the direction of the incident photon. Determine (a) the energy E_3 of the scattered photon; (b) the frequency of the scattered photon; (c) the (total) energy E_4 of the recoiling electron; (d) the angle ϕ that the path of the recoiling electron makes with the direction of the incident photon. [*Hint*: Use conservation of momentum and of energy. Check your value of E_3 using equation (6.33).]

6.6 Calculate the total binding energy, expressed in MeV, of a helium nucleus of mass $6.645\,64 \times 10^{-27}$ kg consisting of two protons, each of mass $1.672\,39 \times 10^{-27}$ kg and two neutrons, each of mass $1.674\,70 \times 10^{-27}$ kg.

6.7 Calculate the energy "released" in the reaction

$$^{14}_{7}\mathrm{N} + {}^{4}_{2}\mathrm{He} \rightarrow {}^{17}_{8}\mathrm{O} + {}^{1}_{1}\mathrm{H}.$$

The atomic masses in atomic mass units (1 u ≡ 931.50 MeV/c^2) are

$$^{14}_{7}\mathrm{N} = 14.003\,074\ \mathrm{u}, \qquad {}^{4}_{2}\mathrm{He} = 4.002\,603\ \mathrm{u},$$

$$^{17}_{8}\mathrm{O} = 16.999\,134\ \mathrm{u}, \qquad {}^{1}_{1}\mathrm{H} = 1.007\,825\ \mathrm{u}.$$

6.8 The Earth receives solar energy at the rate of 1.35×10^3 W m^{-2}. If the distance from the Earth to the Sun is 1.5×10^{11} m, find the decrease in the mass of the Sun per second due to the emission of radiation.

6.9 Show that if (total) energy and momentum are to be conserved in pair production by a photon, the process cannot take place spontaneously in a vacuum. [*Hint:* Assume for the purposes of the discussion that pair production takes place in a vacuum. Let the electron momenta be $h\nu/2c$, with both electrons being emitted in the direction of the incident photon. In this case the sums of the total energies of the electrons would be $2\,[\,m^2c^4 + (\,h\nu/2c\,)^2c^2\,]^{1/2}$, which would be greater than $h\nu$, so that energy would not be conserved.]

6.10 A K^+ meson (of mass 497.7 MeV/c^2) decays at rest into a π^+ (of mass 139.6 MeV/c^2) and a π^0 (of mass 135.0 MeV/c^2). Determine the kinetic energy and velocity of the π^+. [*Hint*: Use equation (6.50).]

6.11 A moving unstable neutral particle decays into two charged particles of kinetic energies 190 and 30 MeV and momenta 300 and 240 MeV/c respectively. Determine the masses of the decay products. If the angle between the decay particles is 45°, determine (a) the mass of the neutral particle; (b) its momentum; (c) its kinetic energy. [*Hint*: Use $m = (p^2c^2 - T^2)/2Tc^2$ to determine the masses of the decay products. Apply conservation of energy and momentum in the laboratory frame to determine the energy and momentum of the parent particle; hence determine its mass.]

6.12 Determine the maximum kinetic energy of the electron arising from the decay of a muon (μ), into an electron (e) a neutrino (ν) and an antineutrino ($\bar{\nu}$):

$$\mu \rightarrow e + \nu + \bar{\nu}.$$

The mass of the muon is 105.7 MeV/c^2, while the masses of the neutrino and antineutrino may be taken to be zero. The mass of the electron is 0.511 MeV/c^2.

6.13 An unstable particle of mass M_0 and momentum $\boldsymbol{p}_0$ decays in flight into two particles of masses M_1 and M_2, momenta $\boldsymbol{p}_1$ and $\boldsymbol{p}_2$ and (total) energies, E_1 and E_2 respectively. Show that

$$M_0^2 c^4 = (M_1 + M_2)^2 c^4 + 2E_1E_2 - 2M_1M_2c^4 - 2p_1p_2c^2 \cos\theta,$$

where θ is the angle between the two decay particles. [*Hint*: Apply conservation of momentum and energy in the laboratory frame.]

6.14 A neutral pion (π^0) of mass m, velocity u, decays in flight into two photons. If one of the photons is emitted at an angle θ to the direction of motion of the π^0 in the laboratory frame, show that its energy $h\nu$ is given by

$$h\nu = \frac{mc^2}{2\alpha\,[1 - (u/c)\cos\theta]},$$

where $\alpha = (1 - u^2/c^2)^{-1/2}$.

6.15 A stationary atomic nucleus has mass M when it is in an excited state of excitation energy E_0. It emits a photon in a transition to the ground state. Show, using relativistic equations, that the frequency of the photon, after allowing for recoil, is

$$\nu = \frac{E_0}{h}\left(1 - \frac{E_0}{2Mc^2}\right).$$

[*Hint:* Derive and use equation (6.50).]

6.16 An electron (of mass 0.511 MeV/c^2) is moving along the x axis of the inertial reference frame Σ with velocity $u = 0.8c$.

(a) Determine the momentum and (total) energy of the electron in Σ.

(b) Use the momentum and energy transformations to determine the momentum and (total) energy of the electron in the inertial frame Σ' that is moving with uniform velocity $v = 0.6c$ along the x axis of Σ.

6.17 If the electron in Problem 6.16 is moving at an angle of 45° to the x axis of Σ, what are the magnitudes of its momentum and (total) energy in Σ'?

6.18 A photon of energy 2 MeV is moving at an angle of 30° to the x axis of the inertial frame Σ. Use the momentum and energy transformations to determine the energy and direction of the photon in an inertial frame Σ' moving with uniform velocity $0.8c$ along the x axis of Σ. Check your results using the inverses of equations (6.69) and (6.72).

6.19 Show that if a proton of velocity $\boldsymbol{u}$ collides with another proton, the velocity of the zero-momentum frame is

$$u_{zm} = \frac{\alpha u}{\alpha + 1},$$

where $\alpha = (1 - u^2/c^2)^{-1/2}$. Show also that if $\gamma_{zm} = (1 - u_{zm}^2/c^2)^{-1/2}$, then

$$\gamma_{zm} = \left[\frac{1}{2}(\alpha + 1)\right]^{1/2}.$$

Determine u_{zm} if the (total) energy of the incident proton of mass 938 MeV/c^2 is 10 GeV. [*Hint*: Use equation (6.86).]

6.20 A particle of mass m_1, velocity $\boldsymbol{u}_1$ and (total) energy E_1 collides with a stationary particle of mass m_2. Show that in the extreme relativistic case, when $u_1 \approx c$ the velocity of the zero-momentum frame is given, to a good approximation, by $c(1 - m_2c^2/E_1)$. [*Hint*: Start from equation (6.89).]

6.21 A photon of energy 500 MeV is incident on a stationary proton of mass 938 MeV/c^2. Determine the velocity of the zero-momentum frame.

6.22 In the collision of a very high-energy proton, of velocity u and $\alpha = (1 - u^2/c^2)^{-1/2} \gg 1$ with a stationary proton, pions are emitted symmetrically in the zero-momentum frame with velocity u^0, so that the number of pions in the backward direction is equal to the number in the forward direction in the zero-momentum frame. Show that the angle in the laboratory frame containing half the emitted pions, $\theta_{1/2}$, is given by

$$\tan\theta_{1/2} = \frac{u^0}{\gamma_{zm}u_{zm}} \approx \frac{2^{1/2}u^0}{\alpha^{1/2}c},$$

where u_{zm} and γ_{zm} are given in Problem 6.19. [*Hint*: Consider a pion emitted at 90° in the zero-momentum frame and find its direction in the laboratory frame.]

6.23 An electron (of mass 0.511 MeV/c^2) collides with a stationary electron, producing a neutral pion (of mass 135 MeV/c^2) according to the reaction

$$\mathrm{e} + \mathrm{e} = \mathrm{e} + \mathrm{e} + \pi^0.$$

Determine the minimum (total) energy that the incident electron must have for the process to take place. [*Hint*: Use equation (6.98).]

6.24 A rocket propels itself rectilinearly through empty space by emitting pure radiation in the direction opposite to its motion. If V is its final velocity relative to its initial rest frame, prove that the ratio of the initial to the final masses of the rocket is given by

$$\frac{M_i}{M_f} = \left(\frac{c + V}{c - V}\right)^{1/2}.$$

[*Hint*: Since the speed of the radiation emitted from the rocket is always equal to c in the laboratory frame, one can equate energies and momenta at the beginning and at the end of the acceleration, writing $\Sigma h\nu$ and $\Sigma h\nu/c$ for the total energy and momentum respectively of the emitted photons.]

7

★ Four-Vector Methods

★ 7.1 INTRODUCTION

According to equation (4.59), the interval δs between two events recorded at (x, y, z, t) and $(x+\delta x, y+\delta y, z+\delta z, t+\delta t)$ in the inertial reference frame Σ, and defined by the relation

$$\delta s^2 = \delta x^2 + \delta y^2 + \delta z^2 - c^2\,\delta t^2 \tag{7.1}$$

is an invariant. This invariant property can be used to develop the theory of special relativity using 4-vectors. At this point there is a choice of approaches. One could continue to use only the real variables $w = ct$, x, y and z introduced in Chapter 4. Alternatively, one can use complex numbers using the variables

$$X_1 = x, \qquad X_2 = y, \qquad X_3 = z, \qquad X_4 = \mathrm{i}ct \tag{7.2}$$

in Σ, where $\mathrm{i} = \sqrt{(-1)}$. Many writers, particularly those going on to consider general relativity, prefer to use the real-variable form for 4-vectors, as they feel that this is an appropriate place to introduce the reader to some tensor analysis. The methods developed can then be used when they go on to general relativity, where the complex-number approach cannot be used. The author has found that the average experimental physicist, who is not likely to go on to do much general relativity, finds the 4-vector methods easier to understand if one starts with the complex-number method, since one can then employ formal analogies with two-dimensional vectors in coordinate space and use "Euclidean" geometry. Such readers are no more worried about the use of complex numbers in this context than they are of the use of complex numbers in alternating current theory. A number of writers on advanced electromagnetism and high-energy physics use the complex form for 4-vectors, so that all readers should have some familiarity with the method. Hence we shall start with the complex-number approach and then go on later to outline the 4-vector method based on real variables.

★ 7.2 DEFINITION OF A VECTOR IN TWO DIMENSIONS

In elementary textbooks a vector is defined as a quantity that has both magnitude and direction. The most convenient way of representing a vector is to use a

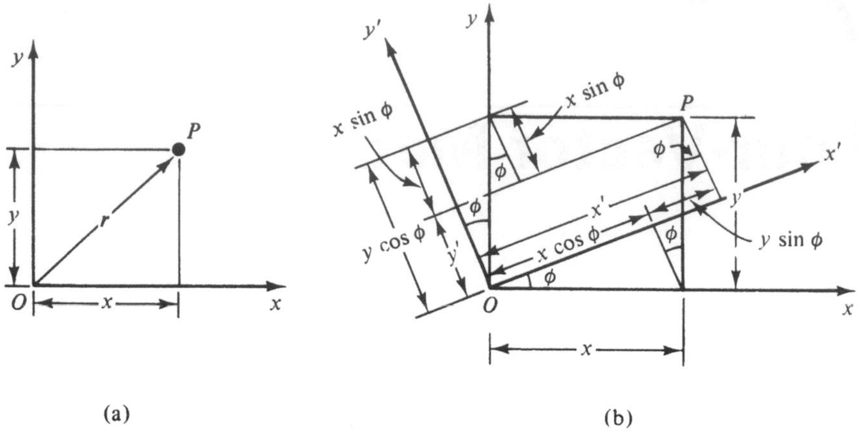

Fig. 7.1 (a) The position vector $\boldsymbol{r}$ is a vector from the origin O to the point P, which has coordinates x and y. (b) The (x, y) axes are rotated through an angle ϕ into the (x', y') coordinate system.

coordinate system fixed in space. The directions of the axes of the coordinate system represent fixed directions in space. The vector can then be represented by its components along the various axes. If the components of a vector are given then its magnitude and direction in space can be calculated. If the axes of the coordinate system are rotated in space, the vector itself must not be changed in either magnitude or direction. As a two-dimensional example, consider the position vector $\boldsymbol{r}$ from the origin O to the point P, which is in the (x, y) plane in Fig. 7.1(a). The vector from O to P has the component (x, y) in the rectangular (Ox, Oy) coordinate system shown in Fig. 7.1(a). The length of the vector $\boldsymbol{r}$ is equal to $(x^2 + y^2)^{1/2}$. If the x and y axes are rotated through an angle ϕ in the (x, y) plane to give new axes x' and y', as shown in Fig. 7.1(b), the position vector $\boldsymbol{r}$ from the origin O to the point P is unchanged in magnitude and direction. It can be seen from Fig. 7.1(b) that if the point P is fixed, the components (x', y') in the rotated coordinate system of the vector from O to P are given in terms of x and y by

$$x' = x\cos\phi + y\sin\phi\,, \tag{7.3}$$

$$y' = -x\sin\phi + y\cos\phi\,. \tag{7.4}$$

In matrix form, we have

$$\begin{bmatrix} x' \\ y' \end{bmatrix} = \begin{bmatrix} \cos\phi & \sin\phi \\ -\sin\phi & \cos\phi \end{bmatrix} \begin{bmatrix} x \\ y \end{bmatrix}. \tag{7.5}$$

The length of the position vector from O to P in Fig. 7.1(b), measured in the rotated coordinate system, is given by

$$(x'^2 + y'^2)^{1/2} = [(x\cos\phi + y\sin\phi)^2 + (-x\sin\phi + y\cos\phi)^2]^{1/2}$$

$$= (x^2 + y^2)^{1/2} = r, \tag{7.6}$$

where we have used $\sin^2 \phi + \cos^2 \phi = 1$. This result confirms that the length of the position vector is unchanged when the coordinate axes are rotated.

Rather than define a vector as a quantity that has magnitude and direction, it is convenient to define it as a quantity that transforms in the same way as the position vector $\boldsymbol{r}$ when the coordinate axes are rotated. By comparison with equation (7.5), we shall define a quantity $\boldsymbol{A}$ to be a vector in two-dimensional space if its components A_x and A_y transform according to the equation

$$\begin{bmatrix} A_x' \\ A_y' \end{bmatrix} = \begin{bmatrix} \cos\phi & \sin\phi \\ -\sin\phi & \cos\phi \end{bmatrix} \begin{bmatrix} A_x \\ A_y \end{bmatrix} \tag{7.7}$$

when the coordinate axes are rotated through an angle ϕ.

If a vector $\boldsymbol{A}$ is multiplied by a scalar ψ, we obtain another vector of magnitude $\psi \boldsymbol{A}$ pointing in the same direction as the vector $\boldsymbol{A}$.

★ 7.3 DEFINITION OF A 4-VECTOR USING COMPLEX NUMBERS

Introduce new variables such that in the inertial reference frame Σ

$$X_1 = x, \qquad X_2 = y, \qquad X_3 = z, \qquad X_4 = \mathrm{i}ct\,. \tag{7.2}$$

Let the corresponding quantities in the inertial reference frame Σ' that is moving with uniform velocity v relative to Σ along their common x axis and whose origin coincides with the origin of Σ at $t = t' = 0$ be

$$X_1' = x', \qquad X_2' = y', \qquad X_3' = z', \qquad X_4' = \mathrm{i}ct', \tag{7.8}$$

From the Lorentz transformations,

$$x' = \gamma(x - vt)\,.$$

Hence

$$X_1' = \gamma\left(X_1 - \frac{vX_4}{\mathrm{i}c}\right) = \gamma\left(X_1 + \mathrm{i}\frac{v}{c}X_4\right), \tag{7.9}$$

$$X_2' = X_2\,, \tag{7.10}$$

$$X_3' = X_3\,. \tag{7.11}$$

Now

$$t' = \gamma\left(t - \frac{vx}{c^2}\right), \qquad \text{or} \qquad \mathrm{i}ct' = \gamma\left(\mathrm{i}ct - \mathrm{i}\frac{v}{c}x\right).$$

Hence

$$X_4' = \gamma\left(X_4 - \mathrm{i}\frac{v}{c}X_1\right). \tag{7.12}$$

Equations (7.9) and (7.12) can be rewritten in the form

$$X_1' = X_1 \cos\phi + X_4 \sin\phi\,, \tag{7.13}$$

$$X_4' = -X_1 \sin\phi + X_4 \cos\phi\,, \tag{7.14}$$

where

$$\cos\phi = \gamma, \tag{7.15}$$

$$\sin\phi = (1-\cos^2\phi)^{1/2} = \left(1 - \frac{1}{1-v^2/c^2}\right)^{1/2}$$

$$= \left(\frac{-v^2/c^2}{1-v^2/c^2}\right)^{1/2} = \frac{\mathrm{i}\gamma v}{c}\,. \tag{7.16}$$

Equations (7.13) and (7.14) have the same mathematical form as equations (7.3) and (7.4). Consider first motion along the x axis of Σ only. The variables X_1 and X_4 suffice in this case, and the Lorentz transformations can be expressed in the matrix form.

$$\begin{bmatrix} X_1' \\ X_4' \end{bmatrix} = \begin{bmatrix} \cos\phi & \sin\phi \\ -\sin\phi & \cos\phi \end{bmatrix} \begin{bmatrix} X_1 \\ X_4 \end{bmatrix}. \tag{7.17}$$

By analogy with equation (7.5), we shall treat X_1 and X_4 as the components of a vector in the (X_1, X_4) plane. The Lorentz transformations can then be treated as a rotation of the X_1 and X_4 axes through an angle ϕ in the (X_1, X_4) plane.

For motion in three dimensions, we need the four variables X_1, X_2, X_3 and X_4. Provided that the directions of the x axis of Σ and the x' axis of Σ' are chosen such that Σ' moves with uniform velocity v along the common x and x' axes, the Lorentz transformations can be written in the matrix form

$$\begin{bmatrix} X_1' \\ X_2' \\ X_3' \\ X_4' \end{bmatrix} = \begin{bmatrix} \gamma & 0 & 0 & \mathrm{i}\frac{v}{c}\gamma \\ 0 & 1 & 0 & 0 \\ 0 & 0 & 1 & 0 \\ -\mathrm{i}\frac{v}{c}\gamma & 0 & 0 & \gamma \end{bmatrix} \begin{bmatrix} X_1 \\ X_2 \\ X_3 \\ X_4 \end{bmatrix}$$

$$= \begin{bmatrix} \cos\phi & 0 & 0 & \sin\phi \\ 0 & 1 & 0 & 0 \\ 0 & 0 & 1 & 0 \\ -\sin\phi & 0 & 0 & \cos\phi \end{bmatrix} \begin{bmatrix} X_1 \\ X_2 \\ X_3 \\ X_4 \end{bmatrix}. \tag{7.18}$$

The variables X_1, X_2, X_3 and X_4 behave like the components of a vector in the 4-dimensional (X_1, X_2, X_3 and X_4) "Euclidean" space, and the Lorentz transformations can be treated as a rotation of axes in the (X_1, X_4) plane.

A 4-vector in the (X_1, X_2, X_3, X_4) space will now be defined as a quantity that transforms under a Lorentz transformation in the same way as the (X_1, X_2, X_3, X_4) coordinates of a point in the four-dimensional space. For example, $\boldsymbol{A}$ is defined to be a 4-vector if, under a Lorentz transformation, by analogy with equations (7.9)–(7.12), its components transform according to the relations

$$\left.\begin{aligned} A_1' &= \gamma\left(A_1 + \mathrm{i}\frac{v}{c}A_4\right), \\ A_2' &= A_2, \qquad A_3' = A_3, \\ A_4' &= \gamma\left(A_4 - \mathrm{i}\frac{v}{c}A_1\right), \end{aligned}\right\} \tag{7.19a}$$

or in matrix form

$$\begin{bmatrix} A_1' \\ A_2' \\ A_3' \\ A_4' \end{bmatrix} = \begin{bmatrix} \gamma & 0 & 0 & \mathrm{i}\frac{v}{c}\gamma \\ 0 & 1 & 0 & 0 \\ 0 & 0 & 1 & 0 \\ -\mathrm{i}\frac{v}{c}\gamma & 0 & 0 & \gamma \end{bmatrix} \begin{bmatrix} A_1 \\ A_2 \\ A_3 \\ A_4 \end{bmatrix}$$

$$= \begin{bmatrix} \cos\phi & 0 & 0 & \sin\phi \\ 0 & 1 & 0 & 0 \\ 0 & 0 & 1 & 0 \\ -\sin\phi & 0 & 0 & \cos\phi \end{bmatrix} \begin{bmatrix} A_1 \\ A_2 \\ A_3 \\ A_4 \end{bmatrix}. \tag{7.19b}$$

Now

$$A_1'^2 + A_2'^2 + A_3'^2 + A_4'^2 = \gamma^2\left(A_1 + \mathrm{i}\frac{v}{c}A_4\right)^2 + A_2^2 + A_3^2 + \gamma^2\left(A_4 - \mathrm{i}\frac{v}{c}A_1\right)^2$$

$$= A_1^2 + A_2^2 + A_3^2 + A_4^2. \tag{7.20}$$

Hence the length of a 4-vector is unchanged under rotation of axes (i.e. by a Lorentz transformation). It is straightforward for a reader familiar with matrix theory to show that the transformation matrix in equations (7.18) and (7.19b) is an orthogonal matrix that leaves the length of the 4-vector unchanged, in agreement with equation (7.20).

Returning to the position 4-vector, its four components can be represented by $(X_1, X_2, X_3, X_4) = (\boldsymbol{r}, \mathrm{i}ct)$. The three components X_1, X_2 and X_3 are the components of a vector in ordinary three-dimensional space; the fourth component is equal to a scalar times $\sqrt{(-1)}$. All 4-vectors have this property.

★ 7.4 THE 4-VELOCITY

Consider a particle moving in three dimensions, as shown previously in Fig. 5.1. Let the particle be at the point (x, y, z) at the time t in Σ. The components of the position 4-vector of this event in Σ are $(X_1 = x, X_2 = y, X_3 = z, X_4 = ict)$. Let the particle be at $(x + dx, y + dy, z + dz)$ an infinitesimal time dt later in Σ. The position vector of the second event has the components $(X_1 + dX_1,\ X_2 + dX_2,\ X_3 + dX_3,\ X_4 + dX_4)$ in Σ. Since $\boldsymbol{X} + d\boldsymbol{X}$ is a 4-vector, it follows from equations (7.9)–(7.12) that

$$X_1' + dX_1' = \gamma\left[X_1 + dX_1 + \frac{iv}{c}(X_4 + dX_4) \right],$$

$$X_2' + dX_2' = X_2 + dX_2\,,$$

$$X_3' + dX_3' = X_3 + dX_3\,,$$

$$X_4' + dX_4' = \gamma\left[X_4 + dX_4 - i\frac{v}{c}(X_1 + dX_1) \right].$$

Subtracting the transformations for the 4-vector (X_1, X_2, X_3, X_4) given by equations (7.9)–(7.12) gives

$$\left.\begin{aligned} dX_1' &= \gamma\left(dX_1 + i\frac{v}{c}\, dX_4 \right), \\ dX_2' &= dX_2, \qquad dX_3' = dX_3, \\ dX_4' &= \gamma\left(dX_4 - i\frac{v}{c}\, dX_1 \right), \end{aligned}\right\} \tag{7.21}$$

Hence the increments in the components of the position 4-vector form the 4-vector

$$d\boldsymbol{X} = (dX_1, dX_2, dX_3, dX_4)\,. \tag{7.22}$$

Since the length of a 4-vector is an invariant, we have

$$dX_1'^2 + dX_2'^2 + dX_3'^2 + dX_4'^2 = dX_1^2 + dX_2^2 + dX_3^2 + dX_4^2\,. \tag{7.23}$$

This corresponds to the relation

$$ds^2 = dx^2 + dy^2 + dz^2 - c^2\, dt^2 = \text{invariant.} \tag{7.24}$$

The above relation was derived in Section 4.10, where it was shown that the interval between two events is an invariant. Note that the presence of $i = \sqrt{(-1)}$ in the definition $X_4 = ict$ converts the negative sign in equation (7.24) into a positive sign in equation (7.23).

In the case of a particle that has a three-dimensional velocity $\boldsymbol{u}$ in Σ, which is always less than c, it is always possible to find an inertial reference frame Σ^0, moving with velocity $\boldsymbol{u}$ relative to Σ, in which the moving particle is instantaneously at rest. In Σ^0 the spatial separation of the two events is zero and the time interval $\mathrm{d}\tau$ between the two events is the *proper time interval* between them, which can be measured by one clock at rest in Σ^0. In Σ^0 the interval $\mathrm{d}s$ between the two events is given by

$$\mathrm{d}s^2 = -c^2\,\mathrm{d}\tau^2. \tag{7.25}$$

Hence equation (7.24) can be written in the form

$$\mathrm{d}s^2 = \mathrm{d}x^2 + \mathrm{d}y^2 + \mathrm{d}z^2 - c^2\,\mathrm{d}t^2 = -c^2\,\mathrm{d}\tau^2.$$

Dividing by $\mathrm{d}t^2$, we have

$$\left(\frac{\mathrm{d}x}{\mathrm{d}t}\right)^2 + \left(\frac{\mathrm{d}y}{\mathrm{d}t}\right)^2 + \left(\frac{\mathrm{d}z}{\mathrm{d}t}\right)^2 - c^2 = -c^2\left(\frac{\mathrm{d}\tau}{\mathrm{d}t}\right)^2.$$

Now $\mathrm{d}x/\mathrm{d}t = u_x$ etc., and $u_x^2 + u_y^2 + u_z^2 = u^2$. Hence

$$u^2 - c^2 = -c^2\left(\frac{\mathrm{d}\tau}{\mathrm{d}t}\right)^2,$$

or

$$\mathrm{d}\tau^2 = \mathrm{d}t^2(1 - u^2/c^2).$$

Therefore

$$\mathrm{d}\tau = \mathrm{d}t\,(1 - u^2/c^2)^{1/2} = \left(\frac{\mathrm{d}t}{\alpha}\right), \tag{7.26}$$

where

$$\alpha = \frac{1}{(1 - u^2/c^2)^{1/2}}. \tag{7.27}$$

It follows from equation (7.25) that, since the interval $\mathrm{d}s$ is an invariant (i.e. it has the same value in all inertial reference frames), the proper time interval $\mathrm{d}\tau$ is an invariant. If all the components of a vector are multiplied by an invariant scalar ψ, one obtains a new vector of length ψ times the length of the original vector. Multiplying the 4-vector $\mathrm{d}\boldsymbol{X}$, given by equation (7.22), by the invariant $1/\mathrm{d}\tau$, which, according to equation (7.26), is equal to $\alpha/\mathrm{d}t$, one obtains a quantity that is also a 4-vector. Let this be denoted by $\boldsymbol{U}$. We have

$$\boldsymbol{U} = \left(\frac{\mathrm{d}X_1}{\mathrm{d}\tau}, \frac{\mathrm{d}X_2}{\mathrm{d}\tau}, \frac{\mathrm{d}X_3}{\mathrm{d}\tau}, \frac{\mathrm{d}X_4}{\mathrm{d}\tau}\right)$$

$$= \left(\alpha\frac{\mathrm{d}x}{\mathrm{d}t}, \alpha\frac{\mathrm{d}y}{\mathrm{d}t}, \alpha\frac{\mathrm{d}z}{\mathrm{d}t}, ic\alpha\right).$$

But $\mathrm{d}x/\mathrm{d}t = u_x$ etc., where $\boldsymbol{u}$ is the ordinary three-dimensional velocity of the particle. Hence

$$\boldsymbol{U} = (\alpha u_x, \alpha u_y, \alpha u_z, \mathrm{i}c\alpha) = (\alpha \boldsymbol{u}, \mathrm{i}c\alpha)\,. \tag{7.28}$$

The 4-vector $\boldsymbol{U}$ is called the 4-velocity of the particle. The corresponding 4-velocity of the particle in Σ' is

$$\boldsymbol{U}' = (\alpha' \boldsymbol{u}', \mathrm{i}c\alpha')\,, \tag{7.29}$$

where $\boldsymbol{u}'$ is the three-dimensional velocity of the particle measured in Σ' and $\alpha' = (1 - u'^2/c^2)^{-1/2}$. Since $\boldsymbol{U}$ is a 4-vector, it must transform according to equations (7.19). Hence

$$U_4' = \gamma\left(U_4 - \mathrm{i}\frac{v}{c}U_1\right) = \gamma\left(\mathrm{i}c\alpha - \mathrm{i}\frac{v}{c}\alpha u_x\right)$$

$$= \gamma \mathrm{i}c\alpha\,(1 - vu_x/c^2)\,.$$

But, from equation (7.28), $U_4' = \mathrm{i}c\alpha'$. Hence

$$\alpha' = \gamma\alpha\,(1 - vu_x/c^2)\,. \tag{7.30}$$

Equation (7.30) is the same as equation (5.14). Using equations (7.19),

$$U_1' = \gamma\left(U_1 + \mathrm{i}\frac{v}{c}U_4\right) = \gamma\left[\alpha u_x + \mathrm{i}\frac{v}{c}(\mathrm{i}c\alpha)\right]$$

$$= \gamma\alpha\,(u_x - v).$$

But, from equation (7.29), $U_1' = \alpha' u_x'$. Hence

$$\alpha' u_x' = \gamma\alpha\,(u_x - v)\,.$$

Using equation (7.30), we obtain

$$u_x' = \frac{u_x - v}{1 - vu_x/c^2}\,.$$

This expression is the transformation for the x component of the three-dimensional velocity of the particle.

From equations (7.19), $U_2' = U_2$ and $U_3' = U_3$. Hence

$$\alpha' U_y' = \alpha U_y\,, \qquad \alpha' U_z' = \alpha U_z\,.$$

Using equation (7.30), we obtain

$$u_y' = \frac{u_y}{\gamma(1 - vu_x/c^2)}\,,$$

$$u_z' = \frac{u_z}{\gamma(1 - vu_x/c^2)}\,.$$

This completes the velocity transformations. The length of a 4-vector must be an invariant. Now

$$U^2 = U_1^2 + U_2^2 + U_3^2 + U_4^2$$

$$= \frac{u^2}{1-u^2/c^2} + \frac{i^2c^2}{1-u^2/c^2}$$

$$= \frac{u^2-c^2}{1-u^2/c^2}$$

$$= -c^2. \qquad (7.31)$$

It follows from the principle of the constancy of the speed of light that $-c^2$ is an invariant, confirming that the length of the 4-velocity is an invariant.

★ 7.5 THE 4-MOMENTUM

If the 4-velocity $\boldsymbol{U}$, given by equation (7.27), is multiplied by the invariant mass m of the particle, we get a new 4-vector called the 4-momentum, which will be denoted by $\boldsymbol{P}$. We have

$$\boldsymbol{P} = m\boldsymbol{U} = (\alpha m\boldsymbol{u}, \mathrm{i}\alpha mc) = (\boldsymbol{p}, \mathrm{i}\alpha mc),$$

where $\boldsymbol{p} = \alpha m\boldsymbol{u}$ is the ordinary 3-dimensional momentum, defined by equation (2.1). According to equation (2.50), $\alpha mc = E/c$, where E is the (total) energy of the particle, which is equal to the sum of its kinetic energy T and its rest energy mc^2. Hence the 4-momentum can be expressed as

$$\boldsymbol{P} = (\boldsymbol{p}, \mathrm{i}E/c). \qquad (7.32)$$

The corresponding 4-vector in Σ' is

$$\boldsymbol{P}' = (\boldsymbol{p}', \alpha' mc) = (\boldsymbol{p}', \mathrm{i}E'/c). \qquad (7.33)$$

The length of a 4-vector is an invariant; therefore

$$p^2 + (\mathrm{i}E/c)^2 = p^2 - E^2/c^2 = \text{constant}. \qquad (7.34)$$

When $p = 0$, $E = mc^2$, so that the constant in equation (7.23) is equal to $-m^2c^2$, which is an invariant since both the mass m and the velocity of light c are constants. Hence equation (7.34) can be rewritten in the form

$$E^2 = p^2c^2 + m^2c^4. \qquad (7.35)$$

Since the length of a 4-vector is an invariant, in Σ' one has

$$p'^2 - E'^2/c^2 = -m^2c^2,$$

giving

$$E'^2 = p'^2c^2 + m^2c^4.$$

To obtain the transformations for the momentum of a single particle, use can be made of the fact that $(\boldsymbol{p}, \mathrm{i}E/c)$ is a 4-vector. For the first component, one has from equations (7.19),

$$P_1' = \gamma\left(P_1 + \frac{iv}{c}P_4\right),$$

or

$$p_x' = \gamma\left(p_x + \frac{iv}{c}\frac{iE}{c}\right)$$
$$= \gamma\,(p_x - vE/c^2)\,.$$

From equations (7.19),

$$P_2' = P_2\,,$$

giving

$$p_y' = p_y\,.$$

Similarly,

$$P_3' = P_3$$

gives

$$p_z' = p_z\,.$$

The relation

$$P_4' = \gamma\left(P_4 - \frac{iv}{c}P_1\right)$$

gives

$$\frac{iE'}{c} = \gamma\left(\frac{iE}{c} - \frac{iv}{c}p_x\right),$$

or

$$E' = \gamma\,(E - vp_x).$$

These transformations are the same as those derived in Section 6.7.1.

★ 7.6 THE MINKOWSKI 4-FORCE ACTING ON A PARTICLE

Since the increments of a 4-vector are the components of another 4-vector (see equation (7.21)), it follows that $d\boldsymbol{P} = (d\boldsymbol{p}, i\,dE/c)$ is a 4-vector. Multiplying by the invariant $1/d\tau$, which, according to equation (7.26), is equal to α/dt, we obtain another 4-vector, called the Minkowski 4-force, defined by the relation

$$\boldsymbol{F} = \frac{d\boldsymbol{P}}{d\tau} = \left(\frac{d\boldsymbol{p}}{d\tau}, \frac{i}{c}\frac{dE}{d\tau}\right) = \left(\alpha\frac{d\boldsymbol{p}}{dt}, \frac{i\alpha}{c}\frac{dE}{dt}\right). \qquad (7.36)$$

In the case of a particle of invariant mass m, this equation can be written in the form

$$\boldsymbol{F} = \frac{\mathrm{d}}{\mathrm{d}\tau}(m\boldsymbol{U}) = m\frac{\mathrm{d}\boldsymbol{U}}{\mathrm{d}\tau} + \boldsymbol{U}\frac{\mathrm{d}m}{\mathrm{d}\tau}$$

$$= m\frac{\mathrm{d}\boldsymbol{U}}{\mathrm{d}\tau}. \tag{7.37}$$

The rate of change of the three-dimensional momentum, $\mathrm{d}\boldsymbol{p}/\mathrm{d}t$, is equal to the three-dimensional force $\boldsymbol{f}$. If the three-dimensional force $\boldsymbol{f}$ acting on the particle gives rise to a displacement $\mathrm{d}\boldsymbol{l}$ of the particle, the work done by the force is given by the scalar product $\boldsymbol{f} \bullet \mathrm{d}\boldsymbol{l}$. If all this work goes into increasing the kinetic energy T of the particle then $\mathrm{d}T = \boldsymbol{f} \bullet \mathrm{d}\boldsymbol{l}$. Dividing by $\mathrm{d}t$ and putting $\mathrm{d}\boldsymbol{l}/\mathrm{d}t = \boldsymbol{u}$, we find that $\mathrm{d}T/\mathrm{d}t = \boldsymbol{f} \bullet \boldsymbol{u}$. Since $E = T + mc^2$, we have for a particle of invariant mass m

$$\frac{\mathrm{d}E}{\mathrm{d}t} = \frac{\mathrm{d}}{\mathrm{d}t}(T + mc^2) = \frac{\mathrm{d}T}{\mathrm{d}t}$$

$$= \boldsymbol{f} \bullet \boldsymbol{u}.$$

Hence the 4-force, given by equation (7.36), can be expressed in the form

$$\boldsymbol{F} = (\alpha \boldsymbol{f}, \mathrm{i}\alpha \boldsymbol{f} \bullet \boldsymbol{u}/c). \tag{7.38}$$

The dynamics of a single particle of invariant mass m is given by equation (7.36), which can now be expressed in the form

$$\boldsymbol{F} = (\alpha \boldsymbol{f}, \mathrm{i}\alpha \boldsymbol{f} \bullet \boldsymbol{u}/c)$$

$$= \left(\alpha\frac{\mathrm{d}\boldsymbol{p}}{\mathrm{d}t}, \frac{\mathrm{i}\alpha}{c}\frac{\mathrm{d}E}{\mathrm{d}t}\right), \tag{7.39}$$

which corresponds to the three-dimensional equations

$$\boldsymbol{f} = \frac{\mathrm{d}\boldsymbol{p}}{\mathrm{d}t}, \tag{7.40}$$

$$\boldsymbol{f} \bullet \boldsymbol{u} = \frac{\mathrm{d}E}{\mathrm{d}t} = \frac{\mathrm{d}T}{\mathrm{d}t}. \tag{7.41}$$

Equation (7.40) is the same as equation (2.4). According to equation (7.41), the rate of increase of the (total) energy and the rate of increase of kinetic energy of a particle are equal to the rate at which the three-dimensional force $\boldsymbol{f}$ is doing work on the particle, which is moving with velocity $\boldsymbol{u}$ in Σ.

The expression for the 4-force in Σ' is

$$\boldsymbol{F}' = (\alpha' \boldsymbol{f}', \mathrm{i}\alpha \boldsymbol{f}' \bullet \boldsymbol{u}'/c). \tag{7.42}$$

For the first component of a 4-vector, from equation (7.19) we have

$$F_1' = \gamma(F_1 + \mathrm{i}vF_4/c).$$

It follows from equations (7.42) and (7.38) that $F_1' = \alpha' f_x'$, $F_1 = \alpha f_x$ and $F_4 = \mathrm{i}\alpha \boldsymbol{f} \bullet \boldsymbol{u}/c$.
Hence

$$\alpha' f_x' = \gamma\left(\alpha f_x + \mathrm{i}\frac{v}{c}\frac{\mathrm{i}}{c}\alpha \boldsymbol{f} \bullet \boldsymbol{u}\right),$$

and therefore

$$f_x' = \frac{\gamma\alpha}{\alpha'}\left[f_x - \frac{v}{c^2}(f_x u_x + f_y u_y + f_z u_z)\right],$$

where we have used the standard relation for the scalar product, namely

$$\boldsymbol{f} \bullet \boldsymbol{u} = f_x u_x + f_y u_y + f_z u_z \,.$$

Using equation (7.30) to eliminate $\gamma\alpha/\alpha'$, we have

$$f_x' = \frac{1}{1 - v u_x/c^2}\left[f_x\left(1 - \frac{v u_x}{c^2}\right) - \frac{v}{c^2}(f_y u_y + f_z u_z)\right]$$

$$= f_x - \frac{v u_y}{c^2(1 - v u_x/c^2)} f_y - \frac{v u_z}{c^2(1 - v u_x/c^2)} f_z \,. \tag{7.43}$$

Since from equation (7.19), $F_2' = F_2$ and $F_3' = F_3$, using equations (7.38) and (7.42) we find that

$$\alpha' f_y' = \alpha f_y \,, \qquad \alpha' f_z' = \alpha f_z \,.$$

Using equation (7.30) to eliminate α'/α, we finally obtain

$$\left.\begin{aligned} f_y' &= \frac{f_y}{\gamma(1 - v u_x/c^2)} \,, \\ f_z' &= \frac{f_z}{\gamma(1 - v u_x/c^2)} \,. \end{aligned}\right\} \tag{7.44}$$

This completes the derivations of the transformations for the three-dimensional force. We have now re-derived, using 4-vector methods, all the relations and transformations developed previously in Chapters 5 and 6. The reader could have gone directly from Section 4.10 to this chapter and then returned to consider the applications given in Chapters 5 and 6.

★ 7.7 SOME MANIPULATIONS WITH 4-VECTORS

By analogy with two-dimensional vectors, it will be assumed that $\boldsymbol{C}$ the sum of two 4-vectors $\boldsymbol{A}$ and $\boldsymbol{B}$ is a 4-vector having components

$$\left.\begin{aligned} C_1 &= A_1 + B_1, \qquad & C_2 &= A_2 + B_2, \\ C_3 &= A_3 + B_3, \qquad & C_4 &= A_4 + B_4 \,. \end{aligned}\right\} \tag{7.45}$$

As a check, using equations (7.19), we have

$$C_1' = A_1' + B_1' = \gamma\left(A_1 + \mathrm{i}\frac{v}{c}A_4\right) + \gamma\left(B_1 + \mathrm{i}\frac{v}{c}B_4\right)$$

$$= \gamma\left[A_1 + B_1 + \mathrm{i}\frac{v}{c}(A_4 + B_4)\right] = \gamma\left(C_1 + \mathrm{i}\frac{v}{c}C_4\right),$$

confirming that the components of $\boldsymbol{C}$ transform as those of a 4-vector.

By analogy with the derivation of equation (7.20), we have

$$C^2 = C_1^2 + C_2^2 + C_3^2 + C_4^2 = C_1'^2 + C_2'^2 + C_3'^2 + C_4'^2$$

$$= C'^2, \tag{7.46}$$

showing that the length of $\boldsymbol{A} + \boldsymbol{B}$ is an invariant; that is, it has the same value in all inertial reference frames. The above analysis can be extended to the sum of more than two 4-vectors.

It is left as an exercise (Problem 7.1) for the reader to use equations (7.19) to show by substitution that the scalar product $\boldsymbol{A} \bullet \boldsymbol{B}$ of two 4-vectors A and $\boldsymbol{B}$, defined by the relation

$$\boldsymbol{A} \bullet \boldsymbol{B} = A_1B_1 + A_2B_2 + A_3B_3 + A_4B_4\,, \tag{7.47}$$

is an invariant; that is, it has the same value in Σ and Σ'.

★ 7.8 CONSERVATION OF 4-MOMENTUM

Consider the interaction of two or more particles having three-dimensional momenta $\boldsymbol{p}_1, \boldsymbol{p}_2, \ldots, \boldsymbol{p}_i, \ldots$ and (total) energies $E_1, E_2, \ldots, E_i, \ldots$ in Σ. Suppose that the particles present after the event have momenta $\bar{\boldsymbol{p}}_1, \bar{\boldsymbol{p}}_2, \ldots, \bar{\boldsymbol{p}}_j, \ldots$ and total energies $\bar{E}_1, \bar{E}_2, \ldots, \bar{E}_j, \ldots$ in Σ. It follows from the generalization of equations (7.45) that the sum of the 4-momenta of the particles present before the event is a 4-vector having components $(\Sigma_i\, p_{ix}, \Sigma_i\, p_{iy}, \Sigma_i\, p_{iz}, \mathrm{i}\Sigma_i\, E_i/c)$. The sum of the 4-momenta of the particles present after the event is a 4-vector having components $\Sigma_j\, \bar{p}_{jx}, \Sigma_j\, \bar{p}_{jy}, \Sigma_j\, \bar{p}_{jz}, \mathrm{i}\Sigma_j\, \bar{E}_j/c)$ in Σ. If it is assumed that the total 4-momentum in Σ is conserved in the event, this implies that

$$\sum_i p_{ix} = \sum_j \bar{p}_{jx}\,, \qquad \sum_i p_{iy} = \sum_j \bar{p}_{jy}\,, \qquad \sum_i p_{iz} = \sum_j \bar{p}_{jz}\,, \tag{7.48}$$

$$\sum_i E_i = \sum_j \bar{E}_j\,. \tag{7.49}$$

Equations (7.48) express the law of conservation of 3-momentum, and equation (7.49) is the law of conservation of energy, showing that both laws are incorporated in the single statement that the total 4-momentum is conserved in point events. Since the length of a 4-vector is the same in all inertial reference frames and since the total 4-momentum is conserved, it follows that the lengths of the total 4-momentum before and after the event in Σ are equal to each other and to the lengths of the total 4-momentum in Σ' before and after the event.

★ 7.9 ENERGY AVAILABLE IN THE ZERO-MOMENTUM FRAME

To illustrate the power of 4-vector methods, we shall now re-derive the results presented in Sections 6.8.1 and 6.8.4. Consider a particle of mass m_1, 3-momentum $\boldsymbol{p}_1$ and (total) energy E_1 that is moving along the $+x$ axis of the laboratory frame Σ when it collides with a particle of mass m_2 that is at rest in Σ, so that $\boldsymbol{p}_2 = 0$ and $E_2 = m_2c^2$. The total 4-momentum in Σ before the collision is $(p_1, 0, 0, \mathrm{i}\,(E_1 + m_2c^2)/c)$. According to equation (7.46), the square of the length of the total 4-momentum in Σ before the collision is $p_1^2 - (E_1 + m_2c^2)^2/c^2$.

Consider the zero-momentum frame Σ^0, which is moving with a uniform velocity v relative to Σ such that the total 3-momentum is zero in Σ^0 before the collision. Using equations (7.19) to transform the resultant 4-momentum from Σ to Σ^0, we have for the first component of the total 4-momentum in Σ^0 before the collision

$$P_1^0 = (p_{1x}^0 + p_{2x}^0) = \gamma\left[p_1 + \mathrm{i}\frac{v}{c}\frac{\mathrm{i}\,(E_1 + m_2c^2)}{c}\right] = \gamma\left[p_1 - \frac{v}{c^2}(E_1 + m_2c^2)\right].$$

Since $p_{1x}^0 + p_{2x}^0$ is zero in the zero-momentum frame Σ^0, we must have

$$v = \frac{c^2p_1}{E_1 + m_2c^2}. \tag{7.50}$$

This is the same as equation (6.89).

It will be assumed that after the event there are particles of masses $\overline{m}_1$, $\overline{m}_2$, $\ldots$, $\overline{m}_j$, $\ldots$ that are all at rest in the zero-momentum frame Σ^0, as shown previously in Fig. 6.9. The total 4-momentum in Σ^0 after the event is $(0, 0, 0, \mathrm{i}\,(\Sigma_j\,\overline{m}_jc^2)/c)$. According to equation (7.46), the square of the length of the total 4-momentum in Σ^0 after the event is $-(\Sigma_j\,\overline{m}_j\,c^2)^2/c^2$. Since the total 4-momentum is conserved in the event in all inertial reference frames, and since the length of a 4-vector is the same in all such frames, it follows that the length of the total 4-momentum in the laboratory frame Σ before the collision is equal to the length of the total 4-momentum in the zero-momentum frame Σ^0 after the collision, so that

$$p_1^2 - (E_1 + m_2c^2)^2/c^2 = -\frac{1}{c^2}\left(\sum_j \overline{m}_j\,c^2\right)^2.$$

Multiplying by $-c^2$ and using equation (7.35) in the form $E_1^2 - c^2p_1^2 = m_1^2c^4$, we obtain the following relation for the threshold energy $E_1 = T_1 + m_1c^2$ needed to produce particles of masses $\overline{m}_1$, $\overline{m}_2$, $\ldots$, $\overline{m}_j$, $\ldots$:

$$2E_1 m_2 c^2 + m_1^2 c^4 + m_2^2 c^4 = \left(\sum_j \overline{m}_j\, c^2 \right)^2. \tag{7.51}$$

If we put $m_1 = m_2 = m$ and $E^0 = \Sigma_j \overline{m}_j c^2$ then this equation reduces to equation (6.98).

As an example, consider again the reaction

$$p^+ + p^+ \to p^+ + p^+ + p^+ + p^- .$$

In this case $m_1 = m_2 = m_p$ and $\Sigma_j \overline{m}_j c^2 = 4m_p c^2$, where m_p is the mass of a proton. Equation (7.51) gives

$$2E_1 m_p c^2 + 2m_p^2 c^4 = 16 m_p^2 c^4,$$

and therefore

$$E_1 = 7 m_p c^2.$$

This is in agreement with the result derived in Section 6.8.4.

As a further example, consider the production of an electron–positron pair by a γ-ray of energy E_γ in the electric field of a particle of mass M. In this example $m_1 = 0$, $m_2 = M$, $E_1 = E_\gamma$ and $\Sigma_j \overline{m}_j c^2 = 2m_e c^2 + Mc^2$, where m_e is the mass of an electron. Substituting in equation (7.51), we have

$$2E_\gamma M c^2 + M^2 c^4 = (Mc^2 + 2m_e c^2)^2,$$

and therefore

$$E_\gamma = 2m_e c^2 \left(1 + \frac{m_e}{M} \right). \tag{7.52}$$

If pair production takes place in the electric field of an atomic nucleus, $M \gg m_e$ and the threshold γ ray energy for pair production is just over $2m_e c^2 =$ 1.022 MeV. If pair production takes place in the electric field of an electron, $M = m_e$ and, according to equation (7.52), the threshold γ-ray energy is $4m_e c^2 =$ 2.044 MeV.

★ 7.10 TREATMENT OF 4-VECTORS USING REAL VARIABLES ONLY

This approach is a natural extension of the use of the real variables $w = ct$, x, y and z in the space–time diagrams and Minkowski diagrams introduced in Chapter 4. Let $X_0 = w = ct$, $X_1 = x$, $X_2 = y$ and $X_3 = z$, or, in vector form, $\boldsymbol{X} = (ct, \boldsymbol{r})$. The Lorentz transformations can be written in the form

$$\left.\begin{aligned} X_0' &= \gamma\left(X_0 - \frac{v}{c} X_1 \right), \\ X_1' &= \gamma\left(X_1 - \frac{v}{c} X_0 \right), \\ X_2' &= X_2, \qquad X_3' = X_3 . \end{aligned}\right\} \tag{7.53}$$

In this section a 4-vector will be *defined* as a quantity that transforms in the same way as the coordinates X_0, X_1, X_2 and X_3 of an event. For example, $\boldsymbol{A}$ is a 4-vector if it transforms from Σ to Σ' according to

$$\left.\begin{aligned} A_0' &= \gamma\left(A_0 - \frac{v}{c}A_1\right), \\ A_1' &= \gamma\left(A_1 - \frac{v}{c}A_0\right), \\ A_2' &= A_2, \qquad A_3' = A_3 . \end{aligned}\right\} \tag{7.54}$$

The square of the length of the 4-vector $\boldsymbol{A}$ will be *defined* as†

$$A^2 = A_0^2 - A_1^2 - A_2^2 - A_3^2 . \tag{7.55}$$

When the length of a 4-vector is defined using this equation, the invariant interval δs between two events is defined using equation (4.60), which is

$$\delta s^2 = c^2\,\delta t^2 - \delta x^2 - \delta y^2 - \delta z^2. \tag{7.56}$$

In the case of the position 4-vector ($X_0 = ct, X_1 = x, X_2 = y, X_3 = z$) the length of the position 4-vector $\boldsymbol{X}$ is

$$X^2 = X_0^2 - X_1^2 - X_2^2 - X_3^2 = c^2t^2 - x^2 - y^2 - z^2. \tag{7.57}$$

According to equation (7.56), the interval between the event at the origin of Σ at $t = 0$ and the event at $X_0 = ct$, $X_1 = x$, $X_2 = y$, $X_3 = z$ is an invariant, so that $c^2t^2 - x^2 - y^2 - z^2$ is an invariant. Hence the length of the position 4-vector is an invariant; that is it has the same value in all inertial reference frames.

The reader can now repeat the developments given in Sections 7.4–7.7 using real variables only, remembering that the proper time interval $d\tau$ is again an invariant, given by equation (7.26). The reader should derive the following relations:

$$\boldsymbol{X} = (ct, x, y, z) = (ct, \boldsymbol{r}), \tag{7.58}$$

$$d\boldsymbol{X} = (c\,dt, d\boldsymbol{r}), \tag{7.59}$$

$$\boldsymbol{U} = \frac{d\boldsymbol{X}}{d\tau} = (c\alpha, \alpha u_x, \alpha u_y, \alpha u_z) = (c\alpha, \alpha\boldsymbol{u}), \tag{7.60}$$

where

$$\alpha = \frac{1}{(1 - u^2/c^2)^{1/2}}, \tag{7.61}$$

† Some books define A^2 as $-A_0^2 + A_1^2 + A_2^2 + A_3^2$, corresponding to the definition of the interval δs by equation (4.59). The reader should always check carefully what definition of A^2 is being used by other authors.

$$\boldsymbol{P} = m\boldsymbol{U} = (mc, p_x, p_y, p_z) = (mc, \boldsymbol{p})$$

$$= (E/c, \boldsymbol{p})\,, \tag{7.62}$$

$$P^2 = E^2/c^2 - p^2 = m^2c^2, \tag{7.63}$$

$$\boldsymbol{F} = \frac{\mathrm{d}\boldsymbol{P}}{\mathrm{d}\tau} = m\frac{\mathrm{d}\boldsymbol{U}}{\mathrm{d}\tau} = \left(\frac{\alpha}{c}\boldsymbol{f}\bullet\boldsymbol{u}, \alpha\boldsymbol{f}\right)$$

$$= \left(\frac{\alpha}{c}\frac{\mathrm{d}E}{\mathrm{d}t}, \alpha\frac{\mathrm{d}\boldsymbol{p}}{\mathrm{d}t}\right). \tag{7.64}$$

The law of conservation of 4-momentum, given by equation (7.62), again leads to both the law of conservation of energy and the law of conservation of 3-momentum.

The above simplified approach avoids the need to introduce tensors by using equation (7.55) to *define* the length of a 4-vector. It is at this stage that some authors like to introduce tensors. Let the variables X_0, X_1, X_2 and X_3 now be denoted by x^0, x^1, x^2 and x^3 respectively. (Superscripts are used to denote *contravariant* vectors.) We can get over the difficulty due to the negative signs in equations (7.56) and (7.55) by putting

$$\mathrm{d}s^2 = \sum_{\mu}\sum_{\nu} g_{\mu\nu}\, x^{\mu}x^{\nu}, \tag{7.65}$$

$$A^2 = \sum_{\mu}\sum_{\nu} g_{\mu\nu}\, A^{\mu}A^{\nu}\,. \tag{7.66}$$

It is conventional to use Greek letters when the summation is from 0 to 3 with 4-vectors. It is also conventional to simplify equations by assuming that when a letter index appears twice, once as a lower and once as an upper index, it is understood that the expression is summed over all possible values of that index (Einstein summation convention). For example, equations (7.65) and (7.66) can be written in the simplified forms

$$\mathrm{d}s^2 = g_{\mu\nu}\, x^{\mu}x^{\nu}, \qquad A^2 = g_{\mu\nu}\, A^{\mu}A^{\nu}.$$

It is assumed that $g_{\mu\nu} = 0$ if μ is not equal to ν and that $g_{00} = 1$ and $g_{11} = g_{22} = g_{33} = -1$, or in matrix form

$$g_{\mu\nu} = \begin{bmatrix} 1 & 0 & 0 & 0 \\ 0 & -1 & 0 & 0 \\ 0 & 0 & -1 & 0 \\ 0 & 0 & 0 & -1 \end{bmatrix}. \tag{7.67}$$

To determine A^2 from equation (7.66), we keep μ fixed and sum first over ν by putting $\nu = 0, 1, 2, 3$ to give

$$A^2 = \sum_{\mu} (g_{\mu 0} A^{\mu} A^0 + g_{\mu 1} A^{\mu} A^1 + g_{\mu 2} A^{\mu} A^2 + g_{\mu 3} A^{\mu} A^3) .$$

All the coefficients $g_{\mu\nu}$ are zero except when $\mu = \nu$. Hence

$$\begin{aligned} A^2 &= g_{00} A^0 A^0 + g_{11} A^1 A^1 + g_{22} A^2 A^2 + g_{33} A^3 A^3 \\ &= (A^0)^2 - (A^1)^2 - (A^2)^2 - (A^3)^2. \end{aligned}$$

The quantity $g_{\mu\nu}$ given by equation (7.67) is called the *metric tensor*. In special relativity the components $g_{\mu\nu}$ of the metric tensor are all constants, but in general relativity the metric varies from point to point, the values of $g_{\mu\nu}$ at a point depending on the distribution of energy and matter. In this approach special relativity is a special case of general relativity corresponding to flat space–time, or Minkowski space, in which the metric tensor is given by equation (7.67).

★ PROBLEMS

7.1 The scalar product of two 4-vectors can be defined by

$$\boldsymbol{A} \bullet \boldsymbol{B} = A_1 B_1 + A_2 B_2 + A_3 B_3 + A_4 B_4 .$$

Show that the scalar product so defined is an invariant; that is, show that $\boldsymbol{A}' \bullet \boldsymbol{B}' = \boldsymbol{A} \bullet \boldsymbol{B}$. [*Hint*: Use equations (7.19).]

7.2 Show that the 4-acceleration $\boldsymbol{A}$ of a particle is given by

$$\boldsymbol{A} = \frac{\mathrm{d}\boldsymbol{U}}{\mathrm{d}\tau} = \left(\frac{c^2}{c^2 - u^2} \boldsymbol{a} + \boldsymbol{u} \frac{c^2 \boldsymbol{u} \bullet \boldsymbol{a}}{(c^2 - u^2)^2} , \frac{\mathrm{i}c^3 \boldsymbol{u} \bullet \boldsymbol{a}}{(c^2 - u^2)^2} \right),$$

where $\boldsymbol{a}$ is the three-dimensional acceleration of the particle.

7.3 Show that the 4-velocity and 4-acceleration of a particle are orthogonal, that is, show that $\boldsymbol{U} \bullet \boldsymbol{A}$ is zero. [*Hint*: Use direct substitution, or start from the fact that $\boldsymbol{U} \bullet \boldsymbol{U} = -c^2$ and differentiate with respect to τ, giving $\boldsymbol{U} \bullet \mathrm{d}\boldsymbol{U}/\mathrm{d}\tau = 0$.]

7.4 The Minkowski 4-force on a particle is defined as $\boldsymbol{F} = \mathrm{d}\boldsymbol{P}/\mathrm{d}\tau$. Show that if the mass m of the particle is constant then $\boldsymbol{F} \bullet \boldsymbol{U} = 0$, whereas if m is not constant then $\boldsymbol{F} \bullet \boldsymbol{U} = -c^2 \mathrm{d}m/\mathrm{d}\tau$. [*Hint*: $\boldsymbol{F} = \mathrm{d}(m\boldsymbol{U})/\mathrm{d}\tau = \boldsymbol{U}\mathrm{d}m/\mathrm{d}\tau + m\,\mathrm{d}\boldsymbol{U}/\mathrm{d}\tau$. Form $\boldsymbol{F} \bullet \boldsymbol{U}$ and remember that $\boldsymbol{U} \bullet \boldsymbol{U} = -c^2$.]

7.5 Calculate the threshold kinetic energies, in MeV, for the following processes, using the method of Section 7.9:

(a) $$\gamma + \mathrm{p} \rightarrow \mathrm{p} + \pi^0;$$

(b) $$\pi^- + \mathrm{p} \rightarrow \pi^+ + \pi^- + \mathrm{n}.$$

The masses of the p, n, $\pi^{\pm}$ and π^0 are 938, 939.5, 139 and 135 MeV/c^2 respectively.

8

Electromagnetism and Special Relativity

8.1 INTRODUCTION

When students are taught electromagnetism for the first time, it is inevitable that the individual laws such as Coulomb's law, Ampère's circuital law and Faraday's law of electromagnetic induction are introduced separately. As a result, many students tend to think of these laws as describing completely independent phenomena, and very few are aware of even the relative orders of magnitude of the electric and magnetic forces between moving charges, let alone the intimate connection between the electric and magnetic forces between moving charges.

In Section 8.2 the ratio of the electric to the magnetic forces in a simple case will be calculated using formulae familiar to most first-year undergraduates. The example will then be discussed in Section 8.3, taking Coulomb's law and the transformations of the theory of special relativity as axiomatic. This will illustrate the unity of electromagnetism in a vivid way, and show that the theory of classical electromagnetism is consistent with the theory of special relativity. The field transformations for $\boldsymbol{E}$ and $\boldsymbol{B}$ will be derived in Section 8.4 from the assumption that the expression for the Lorentz force, given by equation (2.24), is valid in all inertial reference frames. After deriving the expressions for the electric and magnetic fields due to a charge moving with uniform velocity in Section 8.5, we shall go on in Section 8.6 to show that Maxwell's equations are Lorentz-covariant; that is, they obey the principle of relativity when the coordinates and time are transformed using the Lorentz transformations.

8.2 FORCES BETWEEN TWO PARALLEL CONVECTION CURRENTS

Consider two infinitely long, straight, thin, uniformly charged, non-conducting wires a distance r apart *in vacuum* lying in the (x,y) plane of an inertial reference frame Σ. Let both wires move in the positive x direction with uniform velocity v relative to Σ, as shown in Fig. 8.1. Let the electric charge per unit length on the wires be λ, measured relative to Σ. Since the wires are moving relative to Σ,

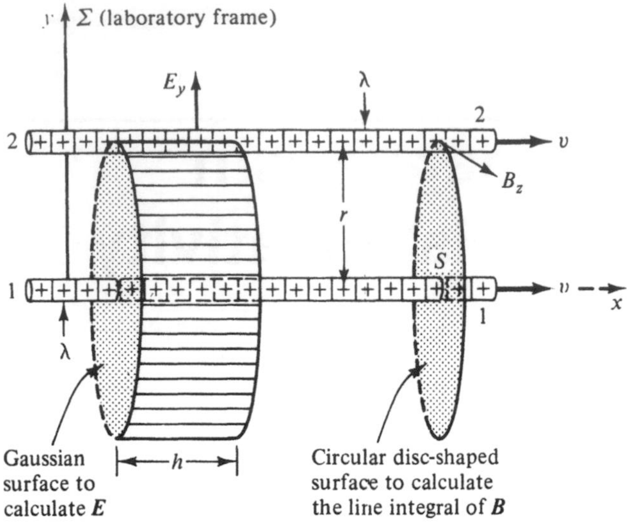

Fig. 8.1 Calculation of the electric and magnetic forces between two parallel convection currents.

the charge distributions give rise to convection currents relative to Σ. The charge passing any point in unit time is λv, so that the magnitude of the convection currents is λv in Σ.

Gauss' law will be used to calculate the electric field. By symmetry, the electric field due to wire 1 must diverge radially from the wire. Considering the cylindrical Gaussian surface of radius r and height h shown in Fig. 8.1 and applying Gauss' law, we have

$$2\pi rhE_r = \lambda h/\epsilon_0 \,,$$

and therefore

$$E_r = E_y = \frac{\lambda}{2\pi\epsilon_0 r}. \tag{8.1}$$

Since the charge per unit length on wire 2 is λ, it follows from equation (2.20) that the electric force on unit length of wire 2 is a repulsive force in the $+y$ direction of magnitude.

$$f_{\text{elec}} = \lambda E_y = \frac{\lambda^2}{2\pi\epsilon_0 r}. \tag{8.2}$$

Consider a circular disc-like surface S of radius r, having wire 1 as centre, as shown in Fig. 8.1. According to Ampère's circuital theorem, the line integral of $\boldsymbol{B}$ around the boundary of the surface S is equal to μ_0 times the total current crossing S, which is λv. Hence

$$\oint \boldsymbol{B} . \mathrm{d}\boldsymbol{l} = \mu_0 \lambda v \,.$$

By symmetry, $\boldsymbol{B}$ has the same value at all points on the circumference of S. Hence

$$2\pi r B = \mu_0 \lambda v \ ,$$

giving

$$B_z = \frac{\mu_0 \lambda v}{2\pi r} \ .$$

The magnetic force on wire 2 is $\int I_2 \, d\boldsymbol{l}_2 \times \boldsymbol{B}_1$, so that the magnetic force per unit length on wire 2 is an *attractive* force of magnitude

$$f_{\text{mag}} = -\frac{\mu_0 \lambda^2 v^2}{2\pi r} \ . \tag{8.3}$$

The resultant force per unit length on wire 2 is given by

$$\begin{aligned} f_y &= f_{\text{elec}} + f_{\text{mag}} \\ &= \frac{\lambda^2}{2\pi\epsilon_0 r} - \frac{\mu_0 \lambda^2 v^2}{2\pi r} = \frac{\lambda^2}{2\pi\epsilon_0 r}(1 - \mu_0 \epsilon_0 v^2). \end{aligned} \tag{8.4}$$

Numerical substitution of 8.85×10^{-12} F m^{-1} for ϵ_0 and $4\pi \times 10^{-7}$ H m^{-1} for μ_0 shows that $\mu_0 \epsilon_0 = 1/c^2$, where $c = 3 \times 10^8$ m s^{-1} is the velocity of light in empty space. Hence the resultant force, electric plus magnetic, on unit length of wire 2 is

$$f_y = \frac{\lambda^2}{2\pi\epsilon_0 r}(1 - v^2/c^2) \ . \tag{8.5}$$

In this simple case, the ratio of the magnetic force to the electric force is v^2/c^2, where v is the velocity of the charges relative to Σ, the inertial reference frame chosen. This illustrates that magnetic forces are of second order in v/c.

Consider a typical case, say a conduction current of 1 A flowing in a copper wire of cross-sectional area, $a = 1$ mm^2 (10^{-6} m^2). Let N be the number of charge carriers per unit volume (approximately 8.5×10^{28} m^{-3} for copper), and let $q = -1.6 \times 10^{-19}$ C be the charge on each electron. The current is given by

$$I = qNav \ ,$$

where v is the mean drift velocity of the electrons. Substituting, we find $v \approx 7 \times 10^{-5}$ m s^{-1}. If the charges shown in Fig. 8.1 moved with this speed, the ratio of the magnetic to the electric forces given by equation (8.5) would be about 5×10^{-26}. Why then are the magnetic forces between electric circuits so important? In practice, in a copper conductor there is no resultant electric charge inside the conductor, since the charges on the moving electrons are compensated by the positively charged ions, which are virtually at rest in a stationary conductor, as illustrated in Fig. 8.2. It will be assumed that the total charge per unit

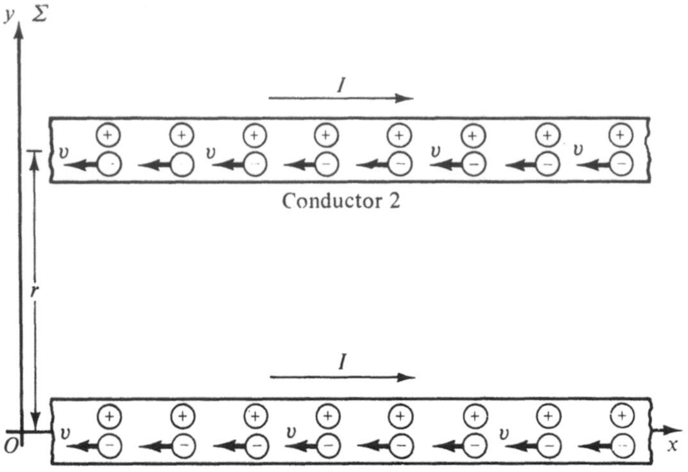

Fig. 8.2 Forces between two conduction currents I. In the simplified model used the positive ions are at rest and the negative electrons all move with the same uniform velocity v. The positive charge per unit length is $+\lambda$ and the negative charge per unit length $-\lambda$. The electric forces cancel, leaving only the second-order magnetic forces between the charges.

length due to the positive ions is λ while that due to the negative electrons is $-\lambda$ in each conductor measured in the laboratory frame Σ.

The force per unit length on the moving electrons in conductor 2 due to the moving electrons in conductor 1 is a repulsive force (electric plus magnetic) given by equation (8.5), namely

$$f_{--} = +\frac{\lambda^2}{2\pi\epsilon_0 r}(1 - v^2/c^2)\,. \tag{8.6}$$

Since the positive ions in conductor 2 are at rest, there is no magnetic force on them, so that the total force per unit length on the positive ions in conductor 2 due to the moving electrons in conductor 1 is an attractive electric force given by

$$f_{-+} = -\frac{\lambda^2}{2\pi\epsilon_0 r}\,. \tag{8.7}$$

The force per unit length on the moving electrons in conductor 2 due to the stationary positive charges in conductor 1 is an attractive electric force given by

$$f_{+-} = -\frac{\lambda^2}{2\pi\epsilon_0 r}\,. \tag{8.8}$$

The force per unit length on the stationary positive ions in conductor 2 due to the stationary positive ions in conductor 1 is a repulsive electric force given by

$$f_{++} = +\frac{\lambda^2}{2\pi\epsilon_0 r}. \tag{8.9}$$

Adding equations (8.6), (8.7), (8.8) and (8.9), we find that the resultant force per unit length of conductor 2 due to all the charges in conductor 1 is an attractive force given by

$$f = -\frac{\lambda^2 v^2}{2\pi\epsilon_0 c^2 r} = -\frac{I^2}{2\pi\epsilon_0 c^2 r}. \tag{8.10}$$

Thus the electric forces between the charges in the two electrically neutral conductors in Fig. 8.2 cancel to give zero, leaving only the second-order magnetic force that is observed experimentally.

8.3 FORCES BETWEEN TWO PARALLEL CONVECTION CURRENTS DERIVED USING THE THEORY OF SPECIAL RELATIVITY

The example illustrated in Fig. 8.1, and discussed in Section 8.2 using the laws of classical electromagnetism, will now be considered from the viewpoint of the theory of special relativity. It will be assumed that λ, the charge per unit length on the moving wires, measured in the laboratory frame Σ, is made up of n discrete charges per unit length of magnitude q each, as shown in Fig. 8.3(b) In Σ,

$$\lambda = nq. \tag{8.11}$$

According to the theory of special relativity, the laws of electromagnetism are the same in all inertial reference systems. Hence the problem can be considered in the reference frame Σ' moving with uniform velocity v relative to Σ along the common x axis. In Σ' the wires are at rest, as shown in Fig. 8.3(a). Let λ', the charge per unit length measured in Σ', be made up of n' charges per unit length of magnitude q each, so that

$$\lambda' = n'q. \tag{8.12}$$

The same value q is used for the total charge on a particle in Σ and Σ' so as to be in accord with the principle of constant electric charge (see Section 2.4). The charge distributions are at rest in Σ'. According to the Lorentz length contraction, equation (4.51), a length l_0 of the wire at rest in Σ' is measured to be $l_0(1 - v^2/c^2)^{1/2}$ in Σ, since the wire is moving with velocity v relative to Σ. The number of charges in a length l_0 of wire in Σ' which is equal to $n'l_0$, is measured to be in a length $l_0(1 - v^2/c^2)^{1/2}$ relative to Σ, as illustrated in Fig. 8.3. Hence the number of charges per unit length in Σ is $n'l_0/l_0(1 - v^2/c^2)^{1/2}$, so that

$$n = \frac{n'}{(1 - v^2/c^2)^{1/2}}, \tag{8.13}$$

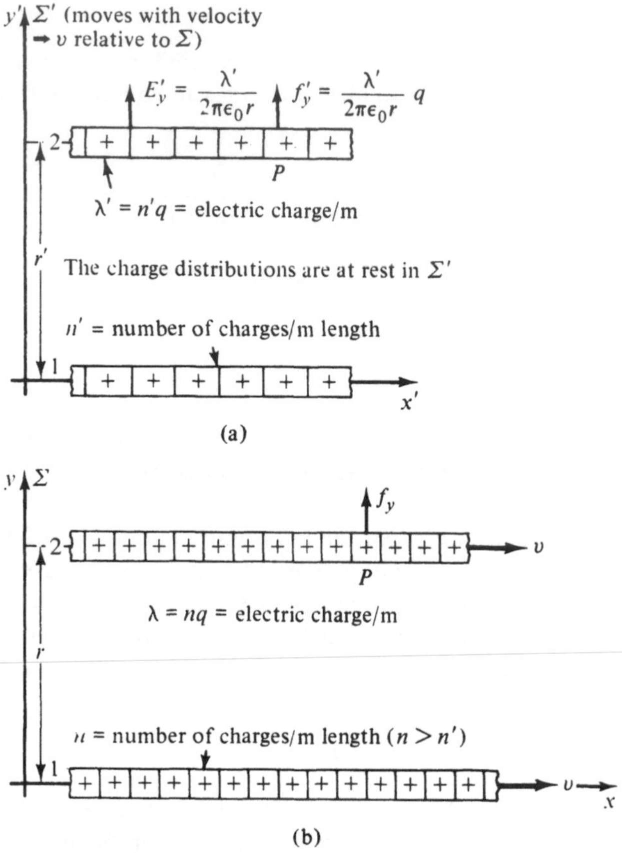

Fig. 8.3 Calculation of the electric and the magnetic forces between two convection currents using the theory of special relativity. (a) The charge distributions are at rest in Σ′; there is only an electric force between the charges in Σ′. (b) In Σ the charge distributions move with uniform velocity v, and there are both electric and magnetic forces between the charges.

showing that the charge per unit length is greater in Σ than Σ′, as illustrated in Fig. 8.3. Hence, using equation (8.11),

$$\lambda = nq = \frac{n'q}{(1 - v^2/c^2)^{1/2}}$$

$$= \frac{\lambda'}{(1 - v^2/c^2)^{1/2}}. \tag{8.14}$$

In Σ′, since the charge distributions are at rest, the laws of classical electromagnetism reduce to Coulomb's law of force between electrostatic charges. Using Gauss' law as in Section 8.2, we find that the electric field at the position of wire 2 due to wire 1 is

$$E_y' = \frac{\lambda'}{2\pi\epsilon_0 r'}. \tag{8.15}$$

According to the Lorentz transformations, the separation of the wires is the same in Σ and Σ', as it is measured in the y direction, so that $r = r'$. The force on one of the charges (labelled P and of magnitude q) of wire 2 in Fig. 8.3(a) has components

$$f_x' = 0, \qquad f_y' = \frac{\lambda' q}{2\pi\epsilon_0 r}, \qquad f_z' = 0 \tag{8.16}$$

in Σ'. The force transformations, derived in Section 6.7.3, will now be applied to determine the force acting on the single charge P, measured in the inertial reference frame Σ shown in Fig. 8.3(b). Since the velocity $\boldsymbol{u}'$ of the charged particle P is zero in Σ', the inverses of equations (6.81), (6.82) and (6.83) can be rewritten as

$$f_x = f_x', \tag{8.17}$$

$$f_y = f_y'/\gamma = f_y'(1 - v^2/c^2)^{1/2}, \tag{8.18}$$

$$f_z = f_z'/\gamma = f_z'(1 - v^2/c^2)^{1/2}. \tag{8.19}$$

Since, according to equations (8.16),

$$f_x' = f_z' = 0,$$

it follows from equations (8.17) and (8.19) that in Σ

$$f_x = f_z = 0.$$

Substituting from equation (8.16) into equation (8.18), we have for the force on the charge P measured in Σ

$$f_y = \frac{\lambda' q}{2\pi\epsilon_0 r}(1 - v^2/c^2)^{1/2}. \tag{8.20}$$

The force per unit length on wire 2, measured in Σ, is equal to the number of charges per unit length, measured in Σ, times the force on each charge, given by equation (8.20). Hence in Σ

$$\text{force/unit length} = nf_y = n\frac{\lambda' q}{2\pi\epsilon_0 r}(1 - v^2/c^2)^{1/2}.$$

But, from equation (8.14), $\lambda' = \lambda(1 - v^2/c^2)^{1/2}$, and, from equation (8.11), $nq = \lambda$. Hence in Σ

$$\text{force/unit length} = \frac{\lambda^2}{2\pi\epsilon_0 r}(1 - v^2/c^2). \tag{8.21}$$

This is in agreement with equation (8.5), derived in Section 8.2 by applying the laws of electromagnetism in the inertial frame Σ. This example illustrates how the magnetic forces produced by electric currents can be calculated, in some cases, from Coulomb's law for the forces between electrostatic charges, if the principle of constant electric charge and the force transformations of the theory of special relativity are taken as axiomatic. One has to include all second-order effects, since the magnetic forces between moving charges are themselves of second order. For example, in the present case the effects associated with the Lorentz contraction had to be included.

Equation (8.21) was derived from Coulomb's law and the transformations of the theory of special relativity. Comparing equations (8.4) and (8.21), it can be seen that, in order to be consistent with the transformations of special relativity, $\mu_0\epsilon_0$ must be exactly equal to $1/c^2$.

According to Newtonian mechanics, force should be absolute; that is, it should have the same numerical value in all inertial reference frames. If equation (1.26) were used instead of equation (8.18), we would have $f_y = f_y'$. Using the value of f_y' given by equation (8.16), which was derived from Coulomb's law, we would find that, according to Newtonian mechanics, the total force acting on the single charge P in Fig. 8.3(b), measured in Σ, would be

$$f_y = f_y' = \frac{\lambda' q}{2\pi\epsilon_0 r}. \tag{8.22}$$

According to the Galilean transformations, there is no length contraction (see Problem 1.4). Hence the number of charges per unit length should be the same in Σ and Σ', so that $n = n'$ and so $\lambda = nq = n'q = \lambda'$. Therefore according to Newtonian mechanics and the Galilean transformations, equation (8.22) giving the force on the particle P in Fig. 8.3(a), measured in Σ, should be

$$f_y = \frac{\lambda q}{2\pi\epsilon_0 r}.$$

The total force per unit length of wire 2, measured in Σ, would, according to Newtonian mechanics and the Galilean transformations, be

$$\text{force/unit length} = nf_y = \frac{n\lambda q}{2\pi\epsilon_0 r} = \frac{\lambda^2}{2\pi\epsilon_0 r}. \tag{8.23}$$

This equation is the same as equation (8.2), which was derived from Coulomb's law. This result shows that, according to Newtonian mechanics and the Galilean transformations, there would be no magnetic forces between the two convection currents in Fig. 8.1, and hence no resultant force between the two conduction currents in Fig. 8.2. The above example illustrates how the magnetic forces between moving charges can be interpreted as second-order relativistic effects, which arise from deviations from Newtonian mechanics. It also illustrates that classical electromagnetism is not consistent with the Galilean transformations, but is entirely consistent with the theory of special relativity.

★ 8.4 TRANSFORMATIONS FOR THE FIELDS *E* AND *B*

In Section 8.3 the electromagnetic forces between two parallel convection currents were determined from Coulomb's law using the force transformations of the theory of special relativity. It is generally more convenient in more complicated cases to use the transformations for the electric field intensity $\boldsymbol{E}$ and the magnetic induction $\boldsymbol{B}$. These will now be derived by assuming that the expression for the Lorentz force acting on a moving charge satisfies the principle of relativity; that is, that it is valid in all inertial reference frames when the transformations of the theory of special relativity are used to transform from one inertial reference frame to another. According to equation (2.24), the Lorentz force acting on a non-radiating test charge of magnitude q that is moving with velocity $\boldsymbol{u}$ in the inertial reference frame Σ is

$$\boldsymbol{f}=\frac{\mathrm{d}}{\mathrm{d}t}\left[\frac{m\boldsymbol{u}}{(1-u^2/c^2)^{1/2}}\right]=q\boldsymbol{E}+q\boldsymbol{u}\times\boldsymbol{B}, \tag{8.24}$$

where $\boldsymbol{E}$ is the electric field intensity and $\boldsymbol{B}$ the magnetic induction, measured in Σ. It was shown in Chapter 2 that equation (8.24) is in excellent agreement with the experimental results obtained in the laboratory. For example, it was shown in Section 2.7 that this equation can be used to design accelerators that can accelerate protons up to $1\ \mathrm{TeV}=10^{12}\ \mathrm{eV}$.

If the Lorentz force law is also valid in the inertial reference frame Σ' that is moving with uniform velocity v relative to Σ along their common x axis then in Σ'

$$\boldsymbol{f}'=q\boldsymbol{E}'+q\boldsymbol{u}'\times\boldsymbol{B}', \tag{8.25}$$

where $\boldsymbol{u}'$, $\boldsymbol{E}'$ and $\boldsymbol{B}'$ are the velocity of the test charge, the electric field intensity and the magnetic induction measured in the inertial frame Σ'. The same value is used for q in equations (8.24) and (8.25) so as to be in accord with the principle of constant charge (see Section 2.4).

It will be assumed that equations (8.24) and (8.25) refer to the same act of measurement of the fields, so that if the fields are measured at (x, y, z) at a time t in Σ and at (x', y', z') at a time t' in Σ', these coordinates and times are related by the Lorentz transformations, and $\boldsymbol{u}$ and $\boldsymbol{u}'$ are related by the relativistic velocity transformations.

From equation (6.82),

$$f_y'=\frac{f_y}{\gamma(1-vu_x/c^2)}.$$

Rearranging, we have

$$f_y=\gamma(1-vu_x/c^2)f_y', \tag{8.26}$$

where

$$\gamma=(1-v^2/c^2)^{-1/2}.$$

The y component of equation (8.24) is

$$f_y=qE_y+q\,(u_zB_x-u_xB_z),$$

while the y' component of equation (8.25) is

$$f_y' = qE_y' + q\,(u_z'B_x' - u_x'B_z') \,.$$

Substituting for f_y and f_y' in equation (8.26), we have

$$qE_y + q(u_zB_x - u_xB_z) = \gamma q\,(1 - vu_x/c^2)\,(E_y' + u_z'B_x' - u_x'B_z') \,. \qquad (8.27)$$

Now, from the relativistic velocity transformations,

$$u_x' = \frac{u_x - v}{1 - vu_x/c^2}\,, \qquad u_z' = \frac{u_z}{\gamma(1 - vu_x/c^2)}\,.$$

Substituting for u_x' and u_z' in equation (8.27), we obtain

$$E_y + u_zB_x - u_xB_z = \gamma\,(1 - vu_x/c^2)\,E_y' + u_zB_x' - \gamma\,(u_x - v)\,B_z'\,,$$

and, rearranging,

$$E_y + u_zB_x - u_xB_z = \gamma\,(E_y' + vB_z') + u_zB_x' - u_x\gamma\,(B_z' + vE_y'/c^2) \,. \qquad (8.28)$$

This equation must be valid whatever the value of the velocity $\boldsymbol{u}$ of the test charge relative to Σ. If, for example, we double u_x, keeping u_z constant, the terms containing u_z and the terms independent of the velocity of the charge in equation (8.28) are unchanged. Equation (8.28) can only remain valid for both values of u_x if the terms containing u_x on the left-hand side are always equal to the terms containing u_x on the right-hand side, whatever the value of u_x; that is, if

$$u_xB_z = u_x\gamma\,(B_z' + vE_y'/c^2)$$

for all u_x. This condition is satisfied if

$$B_z = \gamma\,(B_z' + vE_y'/c^2) \,.$$

Equating the terms containing u_z on both sides of equation (8.28) gives

$$B_x = B_x' \,.$$

Equating the terms independent of u_x and u_z on both sides of equation (8.28) gives

$$E_y = \gamma\,(E_y' + vB_z') \,.$$

Similarly, from equation (6.83),

$$f_z = f_z'\gamma\,(1 - vu_x/c^2) \,.$$

Substituting the f_z and f_z' components of equations (8.24) and (8.25), using the velocity transformations for u_x' and u_y' and comparing coefficients of the terms containing u_x, u_y and the terms independent of the velocity, we find

$$B_y = \gamma\,(B_y' - vE_z'/c^2)\,,$$

$$B_x = B_x'\,,$$

$$E_z = \gamma\,(E_z' - vB_y') \,.$$

Now, from equation (6.81),

$$f_x = f_x' + \frac{vu_y'}{c^2(1+vu_x'/c^2)} f_y' + \frac{vu_z'}{c^2(1+vu_x'/c^2)} f_z'$$

$$\equiv f_x' + K_1 u_y' + K_2 u_z' \, . \tag{8.29}$$

Substituting the x and x' components of equations (8.24) and (8.25) into equation (8.29), we obtain

$$qE_x + q(u_y B_z - u_z B_y) = qE_x' + q(u_y' B_z' - u_z' B_y') + K_1 u_y' + K_2 u_x' \, . \tag{8.30}$$

Since, from the velocity transformations,

$$u_y' = \frac{u_y}{\gamma(1 - vu_x/c^2)} \, , \qquad u_z' = \frac{u_z}{\gamma(1 - vu_x/c^2)} \, ,$$

the only terms in equation (8.30) that are independent of the velocity of the test charge are qE_x and qE_x'. Hence we conclude that

$$E_x = E_x' \, .$$

This completes the derivation of the transformations for $\boldsymbol{E}$ and $\boldsymbol{B}$. Collecting them, we have

$$E_x' = E_x \, , \qquad E_x = E_x' \, , \tag{8.31}$$

$$E_y' = \gamma(E_y - vB_z) \, , \qquad E_y = \gamma(E_y' + vB_z') \, , \tag{8.32}$$

$$E_z' = \gamma(E_z + vB_y) \, , \qquad E_z = \gamma(E_z' - vB_y') \, , \tag{8.33}$$

$$B_x' = B_x \, , \qquad B_x = B_x' \, , \tag{8.34}$$

$$B_y' = \gamma\left(B_y + \frac{v}{c^2} E_z\right) , \qquad B_y = \gamma\left(B_y' - \frac{v}{c^2} E_z'\right) , \tag{8.35}$$

$$B_z' = \gamma\left(B_z - \frac{v}{c^2} E_y\right) , \qquad B_z = \gamma\left(B_z' + \frac{v}{c^2} E_y'\right) , \tag{8.36}$$

These transformations relate the numerical values of the electric and magnetic fields at a point (x, y, z) at time t in Σ with the values of the electric and magnetic fields at a point (x', y', z') at time t' in Σ', where x, y, z and t are related to x', y', z' and t' by the Lorentz transformations.

As an example of the application of the field transformations, the reader can repeat the calculation given in Section 8.3 and illustrated in Figs. 8.3(a,b), by starting with equation (8.15) for the fields in Σ', using the field transformations to determine $\boldsymbol{E}$ and $\boldsymbol{B}$ in Σ and then using the Lorentz force to determine the force on the charge labelled P in Σ, leading to equation (8.20). We shall now apply the field transformations to determine the fields of a charge moving with *uniform* velocity, starting from Coulomb's law in the inertial reference frame in which the charge is at rest.

★ 8.5 ELECTRIC AND MAGNETIC FIELDS DUE TO A CHARGE MOVING WITH UNIFORM VELOCITY DERIVED FROM COULOMB'S LAW

Consider a point charge of magnitude q that is permanently at rest at the origin O' of the inertial reference frame Σ', as shown in Fig. 8.4(b). According to Coulomb's law, the electric field $\boldsymbol{E}'$ measured in Σ' at the field point P' at (x', y', z') at a distance r' from the origin O' of Σ' is given at all times t' in Σ' by

$$\boldsymbol{E}' = \frac{q\boldsymbol{r}'}{4\pi\epsilon_0 r'^3} = \frac{q\,(\boldsymbol{i}'x' + \boldsymbol{j}'y' + \boldsymbol{k}'z')}{4\pi\epsilon_0 r'^3}, \tag{8.37}$$

where $\boldsymbol{i}'$, $\boldsymbol{j}'$ and $\boldsymbol{k}'$ are unit vectors pointing in the $+x'$, $+y'$, and $+z'$ directions of Σ' respectively. The magnetic induction $\boldsymbol{B}'$ due to the stationary charge q is zero in Σ', so that

$$\boldsymbol{B}' = 0. \tag{8.38}$$

If the inertial reference frame Σ' is moving with uniform velocity v along the x axis of the inertial reference frame Σ, as shown in Fig. 8.4(b), and if the origins of Σ and Σ' coincide at $t = t' = 0$, then at the time $t = 0$ in Σ the charge q is at the origin of the Σ and is moving with *uniform* velocity v along the x axis of Σ, as shown in Fig. 8.4(a). Let $\beta = v/c$. According to the principle of constant electric charge, the magnitude q of the charge is the same in Σ as in Σ'. The fields $\boldsymbol{E}$ and $\boldsymbol{B}$ measured in Σ due to the moving charge will be determined at the field point P in Fig. 8.4(a) at (x, y, z) at a distance $\boldsymbol{r}$ from the origin of Σ at the time $t = 0$,

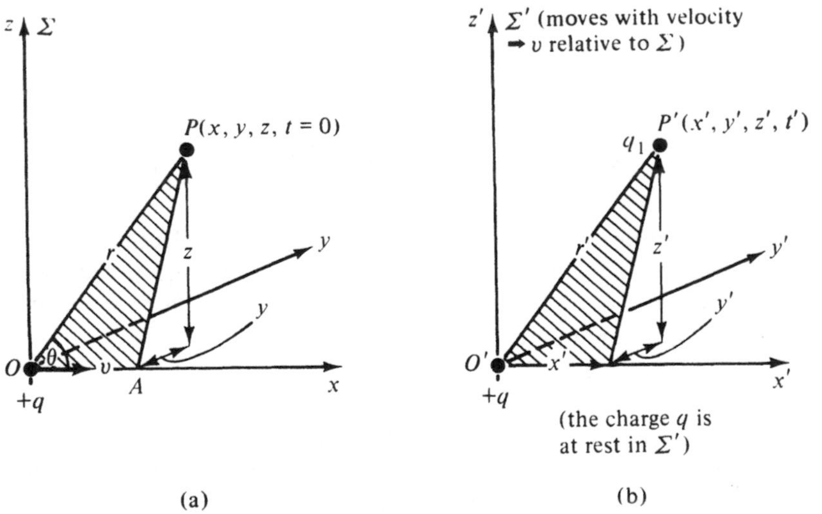

Fig. 8.4 (a) In the inertial reference frame Σ the charge q moves with uniform velocity v. The fields $\boldsymbol{E}$ and $\boldsymbol{B}$ are determined at the point P at $t = 0$, when q is at the origin. (b) In the inertial reference frame Σ' the charge q remains at rest at the origin.

when the moving charge q is at the origin of Σ. The corresponding event of field measurement is at (x', y', z') at the time t' in Σ'. From the Lorentz transformations, when $t = 0$, we have

$$x' = \gamma x, \qquad y' = y, \qquad z' = z, \qquad t' = -\gamma v x/c^2, \tag{8.39}$$

where

$$\gamma = (1 - \beta^2)^{-1/2}.$$

Hence, in Fig. 8.4(b)

$$\begin{aligned}(O'P')^2 = r'^2 = x'^2 + y'^2 + z'^2 &= \gamma^2 x^2 + y^2 + z^2, \\ &= \gamma^2 [x^2 + (1 - \beta^2)(y^2 + z^2)] \\ &= \gamma^2 [x^2 + y^2 + z^2 - \beta^2 (y^2 + z^2)].\end{aligned}$$

But, in Fig. 8.4(a)

$$r^2 = x^2 + y^2 + z^2.$$

Hence

$$r'^2 = \gamma^2 r^2 \left[1 - \frac{\beta^2 (y^2 + z^2)}{r^2} \right]. \tag{8.40}$$

It can be seen from Fig. 8.4(a) that

$$\frac{y^2 + z^2}{r^2} = \frac{(AP)^2}{r^2} = \sin^2 \theta,$$

where θ is the angle between $\boldsymbol{v}$ and $\boldsymbol{r}$ in Σ. Thus equation (8.40) can be rewritten as

$$r' = \gamma r (1 - \beta^2 \sin^2 \theta)^{1/2}. \tag{8.41}$$

Writing equations (8.37) and (8.38) in component form, we have in Σ'

$$E_x' = \frac{qx'}{4\pi\epsilon_0 r'^3}, \qquad E_y' = \frac{qy'}{4\pi\epsilon_0 r'^3}, \qquad E_z' = \frac{qz'}{4\pi\epsilon_0 r'^3},$$

$$B_x' = B_y' = B_z' = 0.$$

Using the field transformations, equations (8.31)–(8.36), substituting for r' using equation (8.41), and using equations (8.39) to substitute for x', y' and z' we obtain the following:

$$\begin{aligned}E_x = E_x' &= \frac{qx'}{4\pi\epsilon_0 r'^3} \\ &= \frac{q\gamma x}{4\pi\epsilon_0 \gamma^3 r^3 (1 - \beta^2 \sin^2 \theta)^{3/2}}\end{aligned}$$

$$= \frac{q(1-\beta^2)x}{4\pi\epsilon_0 r^3(1-\beta^2\sin^2\theta)^{3/2}} \tag{8.42}$$

$$E_y = \gamma(E_y' + vB_z') = \gamma E_y' = \frac{\gamma q y'}{4\pi\epsilon_0 r'^3}$$

$$= \frac{\gamma q y}{4\pi\epsilon_0 r^3\gamma^3(1-\beta^2\sin^2\theta)^{3/2}}$$

$$= \frac{q(1-\beta^2)y}{4\pi\epsilon_0 r^3(1-\beta^2\sin^2\theta)^{3/2}}. \tag{8.43}$$

Similarly,

$$E_z = \frac{q(1-\beta^2)z}{4\pi\epsilon_0 r^3(1-\beta^2\sin^2\theta)^{3/2}}. \tag{8.44}$$

Equations (8.42), (8.43) and (8.44) are the components of the vector equation

$$\boldsymbol{E} = \frac{q(1-\beta^2)\boldsymbol{r}}{4\pi\epsilon_0 r^3(1-\beta^2\sin^2\theta)^{3/2}}, \tag{8.45}$$

where $\boldsymbol{r}$ is a vector from the charge to the field point where the field is required, $\beta = v/c$ and θ is the angle between $\boldsymbol{r}$ and $\boldsymbol{v}$, as shown in Fig. 8.4(a). The position of the charge is its position at the time the field is required, which in the above example is at the time $t = 0$ when the charge q is at the origin of Σ in Fig. 8.4(a)†. Using the transformations for $\boldsymbol{B}$, we have

$$B_x = B_x' = 0$$

$$B_z = \gamma(B_z' + vE_y'/c^2) = \gamma v E_y'/c^2.$$

But, according to equation (8.32), when $B_z' = 0$, $E_y = \gamma E_y'$ Hence

$$B_z = vE_y/c^2.$$

Similarly,

$$B_y = -vE_z/c^2.$$

Therefore

$$\boldsymbol{B} = B_x\boldsymbol{i} + B_y\boldsymbol{j} + B_z\boldsymbol{k} = -(vE_z/c^2)\boldsymbol{j} + (vE_y/c^2)\boldsymbol{k}\,. \tag{8.46}$$

where $\boldsymbol{i}, \boldsymbol{j}$ and $\boldsymbol{k}$ are unit vectors pointing in the $+x$, $+y$ and $+z$ directions of Σ respectively. Now the velocity $\boldsymbol{v}$ of the charge q in Fig. 8.4(a) has components $v_x = v$, $v_y = v_z = 0$.

---

†If the charge q is accelerating then the fields $\boldsymbol{E}$ and $\boldsymbol{B}$ must be related to what is known as the retarded position of the charge. There are also contributions to $\boldsymbol{E}$ and $\boldsymbol{B}$ that depend on the acceleration of the charge at its retarded position. These are the radiation fields (see Panofsky and Phillips 1962). These more complicated formulae reduce to equations (8.45) and (8.49) in the *special* case of a charge moving with *uniform* velocity.

Hence the vector product $\boldsymbol{v} \times \boldsymbol{E}$ is given by

$$\boldsymbol{v} \times \boldsymbol{E} = (v_y E_z - v_z E_y)\,\boldsymbol{i} + (v_z E_x - v_x E_z)\,\boldsymbol{j} + (v_x E_y - v_y E_x)\,\boldsymbol{k}$$
$$= -vE_z\boldsymbol{j} + vE_y\boldsymbol{k}\,. \tag{8.47}$$

Comparing equations (8.46) and (8.47), it can be seen that

$$\boldsymbol{B} = \boldsymbol{v} \times \boldsymbol{E}/c^2. \tag{8.48}$$

Using equation (8.45), we have

$$\boldsymbol{B} = \frac{q\boldsymbol{v} \times \boldsymbol{r}\,(1 - \beta^2)}{4\pi\epsilon_0 c^2 r^3\,(1 - \beta^2 \sin^2\theta)^{3/2}}\,, \tag{8.49}$$

where $\boldsymbol{B}$ is the magnetic induction at a distance $\boldsymbol{r}$ from a charge of magnitude q that is moving with uniform velocity $\boldsymbol{v}$.

When the velocity $\boldsymbol{v}$ of the charge is zero, that is $\beta = 0$, equation (8.45) reduces to

$$E = \frac{q}{4\pi\epsilon_0 r^2}\,, \tag{8.50}$$

in agreement with Coulomb's law. When $\boldsymbol{v} = 0$, the electric field intensity $\boldsymbol{E}$ is the same in all directions, as illustrated in Fig. 8.5(a). The number of lines of $\boldsymbol{E}$ is limited in Fig. 8.5(a) such that the number of lines per unit area perpendicular to $\boldsymbol{E}$ is proportional to the magnitude of $\boldsymbol{E}$. This gives a visual picture of the strength of $\boldsymbol{E}$, the lines of $\boldsymbol{E}$ being closest together in Fig. 8.5(a) where the field strength is greatest.

When $\boldsymbol{v}$ is finite, the electric field is still radial; that is, $\boldsymbol{E}$ diverges radially from the position of the charge, as shown in Fig. 8.5(b). The electric field

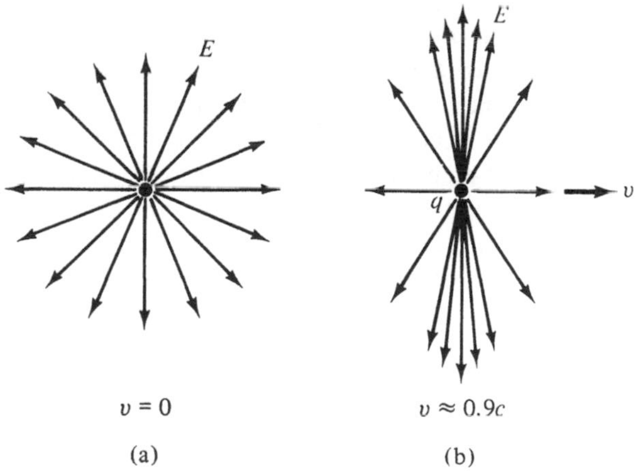

Fig. 8.5 (a) The electric field of a stationary positive charge is spherically symmetric. (b) If the charge is moving with uniform velocity, the electric field diverges radially from the position of the charge. The electric intensity is increased in the direction perpendicular to $\boldsymbol{v}$ but decreased in the directions parallel and antiparallel to $\boldsymbol{v}$.

strength is still proportional to $1/r^2$; however, according to equation (8.45), the magnitude of $\boldsymbol{E}$ is not the same in all directions, although it is still symmetric about $\theta = \frac{1}{2}\pi$. When $\theta = 0$ or $\theta = \pi$, equation (8.45) reduces to

$$E = \frac{q(1-\beta^2)}{4\pi\epsilon_0 r^2}.$$

Thus the electric field intensity is reduced in the direction of $\boldsymbol{v}$, the direction of motion of the charge, and in the direction opposite to $\boldsymbol{v}$, to $1-\beta^2$ times the electrostatic value given by equation (8.50). When $\theta = \frac{1}{2}\pi$,

$$E = \frac{q}{4\pi\epsilon_0 r^2(1-\beta^2)^{1/2}}.$$

Hence the eletric field intensity $\boldsymbol{E}$ is increased in the direction perpendicular to $\boldsymbol{v}$ to $(1-\beta^2)^{-1/2}$ times the electrostatic value given by equation (8.50). It can be shown that the total flux of $\boldsymbol{E}$ remains equal to q/ϵ_0 when the charge is moving. The example of a positive charge moving with uniform velocity of approximately $0.9c$ is shown in Fig. 8.5(b). This illustrates how, although the total number of lines of $\boldsymbol{E}$ is the same as for the electrostatic case, the lines are bunched towards the direction perpendicular to $\boldsymbol{v}$. For $\beta = 0.9$ the electric field is 0.19 times the electrostatic value for $\theta = 0$ and 2.3 times the electrostatic value for $\theta = \frac{1}{2}\pi$. For $\beta = 0.99$ the ratios are 0.02 and 7.1 respectively. Thus in the extreme relativistic case the lines of $\boldsymbol{E}$ are all nearly perpendicular to $\boldsymbol{v}$.

According to equation (8.49), the lines of magnetic induction $\boldsymbol{B}$ are perpendicular to both the velocity $\boldsymbol{v}$ of the charge q, and the lines of $\boldsymbol{E}$, which diverge radially from the position of the charge. The lines of $\boldsymbol{B}$ therefore form closed circles in the plane perpendicular to the direction of motion of q, which is the direction of $\boldsymbol{v}$. The circles are concentric with the direction of $\boldsymbol{v}$. The lines of $\boldsymbol{B}$ in two planes are sketched in Fig. 8.6. The magnitude of $\boldsymbol{B}$ decreases as $1/r^2$. For a given r, the magnitude of $\boldsymbol{B}$ is the same for the positions $\theta = \alpha$ and $\theta = \pi - \alpha$. The sense of rotation of the lines of $\boldsymbol{B}$ about the direction of $\boldsymbol{v}$ is the same in both cases. The direction of $\boldsymbol{B}$ is consistent with the right-handed corkscrew rule, if $q\boldsymbol{v}$ is treated as a current element $I\mathrm{d}\boldsymbol{l}$. According to the right-handed corkscrew rule, the direction of $\boldsymbol{B}$ is the direction in which a right-handed corkscrew would have to be rotated in order that it should advance in the direction of current flow.

When $v \ll c$, $\beta \ll 1$ and equations (8.45) and (8.49) reduce to

$$\boldsymbol{E} \approx \frac{q\boldsymbol{r}}{4\pi\epsilon_0 r^3}, \tag{8.51}$$

$$\boldsymbol{B} \approx \frac{q\boldsymbol{v}\times\boldsymbol{r}}{4\pi\epsilon_0 c^2 r^3}. \tag{8.52}$$

Equation (8.51) is the same as Coulomb's law. Equation (8.52) is sometimes called the Biot–Savart approximation for the magnetic field due to a charge moving with a uniform velocity $v \ll c$.

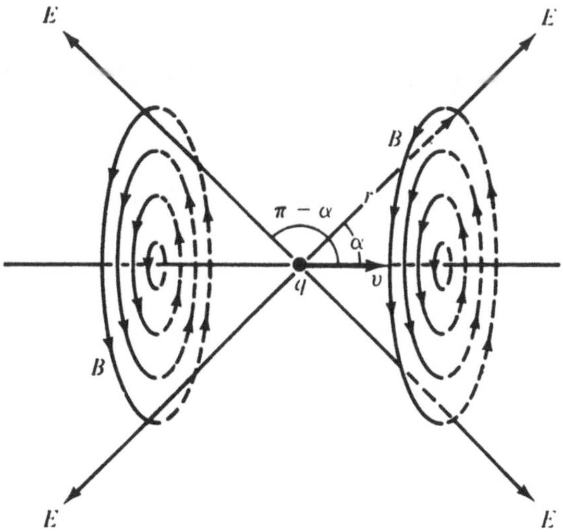

Fig. 8.6 The magnetic field lines of a charge moving with uniform velocity $\boldsymbol{v}$ are circles concentric with the direction of $\boldsymbol{v}$. The direction of $\boldsymbol{B}$ is given by the right-handed corkscrew rule. The field is in the same direction for values of $\theta = \alpha$ and $\theta = \pi - \alpha$, where θ is the angle between $\boldsymbol{r}$ and $\boldsymbol{v}$.

The above example again illustrates the unity of electromagnetism. It shows how, if one has only an electrostatic field in one inertial reference frame then there is both an electric field and a magnetic field in an inertial reference frame moving relative to the first inertial reference frame.

★ 8.6 RELATIVISTIC INVARIANCE OF MAXWELL'S EQUATIONS

★ 8.6.1 Introduction

For the benefit of readers familiar with Maxwell's equations, it will now be shown that Maxwell's equations obey the principle of relativity; that is, they hold in all inertial reference frames if the coordinates and time are transformed using the Lorentz transformations.

In the inertial frame Σ Maxwell's equations are

$$\operatorname{curl} \boldsymbol{E} = -\frac{\partial \boldsymbol{B}}{\partial t}, \tag{8.53}$$

$$\operatorname{div} \boldsymbol{B} = 0, \tag{8.54}$$

$$\operatorname{curl} \boldsymbol{H} = \frac{\partial \boldsymbol{D}}{\partial t} + \boldsymbol{J}, \tag{8.55}$$

$$\operatorname{div} \boldsymbol{D} = \rho, \tag{8.56}$$

where $\boldsymbol{E}$ is the electric intensity, $\boldsymbol{D}$ the electric displacement, $\boldsymbol{B}$ the magnetic induction, $\boldsymbol{H}$ the magnetizing force (magnetic field strength), ρ the true macroscopic charge density and $\boldsymbol{J}$ the true macroscopic conduction current density defined and measured in the inertial reference frame Σ.

If Maxwell's equations obey the principle of relativity then in the inertial frame Σ' moving with uniform velocity v in the positive Ox direction relative to Σ one should have

$$\text{curl}' \boldsymbol{E}' = -\frac{\partial \boldsymbol{B}'}{\partial t'}, \tag{8.57}$$

$$\text{div}' \boldsymbol{B}' = 0, \tag{8.58}$$

$$\text{curl}' \boldsymbol{H}' = \frac{\partial \boldsymbol{D}'}{\partial t'} + \boldsymbol{J}', \tag{8.59}$$

$$\text{div}' \boldsymbol{D}' = \rho', \tag{8.60}$$

where $\boldsymbol{E}', \boldsymbol{B}', \boldsymbol{D}', \boldsymbol{H}', \boldsymbol{J}'$ and ρ' are defined and measured in Σ'. These equations should hold for a point (x', y', z') at a time t' in Σ', where x', y', z' and t' are related to x, y, z and t by the Lorentz transformations. All the derivatives in equations (8.57)–(8.60) are with respect to x', y', z' or t'. For example, equation (8.58) is

$$\frac{\partial B_x'}{\partial x'} + \frac{\partial B_y'}{\partial y'} + \frac{\partial B_z'}{\partial z'} = 0. \tag{8.58}$$

It will be assumed throughout Section 8.6 that Σ' moves with uniform velocity v relative to Σ along their common x axis, so that v and $\gamma = (1 - v^2/c^2)^{-1/2}$ are constants.

★ 8.6.2 Transformation of $\partial/\partial x$, $\partial/\partial y$, $\partial/\partial z$ and $\partial/\partial t$

Consider a function F of x', y', z' and t' in Σ'. The total differential of F is

$$\mathrm{d}F = \frac{\partial F}{\partial x'}\mathrm{d}x' + \frac{\partial F}{\partial y'}\mathrm{d}y' + \frac{\partial F}{\partial z'}\mathrm{d}z' + \frac{\partial F}{\partial t'}\mathrm{d}t'. \tag{8.61}$$

Now for a given event, x', y', z' and t' are all functions of x, y, z and t. The total differential of x' can be expressed as

$$\mathrm{d}x' = \frac{\partial x'}{\partial x}\mathrm{d}x + \frac{\partial x'}{\partial y}\mathrm{d}y + \frac{\partial x'}{\partial z}\mathrm{d}z + \frac{\partial x'}{\partial t}\mathrm{d}t. \tag{8.62}$$

From the Lorentz transformations,

$$x' = \gamma(x - vt),$$

$$y' = y, \qquad z' = z,$$

$$t' = \gamma (t - vx/c^2),$$

where $\gamma = (1 - v^2/c^2)^{-1/2}$. Remembering that v and γ are constants, we have

$$\frac{\partial x'}{\partial x} = \gamma, \qquad \frac{\partial x'}{\partial y} = 0, \qquad \frac{\partial x'}{\partial z} = 0, \qquad \frac{\partial x'}{\partial t} = -\gamma v,$$

and, substituting in equation (8.62),

$$\mathrm{d}x' = \gamma\, \mathrm{d}x - \gamma v\, \mathrm{d}t. \tag{8.63}$$

Similarly,

$$\mathrm{d}y' = \mathrm{d}y\,, \tag{8.64}$$

$$\mathrm{d}z' = \mathrm{d}z\,, \tag{8.65}$$

$$\mathrm{d}t' = \frac{\partial t'}{\partial x}\mathrm{d}x + \frac{\partial t'}{\partial y}\mathrm{d}y + \frac{\partial t'}{\partial z}\mathrm{d}z + \frac{\partial t'}{\partial t}\mathrm{d}t$$

$$= -\frac{\gamma v}{c^2}\mathrm{d}x + \gamma\, \mathrm{d}t\,. \tag{8.66}$$

Substituting for $\mathrm{d}x', \mathrm{d}y', \mathrm{d}z'$ and $\mathrm{d}t'$ from equations (8.63), (8.64), (8.65) and (8.66) respectively into equation (8.61), we have

$$\mathrm{d}F = \frac{\partial F}{\partial x'}(\gamma\, \mathrm{d}x - \gamma v\, \mathrm{d}t) + \frac{\partial F}{\partial y'}\mathrm{d}y + \frac{\partial F}{\partial z'}\mathrm{d}z + \frac{\partial F}{\partial t'}\left(\gamma\, \mathrm{d}t - \frac{\gamma v}{c^2}\mathrm{d}x\right),$$

and, rearranging,

$$\mathrm{d}F = \left(\gamma\frac{\partial F}{\partial x'} - \frac{\gamma v}{c^2}\frac{\partial F}{\partial t'}\right)\mathrm{d}x + \frac{\partial F}{\partial y'}\mathrm{d}y + \frac{\partial F}{\partial z'}\mathrm{d}z + \left(\gamma\frac{\partial F}{\partial t'} - \gamma v\frac{\partial F}{\partial x'}\right)\mathrm{d}t\,. \tag{8.67}$$

But, if F is a function of x, y, z and t, the total differential of F can be written as

$$\mathrm{d}F = \frac{\partial F}{\partial x}\mathrm{d}x + \frac{\partial F}{\partial y}\mathrm{d}y + \frac{\partial F}{\partial z}\mathrm{d}z + \frac{\partial F}{\partial t}\mathrm{d}t\,. \tag{8.68}$$

Comparing the coefficients of $\mathrm{d}x$, $\mathrm{d}y$, $\mathrm{d}z$ and $\mathrm{d}t$ in equations (8.67) and (8.68), we conclude that

$$\frac{\partial}{\partial x} = \gamma\left(\frac{\partial}{\partial x'} - \frac{v}{c^2}\frac{\partial}{\partial t'}\right), \tag{8.69}$$

$$\frac{\partial}{\partial y} = \frac{\partial}{\partial y'}, \tag{8.70}$$

$$\frac{\partial}{\partial z} = \frac{\partial}{\partial z'}, \tag{8.71}$$

$$\frac{\partial}{\partial t} = \gamma\left(\frac{\partial}{\partial t'} - v\frac{\partial}{\partial x'}\right). \tag{8.72}$$

★ 8.6.3 Transformation of the equations curl $E = -\partial B/\partial t$ and div $B = 0$

The y component of equation (8.53) is

$$\frac{\partial E_x}{\partial z} - \frac{\partial E_z}{\partial x} = -\frac{\partial B_y}{\partial t}\,. \tag{8.73}$$

Substituting for $\partial/\partial z$, $\partial/\partial x$ and $\partial/\partial t$ from equations (8.71), (8.69) and (8.72) respectively, we have

$$\frac{\partial E_x}{\partial z'} - \gamma\left(\frac{\partial E_z}{\partial x'} - \frac{v}{c^2}\frac{\partial E_z}{\partial t'}\right) = -\gamma\left(\frac{\partial B_y}{\partial t'} - v\frac{\partial B_y}{\partial x'}\right),$$

and, rearranging

$$\frac{\partial E_x}{\partial z'} - \frac{\partial}{\partial x'}\gamma(E_z + vB_y) = -\frac{\partial}{\partial t'}\gamma\left(B_y + \frac{v}{c^2}E_z\right). \tag{8.74}$$

According to the field transformations, equations (8.31), (8.33) and (8.35), which were derived in Section 8.4 by assuming that the Lorentz force satisfies the principle of relativity when the Lorentz transformations are used to transform from one inertial reference frame to another, we have

$$E_x = E_x'\,,$$

$$\gamma(E_z + vB_y) = E_z'\,,$$

$$\gamma(B_y + vE_z/c^2) = B_y'\,.$$

Substituting in equation (8.74), we obtain

$$\frac{\partial E_x'}{\partial z'} - \frac{\partial E_z'}{\partial x'} = -\frac{\partial B_y'}{\partial t'}\,, \tag{8.75}$$

which is the x' component of equation (8.57).

Using the same method, it can be shown that the z component of equation (8.53) transforms into

$$\frac{\partial E_y'}{\partial x'} - \frac{\partial E_x'}{\partial y'} = -\frac{\partial B_z'}{\partial t'}\,. \tag{8.76}$$

Consider the x component of equation (8.53), namely

$$\frac{\partial E_z}{\partial y} - \frac{\partial E_y}{\partial z} = -\frac{\partial B_x}{\partial t}.$$

Substituting for $\partial/\partial y$, $\partial/\partial z$ and $\partial/\partial t$ from equations (8.70), (8.71) and (8.72) and using the field transformations, equations (8.33), (8.32) and (8.34), to substitute for E_z, E_y, and B_x we have

$$\gamma\frac{\partial}{\partial y'}(E_z' - vB_y') - \gamma\frac{\partial}{\partial z'}(E_y' + vB_z') = -\gamma\left(\frac{\partial}{\partial t'} - v\frac{\partial}{\partial x'}\right)B_x'.$$

Cancelling γ and rearranging, we obtain

$$\frac{\partial E_z'}{\partial y'} - \frac{\partial E_y'}{\partial z'} + \frac{\partial B_x'}{\partial t'} = v\left(\frac{\partial B_x'}{\partial x'} + \frac{\partial B_y'}{\partial y'} + \frac{\partial B_z'}{\partial z'}\right). \tag{8.77}$$

Consider equation (8.54), which in Σ is

$$\frac{\partial B_x}{\partial x} + \frac{\partial B_y}{\partial y} + \frac{\partial B_z}{\partial z} = 0. \tag{8.78}$$

Substituting for $\partial/\partial x$, $\partial/\partial y$ and $\partial/\partial z$ from equations (8.69), (8.70) and (8.71), and using the field transformations, equations (8.34), (8.35) and (8.36), we obtain

$$\gamma\left(\frac{\partial}{\partial x'} - \frac{v}{c^2}\frac{\partial}{\partial t'}\right)B_x' + \gamma\frac{\partial}{\partial y'}\left(B_y' - \frac{vE_z'}{c^2}\right) + \gamma\frac{\partial}{\partial z'}\left(B_z' + \frac{vE_y'}{c^2}\right) = 0.$$

Cancelling γ and rearranging, we obtain

$$\frac{\partial E_z'}{\partial y'} - \frac{\partial E_y'}{\partial z'} + \frac{\partial B_x'}{\partial t'} = \frac{c^2}{v}\left(\frac{\partial B_x'}{\partial x'} + \frac{\partial B_y'}{\partial y'} + \frac{\partial B_z'}{\partial z'}\right). \tag{8.79}$$

Subtracting equation (8.77) from equation (8.79), we have

$$\left(\frac{c^2}{v} - v\right)\operatorname{div}' \boldsymbol{B}' = 0.$$

Since v must be less than c, $c^2/v - v$ cannot be zero, so that we must conclude that

$$\operatorname{div}' \boldsymbol{B}' = 0. \tag{8.80}$$

Putting $\operatorname{div}' \boldsymbol{B}'$ equal to zero in equation (8.79), we obtain

$$\frac{\partial E_z'}{\partial y'} - \frac{\partial E_y'}{\partial z'} = -\frac{\partial B_x'}{\partial t'}. \tag{8.81}$$

Equations (8.81), (8.75) and (8.76) can be combined to give

$$\operatorname{curl}' \boldsymbol{E}' = -\frac{\partial \boldsymbol{B}'}{\partial t'}. \tag{8.82}$$

These results show that, if one starts with the Maxwell equations (8.53) and (8.54) in Σ, applies the Lorentz transformations and the transformations for $\boldsymbol{E}$ and $\boldsymbol{B}$ given by equations (8.31)–(8.36), one obtains equations (8.82) and (8.80), showing that the two Maxwell equations (8.53) and (8.54) obey the principle of relativity when the Lorentz transformations are used to transform from one inertial reference frame to another.

Equations (8.55) and (8.56) include the charge density ρ and the current density $\boldsymbol{J}$. Before showing that equations (8.55) and (8.56) obey the principle of relativity when the Lorentz transformations are used, we shall derive the transformations for ρ and $\boldsymbol{J}$.

★ 8.6.4 Transformation of charge and current densities

We shall start with a simplified model and consider a "uniform" electric charge distribution of volume V in the inertial frame Σ, consisting of n discrete charges per unit volume, of magnitude q each. It will be assumed that all the charges have the same velocity $\boldsymbol{u}$ relative to the inertial frame Σ. For this simplified case the charge density in Σ is

$$\rho = nq. \tag{8.83}$$

The current density, which is the current crossing unit area normal to the direction of current flow, is given in Σ by

$$\boldsymbol{J} = nq\boldsymbol{u}\,, \tag{8.84}$$

or in components from $J_x = nqu_x$, $J_y = nqu_y$, $J_z = nqu_z$. Let the same charge distribution have a volume V' when measured relative to an inertial frame Σ' that is moving with uniform velocity v relative to Σ, and let it consist of n' discrete charges per unit volume moving with velocity $\boldsymbol{u}'$ relative to Σ'. We then have in Σ'

$$\rho' = n'q, \tag{8.85}$$

$$\boldsymbol{J}' = n'q\boldsymbol{u}'. \tag{8.86}$$

The principle of the invariance of total electric charge will be taken as axiomatic in this section. According to this principle, the total charge on a body is independent of the velocity of the body, and the total charge has the same numerical value in all inertial reference frames.

Consider an inertial frame Σ^0 in which the charge distribution is at rest; Σ^0 moves with velocity $\boldsymbol{u}$ relative to Σ. Let V_0 be the proper volume of the charge distribution in Σ^0, and let n_0 be the number of charges per unit volume measured in Σ^0, such that the total number of charges is n_0V_0. According to equation (4.51), owing to length contraction the volume of the charge distribution in Σ should be measured to be

$$V = V_0\,(1 - u^2/c^2)^{1/2}.$$

The total number of charges measured in Σ is nV, which is equal to $nV_0\,(1 - u^2/c^2)^{1/2}$. But the total *number* of charges is an invariant, since it is a pure number. Hence

$$nV = nV_0\,(1 - u^2/c^2)^{1/2} = n_0V_0,$$

so that

$$n = \frac{n_0}{(1 - u^2/c^2)^{1/2}}\,.$$

Similarly, in Σ',

$$n' = \frac{n_0}{(1 - u'^2/c^2)^{1/2}},$$

so that

$$n' = n\,\frac{(1 - u^2/c^2)^{1/2}}{(1 - u'^2/c^2)^{1/2}}. \tag{8.87}$$

From equation (5.13),

$$\frac{(1 - u^2/c^2)^{1/2}}{(1 - u'^2/c^2)^{1/2}} = \gamma\,(1 - vu_x/c^2)\,. \tag{8.88}$$

Substituting in equation (8.87), we obtain

$$n' = \gamma n\,(1 - vu_x/c^2), \tag{8.89}$$

and multiplying by q,

$$qn' = \gamma\,(qn - vqnu_x/c^2)\,.$$

Substituting from equations (8.85), (8.83) and (8.84), we have

$$\rho' = \gamma\,(\rho - vJ_x/c^2)\,.$$

Now, from the x' component of equation (8.86),

$$J_x' = qu_x'n'.$$

But from the velocity transformations, equation (5.6), we have

$$u_x' = \frac{u_x - v}{1 - vu_x/c^2}.$$

Substituting for u_x', and for n' from equation (8.89), we find

$$J_x' = \frac{q\,(u_x - v)}{1 - vu_x/c^2}\,n\gamma\,(1 - vu_x/c^2),$$

and, using equations (8.86), (8.84) and (8.83),

$$J_x' = \gamma\,(J_x - v\rho)\,.$$

Similarly, from equation (8.86),

$$J_y' = qu_y'n'.$$

From the velocity transformations, equation (5.7),

$$u_y' = \frac{u_y}{\gamma\,(1 - vu_x/c^2)}.$$

Substituting for u_y' and n',

$$J_y' = \frac{qu_y}{\gamma(1 - vu_x/c^2)} n\gamma(1 - vu_x/c^2)$$
$$= J_y.$$

Similarly,

$$J_z' = J_z .$$

Collecting the transformations, we have

$$J_x' = \gamma(J_x - v\rho), \qquad J_x = \gamma(J_x' + v\rho'), \tag{8.90}$$

$$J_y' = J_y, \qquad J_y = J_y', \tag{8.91}$$

$$J_z' = J_z, \qquad J_z = J_z', \tag{8.92}$$

$$\rho' = \gamma\left(\rho - \frac{v}{c^2} J_x\right), \qquad \rho = \gamma\left(\rho' + \frac{v}{c^2} J_x'\right). \tag{8.93}$$

It is left as an exercise for the reader to extend the treatment to systems containing both positive and negative charges and to show that equations (8.90)–(8.93) are valid in the general case.

★ 8.6.5 Transformations for the fields *D* and *H*

Consider the y component of the equation

$$\text{curl}\,\boldsymbol{H} = \boldsymbol{J} + \frac{\partial \boldsymbol{D}}{\partial t}, \tag{8.55}$$

which is

$$\frac{\partial H_x}{\partial z} - \frac{\partial H_z}{\partial x} = J_y + \frac{\partial D_y}{\partial t}.$$

Substituting for $\partial/\partial z$, $\partial/\partial x$ and $\partial/\partial t$ from equations (8.71), (8.69) and (8.72) and for J_y from equation (8.91),

$$\frac{\partial H_x}{\partial z'} - \gamma\left(\frac{\partial}{\partial x'} - \frac{v}{c^2}\frac{\partial}{\partial t'}\right)H_z = J_y' + \gamma\left(\frac{\partial}{\partial t'} - v\frac{\partial}{\partial x'}\right)D_y,$$

or

$$\frac{\partial H_x}{\partial z'} - \frac{\partial}{\partial x'}\gamma(H_z - vD_y) = J_y' + \frac{\partial}{\partial t'}\gamma\left(D_y - \frac{v}{c^2}H_z\right). \tag{8.94}$$

If Maxwell's equations are valid in Σ' then one must have in Σ'

$$\frac{\partial H_x'}{\partial z'} - \frac{\partial H_z'}{\partial x'} = J_y' + \frac{\partial D_y'}{\partial t'}. \tag{8.95}$$

Equations (8.94) and (8.95) have the same mathematical form, and if one puts

$$H_x' = H_x, \qquad H_z' = \gamma(H_z - vD_y),$$

$$D_y' = \gamma\left(D_y - \frac{v}{c^2}H_z\right)$$

then they are exactly the same.

It is left as an exercise for the reader to show by similar methods that the x and z components of equation (8.55) transform into the x' and z' components of equation (8.59) and that equation (8.56) transforms into equation (8.60), if $\boldsymbol{J}$ and ρ are transformed using equations (8.90)–(8.93) and if $\boldsymbol{H}$ and $\boldsymbol{D}$ satisfy the transformations:

$$D_x' = D_x, \qquad D_x = D_x', \tag{8.96}$$

$$D_y' = \gamma\left(D_y - \frac{v}{c^2}H_z\right), \qquad D_y = \gamma\left(D_y' + \frac{v}{c^2}H_z'\right), \tag{8.97}$$

$$D_z' = \gamma\left(D_z + \frac{v}{c^2}H_y\right), \qquad D_z = \gamma\left(D_z' - \frac{v}{c^2}H_y'\right), \tag{8.98}$$

$$H_x' = H_x, \qquad H_x = H_x', \tag{8.99}$$

$$H_y' = \gamma(H_y + vD_z), \qquad H_y = \gamma(H_y' - vD_z'), \tag{8.100}$$

$$H_z' = \gamma(H_z - vD_y), \qquad H_z = \gamma(H_z' + vD_y'). \tag{8.101}$$

It has been shown that if Maxwell's equations are valid in Σ then, if the coordinates and time are transformed using the Lorentz transformations, and if one takes the principle of constant charge as axiomatic, the transformed equations have the same mathematical form as Maxwell's equations would have if they were valid in Σ'. This is true whether or not there are material bodies present at the point. Thus Maxwell's equations satisfy the principle of relativity when the coordinates and time are transformed according to the Lorentz transformations.

Only the briefest insight into the relativistic invariance of Maxwell's equations has been given here. For more comprehensive accounts of relativistic electromagnetism, including the transformations for the potentials ϕ and $\boldsymbol{A}$ and the electrodynamics of moving media, the reader is referred to Rosser (1964, 1968).

REFERENCES

Panofsky, W. K. H. and Phillips, M. (1962) *Classical Electricity and Magnetism*, 2nd edn, Chap.20. Addison-Wesley, Reading, Massachusetts.

Rosser, W. G. V. (1964) *An Introduction to the Theory of Relativity*, Chaps. 8–10. Butterworth, London.

Rosser, W. G. V. (1968) *Electromagnetism via Relativity, An Alternative Approach to Maxwell's Equations*, Chap.6. Butterworth, London.

★ PROBLEM

8.1 Show by direct substitution that the following quantities are invariants under a Lorentz transformation: (a) $E^2 - c^2B^2$; (b) $\boldsymbol{B} \bullet \boldsymbol{E}$; (c) $H^2 - c^2D^2$; (d) $\boldsymbol{H} \bullet \boldsymbol{D}$; (e) $\boldsymbol{B} \bullet \boldsymbol{H} - \boldsymbol{E} \bullet \boldsymbol{D}$; (f) $c\boldsymbol{B} \bullet \boldsymbol{D} + c^{-1}\boldsymbol{E} \bullet \boldsymbol{H}$. [*Hint*: Show that $\boldsymbol{B}' \bullet \boldsymbol{E}' = \boldsymbol{B} \bullet \boldsymbol{E}$ etc. by using the field transformations.]

9

Space Travel and the Clock Paradox

9.1 SPACE TRAVEL

So many of the stars in the universe are many millions of light years away that it would seem impossible to reach very many of them in a normal life span. However, this ignores the phenomena of time dilatation and length contraction. As an example, consider a star *at rest* relative to the Earth at a distance of 1000 light years. At first sight, it might appear that it would take a spaceship travelling at the speed of light 1000 years to get there. This, however, is the time measured in the laboratory frame. If k is the number of seconds in a year and c is the speed

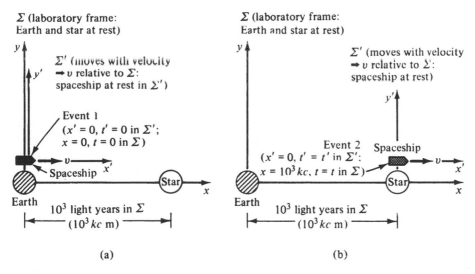

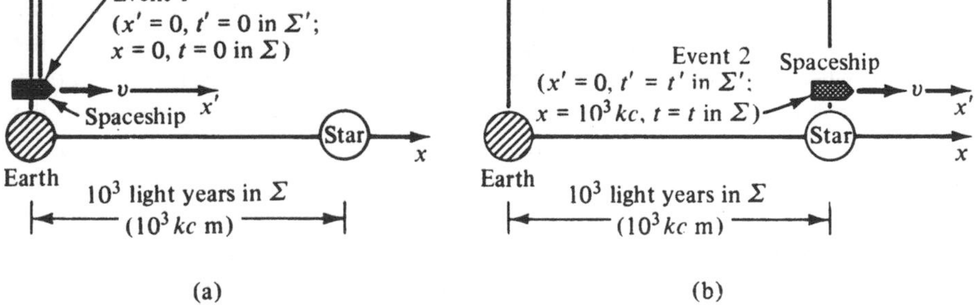

Fig. 9.1 (a) The Earth is at the origin of Σ and the spaceship at the origin of Σ'. The spaceship leaves the Earth at Event 1 at $x = x' = 0$ at $t = t' = 0$ and travels with uniform velocity v relative to the Earth to go to the star 10^3 light years away. (b) The spaceship reaches the star at Event 2 at $x = 10^3kc$ at a time t in Σ. Event 2 is at $x' = 0$ at a time t' in Σ'.

of light (in m s^{-1}) the distance (in m) of the star from the Earth is $10^3\,kc$, measured in the laboratory frame Σ, in which both the Earth and the star are at rest, as shown in Fig. 9.1(a). Suppose that a spaceship leaves the Earth, in a direction opposite to the star, then turns around and accelerates until it reaches a speed v relative to the Earth. Suppose that it continues with uniform velocity v relative to the Earth (Σ) and passes the Earth (Event 1) when the spaceship is moving with uniform velocity v at a time $t = 0$ on a clock on the Earth and a time $t' = 0$ on a clock on board the spaceship. Let the Earth be at the origin of the inertial reference frame Σ and let the spaceship be at rest at the origin of the inertial reference frame Σ' when it is moving away from the Earth towards the star with uniform velocity v, as shown in Fig. 9.1(a). The coordinates of Event 1 in Σ are $x = 0$, $t = 0$, while in Σ' they are $x' = 0$, $t' = 0$. Suppose that the spaceship reaches the star (Event 2) at a time t measured relative to the Earth (Σ) and a time t' measured by the clock on the spaceship (Σ'). Relative to the laboratory frame Σ in which the Earth is at rest, the coordinates of Event 2 (the spaceship's reaching the star) are

$$x = 10^3\,kc, \qquad t = t,$$

while those relative to the spaceship (Σ') are

$$x' = 0, \qquad t' = t',$$

as shown in Fig. 9.1(b). From the Lorentz transformations,

$$t = \gamma\,(t' + vx'/c^2).$$

For Event 2, $x' = 0$. Hence

$$t = \gamma t' = \frac{t'}{(1 - v^2/c^2)^{1/2}}. \tag{9.1}$$

Since $\gamma > 1$, $t' < t$. The time t' of the journey measured by the clock on the spaceship is less by a factor γ than the time t for the journey measured relative to the Earth. Equation (9.1) is the normal expression for time dilatation. The time t' between Events 1 and 2 is the proper time interval between the event of the spaceship's leaving the Earth and the event of its reaching the star, measured by one clock, the clock on the spaceship. The time t for the journey relative to the Earth (Σ) can be measured by the radar methods described in Section 4.2, or by using two synchronized spatially separated clocks, one on the Earth and one on the star.

As a numerical example, it will be assumed that the spaceship reaches the star in a time $t' = 10$ years $= 10k$ s, measured by the clock on the spaceship. If v is the speed of the spaceship relative to the Earth, the spaceship goes with a speed v for a time $t = \gamma t'$ relative to the Earth (Σ) covering a distance $vt = v\gamma t'$ relative to the Earth (Σ), before reaching the star. Since the star is at a distance (in m) from the Earth of $10^3 kc$ measured relative to the laboratory frame (Σ), relative to Σ we have

$$10^3 kc = v\gamma t'. \tag{9.2}$$

If $t' = 10k$ s then

$$10^3 kc = v\gamma\, 10k,$$

or

$$\frac{v}{c} = \frac{10^3}{10\gamma} = \frac{100}{\gamma} = 10^2\left(1 - \frac{v^2}{c^2}\right)^{1/2}, \tag{9.3}$$

and, squaring,

$$\frac{v^2}{c^2} = 10^4 - 10^4\frac{v^2}{c^2};$$

that is

$$\frac{v^2}{c^2} = \frac{10^4}{1 + 10^4} = (1 + 10^{-4})^{-1}.$$

Using the binomial theorem,

$$\frac{v}{c} = (1 + 10^{-4})^{-1/2} = 1 - 0.5 \times 10^{-4},$$

or

$$v = 0.99995\, c.$$

Substituting for v in equation (9.3), we find

$$\gamma = \frac{100}{0.99995} = 100.005\,. \tag{9.4}$$

The above analysis shows that if the spaceship goes at a speed of $0.99995\,c$ relative to the Earth, it will reach the star in 10 years measured by the clock on the spaceship. Relative to the Earth, the spaceship takes a time $t = \gamma t' = 1000.05$ years travelling at a speed $0.99995\,c$ to reach the star. Because the proper time interval t' measured by a clock on the spaceship is less by a factor γ than the time interval t measured in the laboratory frame Σ, it is possible to reach the star, which is 1000 light years away from the Earth, measured relative to the laboratory frame (Σ), in a time of 10 years, measured by the clock on the spaceship (Σ').

Relative to the spaceship, which is at rest at the origin of Σ', the Earth passes the spaceship at Event 1 at $x' = 0, t' = 0$. The Earth and star then move with a velocity $v = 0.99995\,c$ along the *negative* x' axis of Σ' for 10 years or $10k$ s, the star reaching the spaceship at Event 2 at $x' = 0, t' = t'$ relative to Σ'. The astronaut on the spaceship calculates the distance between the Earth and the star to be $l' = vt'$ in Σ', which is only 9.9995 light years. On the other hand, relative to the Earth (Σ), the spaceship goes at a speed v for a time $t = \gamma t'$, covering a distance

$v\gamma t'$ before reaching the star. Hence the *proper* distance between the Earth and the star—that is, the distance between them measured in the inertial reference frame Σ in which they are both at rest—is $l_0 = \gamma v t' = 1000$ light years. Note that $l = l_0/\gamma$. Thus, relative to the moving spaceship, the distance between the Earth and the star is Lorentz-contracted in the direction of motion of the Earth relative to the spaceship (Σ'). This example illustrates how time dilatation and length contraction are closely related.

A proton of mass 938 MeV/c^2 and velocity $0.99995\,c$ such that $\gamma = 100.005$, has a (total) energy of $\gamma mc^2 \approx 100 \times 938$ MeV = 93.8 GeV. There are plenty of cosmic-ray protons of this energy in interstellar space. In fact, protons of energy up to 10^{19} eV have been observed in the cosmic radiation. Provided that the deflection of the protons by galactic magnetic fields can be ignored, the analysis of this section is applicable to cosmic-ray protons. A numerical example was given following Section 4.7.

Now the interesting question is, if, after the time $t' = 10$ years on the spaceship, the spaceship turned around quickly and came back to the Earth with the same uniform speed v, would the time for the complete journey be 20 years, as measured by the clock on the spaceship and $2 \times 1000.05 = 2000.1$ years measured by the clock on the Earth? According to Einstein (1905) this would indeed be the case, and the astronaut should come back to the Earth 20 years older than when the astronaut left, whereas the astronaut's contemporaries on the Earth would have died many centuries earlier. This suggestion is known as the *clock paradox*. It will be shown in Sections 9.2 and 9.3 that the clock paradox is consistent with the postulates of special relativity, and in Section 9.4 that it is consistent with experiments.

Before we rush to look for spaceships in the search for eternal youth, we should consider the fuel requirements to drive such a spaceship. (Incidentally, everything else on the spaceship would go at the slower rate given by equation (9.1), so that the astronaut would not notice that anybody was living longer until the spaceship returned to the Earth.)

Let a rocket of initial mass m_i emit gases at a speed w relative to the rocket until the rocket reaches a speed u, when its final mass is m_f. It is shown in Problem 9.4 that

$$\frac{m_i}{m_f} = \left(\frac{c+u}{c-u}\right)^{c/2w}.$$

With present-day rockets, the exhaust speed w of the gases is of order $10\ \text{km s}^{-1}$, so that $c/2w = 3 \times 10^8/2 \times 10^4 = 1.5 \times 10^4$. If the final speed of the rocket were, say, $u = \frac{1}{2}c$ then

$$\frac{m_i}{m_f} = \left(\frac{c+\frac{1}{2}c}{c-\frac{1}{2}c}\right)^{1.5\times 10^4} = 3^{1.5\times 10^4} \approx 10^{7157}.$$

If the final mass of the rocket is to be large enough to hold an astronaut, say one tonne, one would have to start with a spaceship weighing 10^{7157} tonnes. It can be seen that with present-day rocketry techniques it would be impossible to accelerate a rocket to a speed comparable to the speed of light. If, however, we could use photons (light quanta) or high-speed atomic particles as exhaust "gases" then $w \approx c$, and, for $u = \frac{1}{2}c$, m_i/m_f would only be $\sqrt{3}$. At present, the only particles that can be accelerated to very high speeds are atomic particles. For example, a proton of energy 1 TeV (10^{12} eV) from a proton synchrotron has a speed of 0.999 999 5 c. Hence the predictions of the clock paradox are important for atomic particles accelerated in the laboratory. An experimental check of the clock paradox using atomic particles in the laboratory will be described in Section 9.4.

9.2 IMAGINARY EXPERIMENT ON THE CLOCK PARADOX

A spaceship leaves the Earth at a time $t = 0$ on a clock on the Earth and a time $t' = 0$ on a clock on the spaceship. Suppose that the spaceship travels with uniform velocity v relative to the Earth (laboratory frame Σ), and then turns around at a time $\frac{1}{2}t_L$ relative to the laboratory frame (Σ), when it has covered a distance $\frac{1}{2}vt_L$ relative to the Earth (Σ), as illustrated in the displacement–time curve in Fig. 9.2 for the special case when $v = 0.6\,c$. Let the spaceship turn around quickly, and return with uniform velocity v relative to the Earth, reaching the latter at a time t_L measured by the clock on the Earth, as shown in Fig. 9.2. Suppose that the time of the complete journey is measured to be t_s' by a clock on the spaceship, so that the spaceship turns around at a time $\frac{1}{2}t_s'$ on its clock. Let radio (or light) signals be sent out from the spaceship at time intervals of T_0', measured by the clock on the spaceship, as shown in Fig. 9.2. The frequency of these signals, measured using that clock, is $\nu_0' = 1/T_0'$. The total number of signals transmitted from the spaceship before turning around at the time $\frac{1}{2}t_s'$, relative to the spaceship, is $t_s'/2T_0' = \frac{1}{2}\nu_0' t_s'$. When the spaceship turns around at a time $\frac{1}{2}t_L$ relative to the laboratory frame Σ, it is at a distance $\frac{1}{2}vt_L$ from the Earth, measured relative to the laboratory frame Σ (Fig. 9.2). The radio signal emitted from the spaceship at the time $\frac{1}{2}t_L$ relative to the laboratory frame Σ covers the distance $\frac{1}{2}vt_L$ to the Earth at a speed c in a time $vt_L/2c$, reaching the Earth at a time $\frac{1}{2}t_L + vt_L/2c$ or $\frac{1}{2}t_L(1 + v/c)$ on the Earth clock (Fig. 9.2). According to equation (4.9), when the spaceship is moving away from the Earth, signals emitted at time intervals of T_0' from the spaceship (measured by the spaceship clock) reach the Earth at time intervals of $KT_0' = K/\nu_0'$ (measured by the Earth clock). Since the total number of signals emitted from the spaceship before it turns around is $\frac{1}{2}\nu_0' t_s'$, and the time interval between each signal's reaching the Earth is K/ν_0' on the Earth clock, the signal emitted from the spaceship when it turns around must reach the Earth at a time $\frac{1}{2}\nu_0' t_s' \times K/\nu_0'$ on the Earth clock. Hence

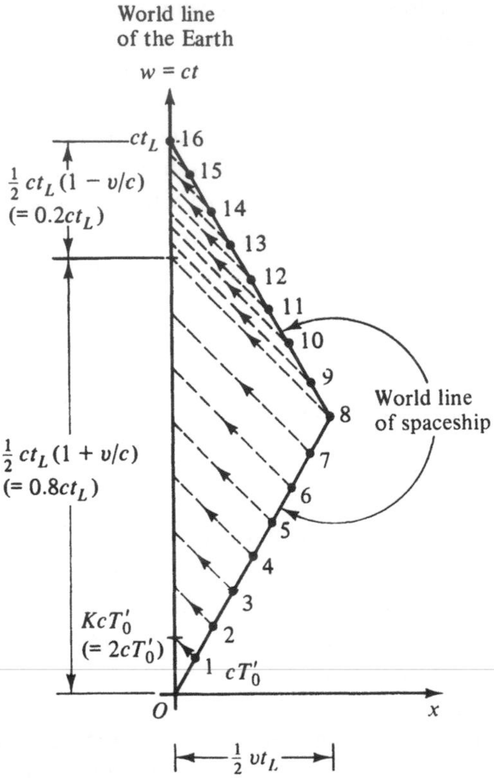

Fig. 9.2 Imaginary experiment on the clock paradox. The speed of the spaceship in this figure is $0{\cdot}6c$ relative to the Earth, so that in this special case $\gamma = (1 - v^2/c^2)^{-1/2} = \frac{5}{4}$, and $K = (1 + v/c)^{1/2}/(1 - v/c)^{1/2} = 2$. From equation (9.7), $t_L = \gamma t'_s = \frac{5}{4} t'_s$. The spaceship transmits 16 radio signals at time intervals of T_0' relative to the spaceship. The first 8 signals reach the Earth at time intervals of $KT_0' = 2T_0'$ on the Earth clock, in a total time of $16\,T_0'$ on the latter. The second 8 signals reach the Earth at time intervals of $T_0'/K = \frac{1}{2}T_0'$ on the Earth clock in a total time interval of $8 \times \frac{1}{2} T_0' = 4\,T_0'$ on the latter. Thus the total time of the journey is $20\,T_0'$ measured by the Earth clock and $16\,T_0'$ measured by the spaceship clock. This is in agreement with equation (9.7).

$$\tfrac{1}{2} t_L \left(1 + \frac{v}{c} \right) = \tfrac{1}{2} \nu_0' \, t_s' \times \frac{K}{\nu_0'} = \tfrac{1}{2} t_s' \, K \, . \qquad (9.5)$$

The $\frac{1}{2}\nu_0' t_s'$ signals emitted from the spaceship between the time it turns around and the time it returns to the Earth reach the Earth between the times $\frac{1}{2} t_L (1 + v/c)$ and t_L on the Earth clock, that is, in a time interval of $\frac{1}{2} t_L (1 - v/c)$ on the Earth clock (Fig. 9.2).

It was shown in Section 4.8 that if the spaceship emits signals of frequency ν_0', and if the frequency of the signals reaching the Earth is ν_0'/K when the spaceship is going away from the Earth with uniform velocity v, then the frequency of the signals reaching the Earth when the spaceship is moving towards the Earth with uniform velocity v is $K\nu_0'$. Hence the time between each successive signal's reaching the Earth is $1/K\nu_0'$, for signals emitted from the spaceship when it is moving towards the Earth. Since the total number of signals emitted from the spaceship between its turning around and returning to the Earth is $\frac{1}{2}\nu_0' t_s'$, and the time interval between each signal's reaching the Earth is $1/K\nu_0'$ on the Earth clock, the total time for which these signals are received on the Earth is $\frac{1}{2}\nu_0' t_s' \times (1/K\nu_0')$ measured by the Earth clock. Hence

$$\tfrac{1}{2} t_L \left(1 - \frac{v}{c}\right) = \tfrac{1}{2}\nu_0' t_s' \times \frac{1}{K\nu_0'} = \frac{t_s'}{2K}. \tag{9.6}$$

Multiplying equations (9.5) and (9.6) gives

$$t_s'^2 = t_L^2 (1 - v^2/c^2),$$

or

$$t_L = \frac{t_s'}{(1 - v^2/c^2)^{1/2}} = \gamma t_s'. \tag{9.7}$$

Equation (9.7) suggests that the time for the journey measured by the clock on the spaceship should be less than the time for the journey measured by a clock at rest on the Earth. The above example illustrates that it is consistent with the principles of special relativity that a person who goes on a journey into outer space and back at a speed comparable to c might age less than a person who stays on the Earth. In the example discussed in Section 9.1 we have $t_s' = 20$ years for the journey to the star and back. If $v = 0.99995c$, such that $\gamma = 100.005$, then t_L would be equal to 2000.1 years.

Dividing equation (9.5) by equation (9.6) gives

$$\frac{1 + v/c}{1 - v/c} = K^2,$$

or

$$K = \left(\frac{1 + v/c}{1 - v/c}\right)^{1/2}.$$

This is in agreement with equation (4.16).

★ 9.3 A CRITIQUE OF THE CLOCK PARADOX

Consider the general case of a clock that is accelerating and moving relative to the inertial reference frame Σ, as shown in Fig. 9.3(a). Let the velocity of the accelerating clock, measured at a time t relative to Σ, be equal to $\boldsymbol{u}$. Choose the direction of the x axis of Σ to be parallel to the direction of $\boldsymbol{u}$ at the time t in Σ

as shown in Fig. 9.3(a). At the time t in Σ the accelerating clock is instantaneously at rest in the inertial reference frame Σ^0 that is moving with uniform velocity u along the x axis of Σ, as shown in Fig. 9.3(b). Let the first tick on the accelerating clock (Event 1) be recorded at (x, y, z) at the time t in Σ, as shown in Fig. 9.3(a), and at (x^0, y^0, z^0) at a time t^0 in Σ^0. The time t^0 in Σ^0 is measured by a *clock that is permanently at rest at* (x^0, y^0, z^0) *in* Σ^0 and that coincides with the position of the accelerating clock when it is instantaneously at rest at (x^0, y^0, z^0) in Σ^0, as shown in Fig. 9.3(b). Let the second tick, at an infinitesimal time later on the accelerating clock (Event 2), be measured to be at $(x + \mathrm{d}x, y + \mathrm{d}y, z + \mathrm{d}z)$ at the time $t + \mathrm{d}t$ in Σ, as shown in Fig. 9.3(a). Since the accelerating clock is instantaneously at rest in Σ^0, the second tick on the accelerating clock (Event 2) is again recorded at (x^0, y^0, z^0) in Σ^0. Let the time of Event 2 be recorded at the time $t^0 + \mathrm{d}t^0$ by the *clock that is permanently at rest at* (x^0, y^0, z^0) *in* Σ^0, as shown in Fig. 9.3(b). The coordinates and times of Events 1 and 2, measured by rulers and synchronized clocks *at rest* in Σ and Σ^0 respectively, are related by the Lorentz transformations. It follows from equation (4.59) that, according to the Lorentz transformations, the interval between Events 1 and 2 is an invariant; that is, it has the same numerical value in Σ and Σ^0. Hence, since $\mathrm{d}x^0 = \mathrm{d}y^0 = \mathrm{d}z^0 = 0$ in Σ^0,

$$\mathrm{d}s^2 = \mathrm{d}x^2 + \mathrm{d}y^2 + \mathrm{d}z^2 - c^2\,\mathrm{d}t^2 = -\,c^2\,(\mathrm{d}t^0)^2. \tag{9.8}$$

Let the time interval between the two ticks on the accelerating clock (Events 1 and 2), *measured by the accelerating clock itself*, be equal to $\mathrm{d}\tau$. Since the rate of an accelerating clock may be affected by its acceleration, one cannot, without making an extra assumption, relate the time interval $\mathrm{d}\tau$ measured by the accelerating clock to the time interval $\mathrm{d}t^0$ measured by the clock permanently at

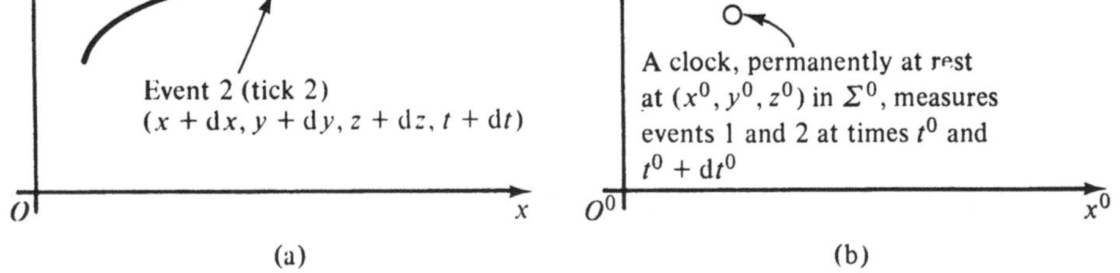

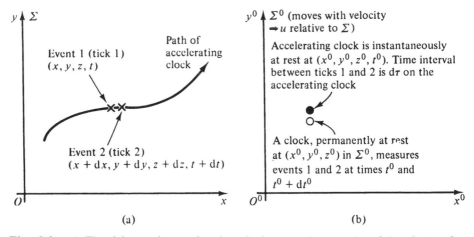

Fig. 9.3 (a) The ticks on the accelerating clock are at (x, y, z, t) and $(x + \mathrm{d}x, y + \mathrm{d}y, z + \mathrm{d}z, t + \mathrm{d}t)$ in Σ. (b) The accelerating clock is instantaneously at rest in Σ^0. The time between the two ticks on the accelerating clock itself is $\mathrm{d}\tau$. The time interval between the two ticks, measured by a clock permanently at rest in Σ^0 is $\mathrm{d}t^0$.

rest at (x^0, y^0, z^0) in Σ^0. At this stage the *clock hypothesis* is introduced. According to this hypothesis, it is *postulated* that the rate of the accelerating clock is independent of its acceleration and is equal to the rate of the clock that is permanently at rest at (x^0, y^0, z^0) in the inertial reference frame Σ^0 in which the accelerating clock is instantaneously at rest, as shown in Fig. 9.3(b). Hence, according to the clock hypothesis, it is *assumed* that $\mathrm{d}\tau$, the time interval between the two ticks measured by the accelerating clock itself, is equal to $\mathrm{d}t^0$, the time interval between the two ticks measured by the clock permanently at rest at (x^0, y^0, z^0) in the inertial reference frame Σ^0 in which the accelerating clock is instantaneously at rest. (The clock hypothesis goes beyond the Lorentz transformations. It is only the time interval $\mathrm{d}t^0$ measured by the clock at rest in Σ^0 that is related by the Lorentz transformations to the time interval $\mathrm{d}t$ measured by two spatially separated, synchronized clocks at rest in Σ.) Putting $\mathrm{d}t^0$ equal to $\mathrm{d}\tau$ in equation (9.8), we have

$$-c^2\,\mathrm{d}\tau^2 = \mathrm{d}x^2 + \mathrm{d}y^2 + \mathrm{d}z^2 - c^2\,\mathrm{d}t^2,$$

and therefore

$$\begin{aligned}\mathrm{d}\tau^2 &= \mathrm{d}t^2\left\{1 - \frac{1}{c^2}\left[\left(\frac{\mathrm{d}x}{\mathrm{d}t}\right)^2 + \left(\frac{\mathrm{d}y}{\mathrm{d}t}\right)^2 + \left(\frac{\mathrm{d}z}{\mathrm{d}t}\right)^2\right]\right\}\\ &= \mathrm{d}t^2\,(1 - u^2/c^2)\,.\end{aligned}$$

Hence

$$\mathrm{d}\tau = \mathrm{d}t\,(1 - u^2/c^2)^{1/2}. \tag{9.9}$$

Integrating this equation over the arbitrary path followed by the accelerating clock in Σ, we have

$$\int \mathrm{d}\tau = \int (1 - u^2/c^2)^{1/2}\,\mathrm{d}t\,. \tag{9.10}$$

Equations (9.9) and (9.10) are the mathematical representations of the clock hypothesis. According to this hypothesis, the time for the journey measured by the accelerating clock itself is $\int \mathrm{d}\tau$, whereas the time for the same journey measured by spacially separated, synchronized clocks at rest in Σ is $\int \mathrm{d}t$.

It was shown in Section 4.10, equation (4.59), that it follows from the Lorentz transformations that the interval $\mathrm{d}s$ is an invariant, so that, applying the clock hypothesis in an inertial reference frame Σ' moving with uniform velocity relative to Σ, we have

$$\begin{aligned}\mathrm{d}s^2 &= \mathrm{d}x'^2 + \mathrm{d}y'^2 + \mathrm{d}z'^2 - c^2\,\mathrm{d}t'^2 = -c^2\,(\mathrm{d}t^0)^2\\ &= -c^2\,\mathrm{d}\tau^2.\end{aligned}$$

Hence in Σ' we have

$$\mathrm{d}\tau = \mathrm{d}t'\,(1 - u'^2/c^2)^{1/2}, \tag{9.11}$$

where dt' is the time interval between the two ticks on the accelerating clock measured using two spatially separated synchronized clocks at rest in Σ', and u' is the instantaneous value of the speed of the accelerating clock relative to Σ'. Integrating equation (9.11), we have in Σ'

$$\int d\tau = \int (1 - u'^2/c^2)^{1/2}\, dt'. \tag{9.12}$$

Comparisons of equations (9.9) and (9.11) shows that the clock hypothesis is Lorentz-covariant; that is, it obeys the principle of relativity when the coordinates and time are changed using the Lorentz transformations. Hence it is consistent with the theory of special relativity to make the clock hypothesis. Whether or not the hypothesis is valid in practice must be decided by experiment. An experimental confirmation of equation (9.10) will be described in Section 9.4.

If the accelerating clock were permanently at rest in Σ^0 then u would be a constant and equation (9.10) would reduce to

$$\int d\tau = \int dt^0 = (1 - u^2/c^2)^{1/2} \int dt,$$

and therefore

$$\Delta t = \frac{\Delta\tau}{(1 - u^2/c^2)^{1/2}}.$$

This is the normal formula for time dilatation. (Compare with equations (4.21) and (4.42), putting $v = u$ and $\Delta t' = \Delta\tau$.)

It is left as an exercise for the reader to apply equation (9.10) to the example shown in Fig. 9.2 to derive equation (9.7).

★ 9.4 EXPERIMENTAL CHECK OF THE CLOCK PARADOX

Consider a spaceship that travels in a circular path with *uniform* speed u relative to the laboratory frame Σ. In this example, according to equation (9.10),

$$\int d\tau = \int (1 - u^2/c^2)^{1/2}\, dt = (1 - u^2/c^2)^{1/2} \int dt.$$

Putting $\int d\tau = t'_{\text{spaceship}}$ and $\int dt = t_{\text{lab}}$, we have

$$t'_{\text{spaceship}} = (1 - u^2/c^2)^{1/2}\, t_{\text{lab}}.$$

Hence

$$t_{\text{lab}} = \alpha t'_{\text{spaceship}}, \tag{9.13}$$

where $\alpha = (1 - u^2/c^2)^{-1/2}$. If one started with N_0 radioactive atoms at rest in a box on the Earth and N_0 identical atoms at rest in a box inside a spaceship when the latter starts on a journey in a circular path with uniform speed u then when the spaceship returns to the Earth, according to the law of radioactive decay, the number of radioactive atoms left in the box that stayed on the Earth (denoted by N_{lab}) should be given by

$$N_{lab} = N_0 \exp\left(-\frac{t_L}{T_0}\right), \tag{9.14}$$

where t_L is the time for the journey measured by the Earth clock. The number of radioactive atoms left in the box in the spaceship at the end of the journey should be

$$N_{spaceship} = N_0 \exp\left(-\frac{t_s'}{T_0}\right), \tag{9.15}$$

where t_s' is the time of the journey measured by the clock on the spaceship, and T_0 is the mean life of the radioactive atoms when they are at rest. If $t_s' < t_L$ as given by equation (9.13), then $N_{spaceship}$ should be greater than N_{lab} and there should be more radioactive atoms left undecayed in the spaceship than on the Earth. Using equation (9.13), equation (9.15) becomes

$$N_{spaceship} = N_0 \exp\left(-\frac{t_s'}{T_0}\right) = N_0 \exp\left(-\frac{t_L}{\alpha T_0}\right). \tag{9.16}$$

An experiment of the above type was carried out by Farley and collaborators using a muon (μ) storage ring at CERN. Many experiments on muons have shown that their mean life is

$$T_0 = 2.20 \times 10^{-6}\ \text{s}$$

when they are at rest, so that for the muons at rest in the laboratory, equation (9.14) becomes

$$N_{lab} = N_0 \exp\left(-\frac{t_L}{2.20 \times 10^{-6}\ \text{s}}\right). \tag{9.17}$$

Instead of putting the muons in a spaceship, in their experiment at CERN, Farley *et al.* (1968) sent them around in circular orbits in a magnetic field in the laboratory, as illustrated in Fig. 9.4(b). They found that the rate of decay of muons of momenta 1.274 GeV/*c* travelling in circles in a magnetic field satisfied the experimental relation

$$N_{spaceship} = N_0 \exp\left[-\frac{t_L}{(2.615 \pm 0.003) \times 10^{-5}\ \text{s}}\right], \tag{9.18}$$

where t_L is the time of the muon journey relative to the laboratory. It can be seen that muons that go on a journey "into outer space" (in this case, move in a circle of 5 m diameter) live longer on the average than "stay-at-home" muons at rest in the laboratory. This is an *experimental* result. To illustrate it, assume that we start with a million muons at rest in the laboratory, and a million muons, each of momentum 1.274 GeV/*c*, that start moving in circular orbits of diameter 5 m. According to equation (9.17), the average number of the muons which remain at

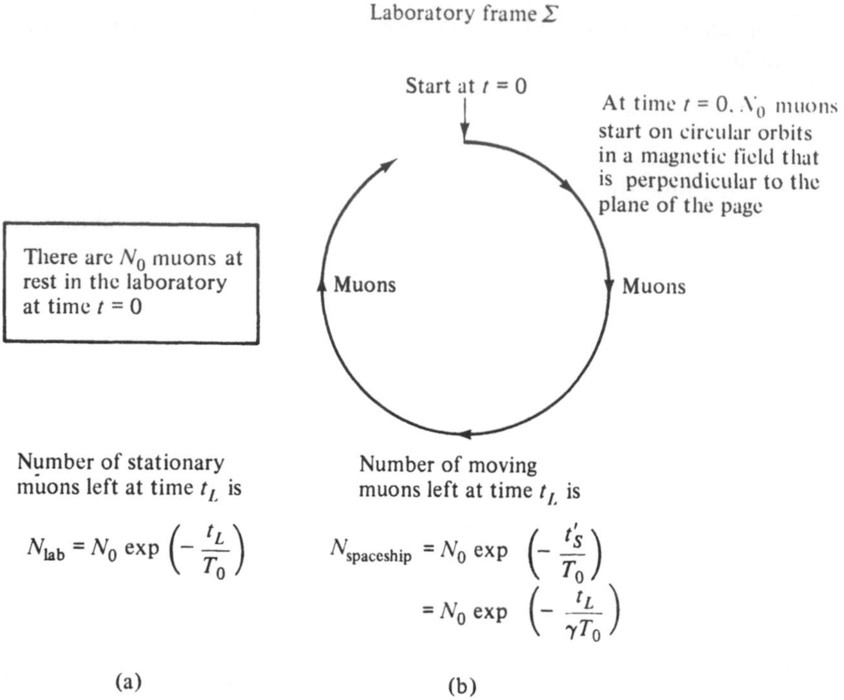

Fig. 9.4 Experimental check of the clock paradox using muons. According to equations (9.14) and (9.16) at a time t_L in the laboratory system, $N_{\text{spaceship}}$ should be greater than N_{lab}. This prediction is confirmed by experiment.

rest in the laboratory that have not decayed in a time of $t_{\text{lab}} = 2.2 \times 10^{-5}$ s is 45.4. According to the experimentally determined equation (9.18), the average number of the moving muons that have not decayed in the time $t_{\text{lab}} = 2.2 \times 10^{-5}$ s measured in the laboratory frame Σ, is 431 150. Thus experiments show that there is a very significant difference in the number of muons that have not decayed in the two cases.

According to equation (9.16), which was derived from the clock hypothesis, equation (9.10), for muons of momentum 1.274 GeV/c, with $T_0 = 2.2 \times 10^{-6}$ s and $\alpha = 12.15$, we have in the laboratory frame:

$$N_{\text{spaceship}} = N_0 \exp\left(-\frac{t_{\text{lab}}}{2.672 \times 10^{-5}\ \text{s}} \right). \tag{9.19}$$

The experimental mean lifetime of 2.615×10^{-5} s in equation (9.18) for muons of momentum 1.274 GeV/c in the laboratory frame is 2% less than the value of 2.672×10^{-5} s predicted using the clock hypothesis and given by equation (9.19). This difference of 2% was interpreted as a loss of muons due to imperfections in the magnetic field. These experimental results confirm the clock hypothesis.

Summarizing, the clock hypothesis goes beyond the Lorentz transformations. The hypothesis is consistent with the theory of special relativity. It is in agreement with experimental results. According to the clock hypothesis, the rate of an accelerating clock is independent of its acceleration. The centripetal acceleration of muons of momentum 1.274 GeV/c moving in circular orbits of radius $\rho = 2.5$ m is $u^2/\rho = 3.6 \times 10^{16}$ m s^{-2}, or $3.7 \times 10^{15}\,g$, where $g = 9.81$ m s^{-2} is the value of the acceleration due to gravity at the surface of the Earth. These results show that the clock hypothesis holds for enormous accelerations.

★ 9.5 RATES OF CLOCKS IN EARTH SATELLITES AND AEROPLANES

When a satellite is in a circular orbit above the Earth, there are two effects to be considered when comparing the rates of clocks on the Earth and in the satellite. First, according to the clock hypothesis, equation (9.10), there is an effect due to the *motion* of the satellite relative to the Earth. According to equation (9.10), the clock on the satellite, which is moving relative to the Earth, should go at a slower rate than an identical clock at rest on the Earth. Secondly, the clock on the satellite is at a higher gravitational potential than the identical clock that is at rest on the Earth. It is shown in Appendix C, using the principle of equivalence, that the rate of a clock is influenced by a gravitational field: according to equation (C.5), the clock at the higher gravitational potential, namely that on the satellite, should go at the faster rate. Both contributions are always present. The net result is that for satellites in circular orbits at altitudes greater than about $\frac{1}{2}R$ above sea level, where $R = 6400$ km is the Earth's radius, the second effect, associated with the difference in gravitational potential, predominates: the clock on the satellite should read more than a similar clock at rest on the Earth, and an astronaut on the satellite should age more than a person at rest on the Earth. If the altitude of the satellite is less than about $\frac{1}{2}R$ above sea level, the first effect, associated with the motion of the satellite relative to the Earth, predominates: the clock on the satellite should read less than the identical clock at rest on the Earth, and an astronaut on the satellite should age less than a person at rest on the Earth. (See Rosser (1964).)

Hafele and Keating (1972) took caesium clocks on aeroplane journeys around the Earth and found that the time differences they observed were in agreement with predictions based on a combination of the effect predicted by equation (9.10) and the effect due to changes in gravitational potential, predicted by equation (C.5).

REFERENCES

Einstein, A. (1905) *Ann. der Phys.* **17**, 891. [English translation in Einstein, A., Lorentz, H.A., Minkowski, H. and Weyl, H. (1952) *The Principle of Relativity*, p.37. Dover, New York.]

Farley, F.J.M., Bailey, J. and Picasso, F. (1968) *Nature* **217**, 17.

Hafele, J.C. and Keating, R.E. (1972) *Science* **177**, 168.

Rosser, W.G.V. (1964) *An Introduction to the Theory of Relativity*, p.449. Butterworth, London.

PROBLEMS

9.1 A spaceship accelerates quickly and then moves with uniform velocity relative to the solar system until it reaches a star eight light years away. A clock on the spaceship records the time for the journey as six years. Calculate (a) the speed of the spaceship relative to the Earth; (b) the time of the journey relative to the Earth. (c) If the spaceship turns around quickly and returns to the Earth with the same velocity as on the outward journey, compare the total times for the journey relative to the Earth and relative to the spaceship.

9.2 An astronaut goes on a journey into space; the spaceship leaves the Earth on the astronaut's 29th birthday and travels at a uniform speed. What is the minimum speed at which the spaceship must move relative to the Earth in order that it may return 10 years later (relative to the Earth) but before the astronaut's 30th birthday (from the astronaut's point of view).

9.3 A total of 2000 charged pions are created at the origin of an inertial frame. Half of the pions remain at rest at the origin, while the other half go on a journey with a uniform speed of $0.995\,c$. After a path length of 15 m, the travelling pions are deflected in a magnetic field so that they return to the origin.

(a) How many of the moving pions should survive the journey?

(b) How many of the stationary pions are left at the origin when the travelling pions return?

Take the mean lifetime of charged pions to be 2.5×10^{-8} s when they are at rest.

9.4 A rocket propels itself forwards by emitting gases in the backwards direction at a constant velocity w relative to the inertial frame in which the rocket is instantaneously at rest. Its final velocity, relative to its initial rest frame, is u. Show that

$$\frac{m_i}{m_f} = \left(\frac{c+u}{c-u}\right)^{c/2w},$$

where m_i and m_f are the initial and final masses of the rocket respectively. [*Hint:* Consider the rocket in the inertial frame Σ' in which it is instantaneously at rest, and which is moving with velocity v relative to the initial rest frame Σ. Let gases of mass dm_g be emitted such that the mass of the rocket changes from m to $m + dm$ and the velocity of the rocket changes from v to $v + dv$ in Σ and from 0 to dv' in Σ'. Apply the law of conservation of energy in Σ' and neglect terms of order $dm\,dv'$ and $(dv')^2$ to show that

$$-d(mc^2) = (1 - w^2/c^2)^{-1/2}\, d(m_g c^2).$$

Apply the law of conservation of momentum in Σ' and neglect terms of order $(dv')^2$ to show that

$$(1 - w^2/c^2)^{-1/2}\, w\, dm_g = m\, dv',$$

and therefore

$$-w\, dm = m\, dv'.$$

From the velocity transformations, $dv' = (1 - v^2/c^2)^{-1}\, dv$. Substitute for dv' and then integrate.]

Appendix A

The Michelson–Morley Experiment

The most famous of the optical experiments to try to determine the velocity of the Earth relative to the hypothetical absolute reference frame in which, it was assumed, the hypothetical luminiferous ether was at rest and in which, it was assumed, the speed of light in empty space was the same in all directions, was carried out by Michelson and Morley in 1887. The principle of the optical interferometer used by Michelson and Morley is illustrated in Fig. A.1. Light from a source S is divided at a half-silvered mirror A, part of the light going in the direction AB and part in the direction AC. The light is reflected by mirrors at B and C. The two pencils of light then combine to produce interference fringes,

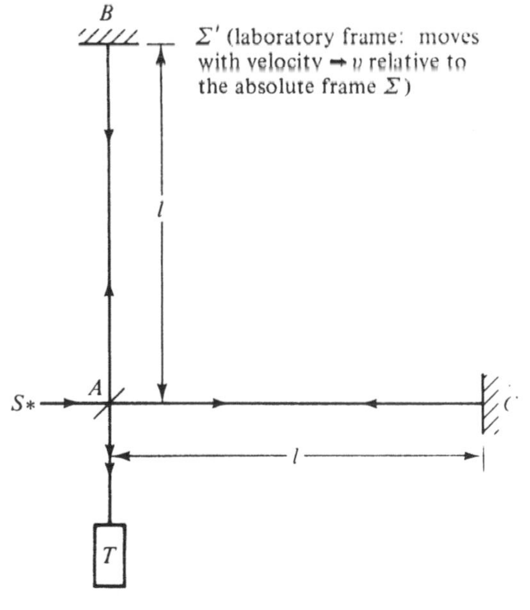

Fig. A.1 Schematic form of the Michelson–Morley experiment.

which are viewed by the telescope T. For a full account of the theory of the Michelson interferometer the reader is referred to a textbook on optics.

For the purposes of discussion it will be assumed that the laboratory frame, denoted by Σ' in this instance, is moving with uniform velocity v in the direction from A to C in Fig. A.1, relative to the absolute reference frame Σ, in which the speed of light in empty space is equal to c in all directions. It will also be assumed, for the purposes of discussion, that the Galilean velocity transformations are correct. It was shown in Section 3.2 (see Figs. 3.1a,b) that, according to the Galilean velocity transformations, the speed of light in the laboratory frame Σ' should be $c - v$ in the direction A to C, $c + v$ in the direction C to A, and $(c^2 - v^2)^{1/2}$ in the directions A to B and B to A. If $AB = AC = l$ then the time t_1' for the light to go from A to B and back to A, and the time t_2' for the light to go from A to C and back to A, are given by

$$t_1' = \frac{2l}{(c^2 - v^2)^{1/2}} = \frac{2l}{c\,(1 - v^2/c^2)^{1/2}}, \tag{A.1}$$

$$t_2' = \frac{l}{c - v} + \frac{l}{c + v} = \frac{2lc}{c^2 - v^2} = \frac{2l}{c\,(1 - v^2/c^2)}. \tag{A.2}$$

Let $\Delta t' = t_2' - t_1'$. Then, using (A.1) and (A.2), we have

$$\Delta t' = t_2' - t_1' = \frac{2l}{c}\,[(1 - v^2/c^2)^{-1} - (1 - v^2/c^2)^{-1/2}].$$

Expanding by the binomial theorem and neglecting terms of order greater than v^2/c^2,

$$\Delta t' = \frac{2l}{c}\,[1 + v^2/c^2 - 1 - \tfrac{1}{2}\,v^2/c^2] = \frac{lv^2}{c^3}.$$

This corresponds to a path difference Δ_1, given by

$$\Delta_1 = c\,\Delta t' = l\,\frac{v^2}{c^2}.$$

If the interferometer is rotated through 90°, such that the interferometer arm AB is parallel to v, the velocity at which the laboratory frame Σ' is moving relative to the absolute system Σ, then the time for light to go from A to C and back to A is less now than the time for light to go from A to B and back to A by an amount lv^2/c^3, and the new path difference is

$$\Delta_2 = -lv^2/c^2 .$$

Hence the total change in path difference on rotating the interferometer through 90° is $2lv^2/c^2$. If λ is the wavelength of the light used then this change in path difference corresponds to a fringe shift of

$$n = \frac{2l}{\lambda}\frac{v^2}{c^2} .$$

Michelson and Morley had a path length of 11 m, and used a wavelength of 5.9×10^{-7} m. If it is assumed that the velocity of the Earth relative to the absolute system is equal to the full velocity of the Earth in its orbit around the Sun, namely 30 km s^{-1}, then $v/c \approx 10^{-4}$. Hence

$$n = \frac{2 \times 11}{5.9 \times 10^{-7}} \times (10^{-4})^2 \approx 0.37.$$

This is a fraction of a fringe. To ensure accuracy and freedom from vibration, particularly when the interferometer was rotated, Michelson and Morley floated the apparatus on mercury and took their readings continuously as the whole apparatus rotated slowly. The maximum displacement that they found was less than one hundredth of a fringe. They could have detected a fringe shift of 0.37 fringe quite easily. Now it might have happened that at the time they performed the experiment the Earth might just have been at rest relative to the absolute system. However, when they performed the experiment at different times of the day and also six months later, when the Earth was moving in the opposite direction relative to the Sun, Michelson and Morley did not find any fringe shift. There have been many repetitions of the Michelson–Morley experiment, but no significant fringe shift has ever been observed. For example, in 1964 Jaseja, Javan, Murray and Townes carried out an experiment using laser beams, which was equivalent to the Michelson–Morley experiment. Their results showed that the velocity of the Earth relative to any absolute system, *if* it exists, is less than 1/1000 of the Earth's orbital velocity around the Sun.

Comment According to the principle of the constancy of the speed of light and the theory of special relativity, the speed of light is equal to c in all directions in *all* inertial reference frames, including the laboratory frame Σ' in Fig. A.1. In this case both the time it takes light to go from A to B and back to A and the time it takes light to go from A to C and back to A are equal to $2l/c$, whatever the orientation of the interferometer in Fig. A.1, and no fringe shift should be observed when the apparatus is rotated.

REFERENCE

Jaseja, T.S., Javan, A., Murray, J. and Townes, C.H. (1964) *Phys. Rev.* **A 133**, 1221.

Appendix B

Simultaneity of Spatially Separated Events

In order to overcome the necessity of measuring time intervals, we shall consider an example due to Einstein. It will be assumed that a long train is moving with uniform velocity v along the x axis of the laboratory frame Σ, as shown in Fig. B.1. It will be assumed that the inertial reference frame Σ' is fixed on the train and that the origins of Σ and Σ' coincide at $t=t'=0$. It will also be assumed that there are inertial observers on the train and on the embankment at the side of the track, and that these observers have devices, such as photomultipliers, to detect

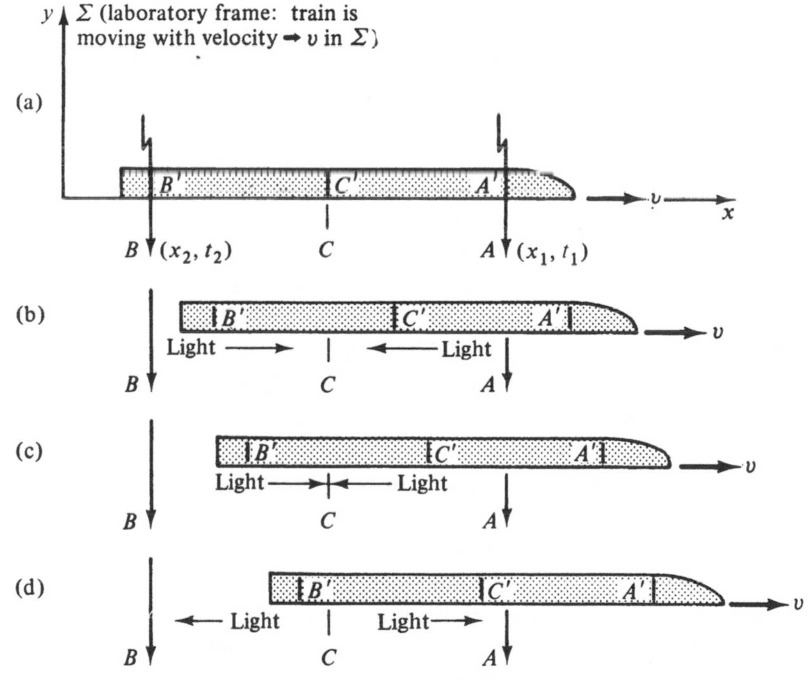

Fig. B.1 Lightning strikes twice, leaving marks on the train at A' and B' and marks on the embankment at A and B. The light signals from A and B reach C at the same time.

light signals coming from different directions as well as measuring rods to measure distances.

Suppose that two lightning bolts strike both the train (Σ') and the embankment (Σ), leaving marks on both. These are two events independent of any particular coordinate system, since the inertial observers on the train and embankment will agree that the lightning did strike on two occasions. Let the positions of the marks on the embankment (Σ) and on the train (Σ') be denoted by A and A' and by B and B' respectively, and let the light detectors on the embankment (Σ) and on the train (Σ') be at C and C' respectively, as shown in Fig. B.1(a). Suppose that both observers find after the events that their light detectors were exactly half-way between the events; that is, $AC = CB$ in the laboratory frame (Σ) and $A'C' = C'B'$ on the train (Σ'). The measuring sticks need not even be calibrated in the same units to show these equalities. The length measurements can be carried out by the observers by walking along the embankment and the train respectively.

It is assumed that the light signals from A and B arrive at the same time at the detectors at C as shown in Fig.B.1(c). If the observer on the embankment (Σ) assumes that the velocity of light in empty space is independent of the velocity of the source and is the same in all directions in space, as predicted by Maxwell's equations and the principle of the constancy of the velocity of light, then the observer on the embankment will conclude that the two lightning flashes struck simultaneously at A and B at a time AC/c before the light signals reach C. The light emitted by the lightning takes a finite time to reach the detectors at C and C', and during this time interval the train (Σ') moves relative to the embankment (Σ) such that C' moves towards A, as illustrated in Fig. B.1(b). Thus the light travelling from A towards C must pass C' before reaching C, as shown in Fig. B.1(b). A little later, relative to the laboratory frame Σ, the light signals from the flashes at A and B reach C simultaneously, as shown in Fig. B.1(c), so that the observer on the embankment (Σ) concludes that the lightning struck at the same time at A and B. Note that the light from B and B' does not reach C' until after it has passed C, as shown in Fig. B.1(d). As a result, the observer on the train (Σ') would record the signals from A and A' before the signals from B and B', say at times t' and $t' + \Delta t'$ respectively. The observer on the train (Σ') measures $A'C'$ to be equal to $C'B'$. If the observer on the train (Σ') assumes that the velocity of light is the same in both directions and is equal to c then that observer estimates that the events at A' and B' occurred at times $t' - A'C'/c$ and $t' + \Delta t' - B'C'/c$ respectively. Hence the observer on the train (Σ') would conclude that the events did not occur simultaneously in the inertial reference frame Σ', but would conclude that the event at A' happened a time $\Delta t'$ before the event at B'. Both observers use exactly the same criteria for the simultaneity of spatially separated events. It is quite possible for the lightning to strike such that the light signals reach C' simultaneously, in which case the observer on the train (Σ') would record the events as simultaneous; but in this case the light from B will pass C before reaching C', while the light from A reaches C at a later time so that the observer on the embankment (Σ) will conclude that the lightning struck at B

before A. When the problem of the simultaneity of spatially separated events is related to measurements that can actually be performed, if the velocity of light in empty space is independent of the velocity of the source relative to the observer, and is the same in all directions, then it can be seen that events measured to be simultaneous in one inertial reference frame are not measured as simultaneous in an inertial reference frame moving with uniform velocity relative to the first.

Note that in Fig. B.1 the velocity of the train is comparable to the speed of light. Normally trains go rather slower than this, so that the differences from absolute simultaneity are normally too small to be noticed in our daily lives.

The Lorentz transformations can be used to fill in the mathematical background to the above qualitative discussion. Let the event of the lightning's striking at A, A' be recorded at (x_1, t_1) in the laboratory frame Σ. Let the same event at A, A' be recorded at (x_1', t_1') relative to Σ', the inertial reference frame in which the train is at rest. Let the event at B, B' be recorded at (x_2, t_2) and (x_2', t_2') relative to Σ and Σ' respectively. From the Lorentz transformations, we have

$$t_2' = \gamma\,(t_2 - v x_2/c^2)\,,$$

$$t_1' = \gamma\,(t_1 - v x_1/c^2)\,.$$

If the events are simultaneous relative to Σ then $t_1 = t_2$. Hence

$$t_2' - t_1' = \gamma v\,(x_1 - x_2)/c^2.$$

In Fig. B.1, $x_1 > x_2$, so that $t_2' > t_1'$. Hence, relative to the inertial frame Σ' in which the train is at rest, the event at B, B' should be measured to occur after the event at A, A', as predicted previously by our qualitative discussion.

Appendix C

The Principle of Equivalence

C.1 INTRODUCTION

Let an astronaut be in a spaceship that is at rest in an inertial reference frame Σ in a region of outer space where there is *no* gravitational field, as shown in Fig. C.1(a). There are no windows in the spaceship, so that the astronaut cannot look out, and it is soundproof, so that the astronaut cannot hear anything. Everything inside the spaceship is in a state of weightlessness. Now let a signal from the Earth fire some small rockets, so that the spaceship starts moving with uniform acceleration $\boldsymbol{a}$ relative to Σ. Consider what happens from the viewpoint of the reference frame A_1, shown in Fig. C.1(b), in which the accelerating spaceship is at rest. The astronaut is no longer in a state of weightlessness. If the astronaut holds up an apple and lets it go, the apple will continue to move with the same velocity, relative to Σ, as the spaceship had when the apple was released. Hence, relative to the spaceship, the apple appears to fall with uniform acceleration $-\boldsymbol{a}$. If two apples of different masses were dropped at the same time, they would have the same acceleration $-\boldsymbol{a}$ relative to the accelerating spaceship, so that all falling bodies would have the same acceleration relative to the spaceship. If a spring balance, with a mass attached to it, were attached to the roof of the spaceship, the mass would remain at rest relative to Σ when the spaceship started accelerating, unless the spring in the spring balance were under sufficient tension to give the mass the same acceleration as the spaceship had relative to Σ. If the acceleration of the spaceship were numerically equal to g, the acceleration due to gravity at the surface of the Earth, the conditions in the spaceship would be similar to the conditions in a spaceship at rest on the surface of the Earth. How should the astronaut interpret the results in the accelerating spaceship if the astronaut did not know that rockets were accelerating the spaceship? The astronaut would be just as justified as Newton was in concluding from the observations on a falling apple that the spaceship was in a gravitational field. The astronaut could interpret the tension in the spring balance in terms of a force of "gravitational" origin on the mass, when it is in a "gravitational" field. In fact, the astronaut inside the spaceship could not tell the difference between times when the spaceship was accelerating relative to the fixed stars, as shown in Fig. C.1(b), and times when it was at rest relative to the fixed stars and a massive

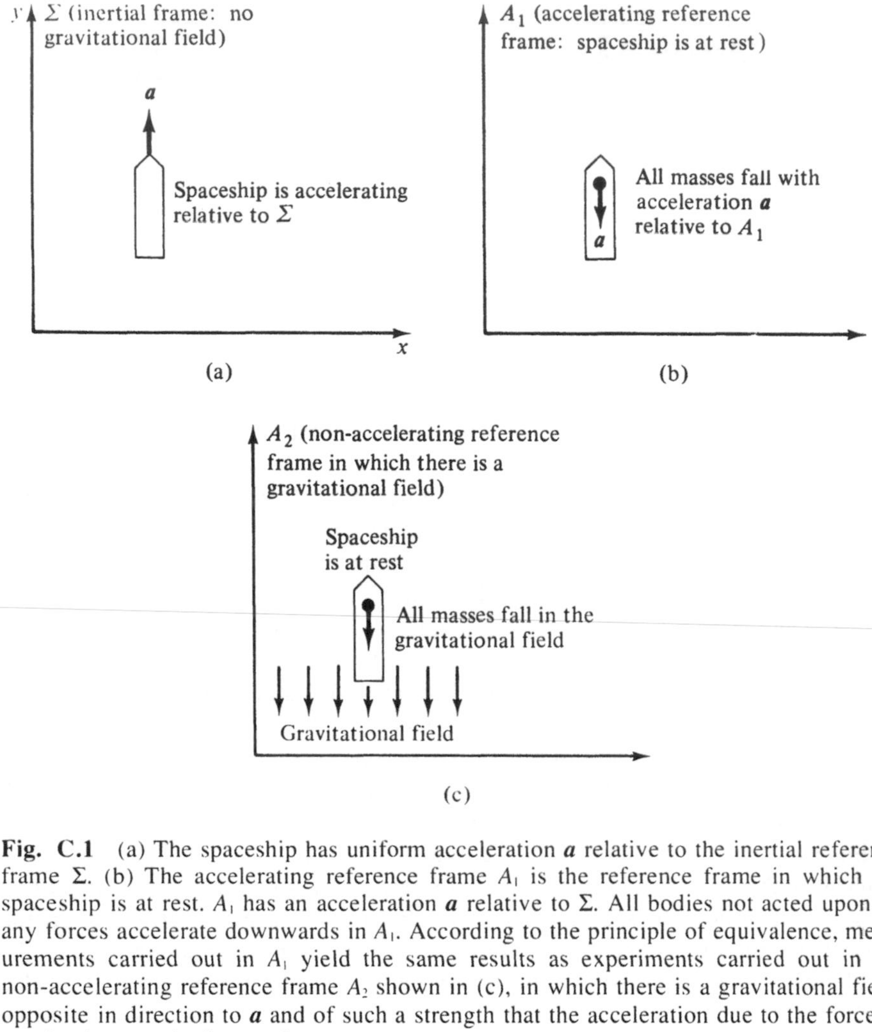

Fig. C.1 (a) The spaceship has uniform acceleration ***a*** relative to the inertial reference frame Σ. (b) The accelerating reference frame A_1 is the reference frame in which the spaceship is at rest. A_1 has an acceleration ***a*** relative to Σ. All bodies not acted upon by any forces accelerate downwards in A_1. According to the principle of equivalence, measurements carried out in A_1 yield the same results as experiments carried out in the non-accelerating reference frame A_2 shown in (c), in which there is a gravitational field, opposite in direction to ***a*** and of such a strength that the acceleration due to the force of gravity is numerically equal to *a*.

body was underneath the spaceship, giving rise to a gravitational field, as shown in Fig. C.1(c). In 1911 Einstein proposed that the coordinate system A_1 shown in Fig. C.1(b), which is accelerating with uniform acceleration ***a*** relative to the fixed stars, and the coordinate system A_2 shown in Fig. C.1(c), which is at rest in a uniform gravitational field, in which the acceleration due to gravity is numerically equal to ***a***, are exactly equivalent. That is, experiments carried out under identical conditions in A_1 and A_2 should give the same numerical results. This is the *principle of equivalence*.

C.2 EQUIVALENCE OF GRAVITATIONAL MASS AND INERTIAL MASS

The *gravitational mass* of a body is the mass that appears in Newton's law of universal gravitation, which is equation (1.9). It is a measure of that property of matter by virtue of which every particle of matter exerts a gravitational force of attraction on every other particle of matter. By analogy with electrostatics, we could call the gravitational mass the gravitational "charge".

Inertial mass is associated with a completely different property of matter, namely the fact that a force, not necessarily of gravitational origin, must be exerted on a particle in order to accelerate it. It is the mass that appears in Newton's second law of motion, equation (1.5). In Newtonian mechanics it is an extra hypothesis, which must be tested by experiment, that the gravitational and inertial masses are proportional to each other. As an analogy, consider the electrostatic forces on a proton and a positron (positive electron) in an electric field. The electric charges on the proton and positron are equal. It follows from the theory developed in Section 2.4 that, since the inertial mass of a proton is 1836 times the inertial mass of a positron, the proton and the positron would not have the same acceleration in the electric field. Since inertial mass is not proportional to electric charge, why should, in the context of Newtonian mechanics, inertial mass be proportional to gravitational "charge" (i.e. gravitational mass)?

Consider a body of gravitational mass m_G falling near the surface of the Earth with a velocity much less than c. The force of gravitational attraction is equal to $G\, m_G M_G/R^2$, where M_G is the gravitational mass of the Earth and R is the radius of the Earth. The acceleration produced by this force is equal to the ratio of the force to the inertial mass of the body on which it is acting. Hence

$$\text{acceleration} = \frac{\text{gravitational mass}}{\text{inertial mass}} \times \frac{GM_G}{R^2}\,.$$

If the gravitational mass of a body were not proportional to its inertial mass then the accelerations of different bodies in the Earth's gravitational field would not always be the same. Experimentally, it is found that this acceleration is the same, within experimental error for all bodies, so that experimental results require the proportionality of gravitational and inertial mass.

It was shown in Section C.1 that all bodies released inside the accelerating spaceship in Fig. C.1 had the same acceleration $-\boldsymbol{a}$ relative to the spaceship, whatever their inertial mass. According to the principle of equivalence, the same result should be true inside the spaceship that is at rest in the uniform gravitational field in Fig. C.1(c). Hence, according to the principle of equivalence, all bodies, whatever their inertial masses, should have the same acceleration in a gravitational field. This shows that the proportionality of gravitational mass and inertial mass follows directly from the principle of equivalence and forms an integral part of the theory of general relativity. If the units are chosen appropriately, the gravitational and inertial masses can be made numerically equal.

The gravitational mass of a body can be determined by weighing, using an equal-arm balance. This is the most convenient way of determining masses experimentally. To use these values in the equation $\boldsymbol{f} = m\boldsymbol{a}$ (Newton's second law) requires the equality of gravitational and inertial mass.

C.3 THE RATES OF CLOCKS IN GRAVITATIONAL FIELDS AND THE GRAVITATIONAL SHIFT

The principle of equivalence will now be used to determine the rates of clocks in gravitational fields. It will be assumed throughout this appendix that all speeds are very much less than the speed of light, and the formulae of Newtonian mechanics will be used.

Consider a spaceship that is moving with uniform acceleration $\boldsymbol{a}$ relative to an inertial reference frame Σ in which there is no gravitational field, as shown in Fig. C.2(a). Let the spaceship be instantaneously at rest in Σ at the time $t = 0$. Let two clocks of identical construction, labelled 1 and 2, be fixed to the spaceship, with Clock 1 at the rear and Clock 2 at the front end of the spaceship at a distance h apart. As examples of suitable clocks, we shall use two light sources that have a frequency ν_0 when they are at rest in an inertial frame in which there is no gravitational field. As an example of real clocks, we can use two caesium clocks. Consider the light emitted from Clock 1 at the time $t = 0$, when the spaceship is instantaneously at rest in the inertial reference frame Σ at the time $t = 0$. Let the light reach Clock 2 at a time t later. By this time, Clock 2 has a speed $v = at$ and has travelled a distance $\frac{1}{2}at^2$ relative to Σ. Since light travels the distance $h + \frac{1}{2}at^2$ from Clock 1 to Clock 2 in time t at speed c,

$$ct = h + \tfrac{1}{2}at^2.$$

If $\frac{1}{2}at^2 << ct$ then

$$t \approx \frac{h}{c}, \qquad v = at \approx \frac{ah}{c}.$$

To an astronaut situated beside Clock 2, it will appear as if the light coming from Clock 1 was emitted by a source moving away from the astronaut with a speed $v = ah/c$. The astronaut will observe the frequency of the light to be Doppler-shifted towards the red end of the spectrum. According to the non-relativistic theory of the Doppler effect (put $v << c$ in equation (4.54)), the frequency ν' observed by the astronaut by Clock 2 is

$$\nu' = \nu_0\left(1 - \frac{v}{c}\right) = \nu_0\left(1 - \frac{ah}{c^2}\right). \tag{C.1}$$

This is a lower frequency than that of the light from Clock 2. Similarly, light emitted from Clock 2 at the instant $t = 0$ in Σ has a frequency $\nu_0\,(1 + ah/c^2)$

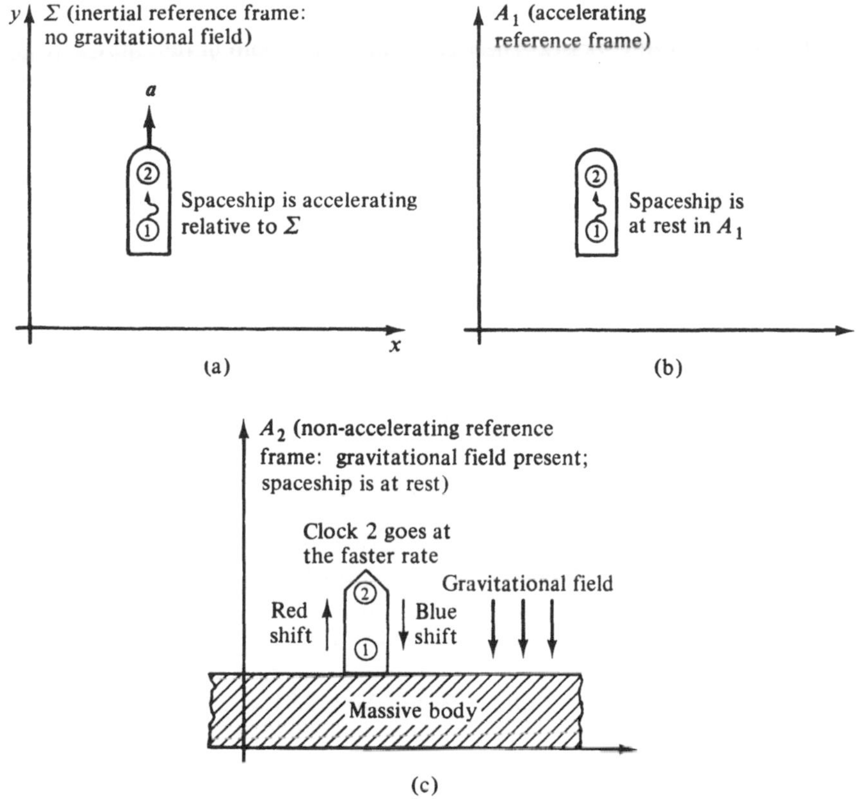

Fig. C.2 (a) In the inertial reference frame Σ the spaceship has an acceleration $\boldsymbol{a}$. Light is emitted from two identical light sources 1 and 2 at the time $t = 0$, when the spaceship is instantaneously at rest in Σ. By the time light from Source 1 reaches Source 2, Source 2 is moving relative to Σ, so that the light from Source 1 will appear Doppler-shifted to an astronaut beside Source 2. (b) In the accelerating reference frame A_1 in which the spaceship is at rest, the light reaching Source 2 from Source 1 has a lower frequency than Source 2. According to the principle of equivalence, measurements in A_1 should be the same as in the non-accelerating reference frame A_2 shown in (c), in which there is a gravitational field. Hence Clock 2 in (c) should go at a faster rate than Clock 1 in A_2, and visible light from Source 1 should be shifted to the red end of the spectrum compared with the frequency of the exactly similar Source 2.

relative to Clock 1 when it reaches Clock 1. To first order in v/c, the difference in both cases is $\nu_0 ah/c^2$, if $ah/c^2 \ll 1$.

According to the principle of equivalence, these experimental results should be exactly the same as when the same experiments are carried out in the non-accelerating coordinate system A_2 shown in Fig. C.2(c), in which the spaceship is at rest and in which there is a gravitational field of such a strength that g, the acceleration due to the gravitational field, is numerically equal, but opposite in direction, to $\boldsymbol{a}$, the acceleration of the spaceship relative to Σ. The light reaching

Clock 2 from Clock 1 should have a frequency lower than Clock 2 by an amount $\nu_0 gh/c^2$. Conversely, light reaching Clock 1 from Clock 2 should have a frequency higher than that of Clock 1. The light sources are both at rest in A_2. No light vibrations should be lost on the way, so that one must conclude that in the presence of a gravitational field the actual frequencies of the stationary clocks differ by an amount $\nu_0 gh/c^2$. This is an actual physical difference associated with the presence of a gravitational field. Now gh is the difference in gravitational potential between the clocks. According to normal convention, Clock 2 is at the higher gravitational potential in A_2 (see Fig. C.2(c)), since work must be done to raise unit mass from Clock 1 to Clock 2 in the gravitational field in A_2. So far, it has been assumed that g is constant between the two clocks. For a varying gravitational field, gh has to be replaced by $\Delta\phi$, the difference in gravitational potential between the clocks, so that the difference between the frequencies is $\Delta\nu = \nu_0\, \Delta\phi/c^2$. Hence the fractional difference in frequency is

$$\frac{\Delta\nu}{\nu_0} \approx \frac{\Delta\phi}{c^2}\,. \tag{C.2}$$

For visible light, light coming from the higher altitude or higher gravitational potential is shifted to the violet end of the spectrum, and light going from the lower to higher altitude undergoes a redshift, as shown in Fig. C.2(c). This prediction was confirmed by Pound and Rebka (1960), who measured the change in frequency of the 14.4 keV γ-ray line emitted by ^{57}Fe, over a vertical distance of 22 m using the Mössbauer effect. Since $\Delta\phi = 9.81 \times 22\ \mathrm{m^2\,s^{-2}}$, the fractional shift in frequency should have been

$$\Delta\phi/c^2 = 9.81 \times 22/\,(3 \times 10^8)^2 = 2.4 \times 10^{-15}.$$

Pound and Rebka's measured value was 0.97 ± 0.04 times the predicted value.

Another experimental check of equation (C.2) is to look at light coming from stars. This light comes from light sources at a lower gravitational potential than the surface of the Earth. (Owing to the larger mass of the star, work would have to be done to take unit mass from the star to the Earth, even though over the last stage of the journey some gravitational potentia' energy would be lost in the gravitational field of the Earth). From equation (C.2), it follows that the light from stars should be shifted to the red end of the spectrum. This is the *gravitational redshift*. The experimental results available are consistent with this prediction.

If ν_1 is the frequency of Clock 1 and ν_2 is the frequency of Clock 2 when the clocks are at rest in a gravitational field (e.g. as in Fig. C.2(c)) then, provided that $\Delta\phi << c^2$,

$$\nu_2 \approx \nu_1 \left(1 + \frac{\Delta\phi}{c^2}\right), \tag{C.3}$$

$$\nu_1 \approx \nu_2 \left(1 - \frac{\Delta\phi}{c^2}\right). \tag{C.4}$$

Equation (C.3) should hold for other processes, such as the successive ticks of clocks. According to this equation, Clock 2 should register more ticks in a given time interval, and so should register a longer time interval than Clock 1. If Δt_1 and Δt_2 are the time intervals registered by Clocks 1 and 2 respectively then since the time interval measured by each clock is proportional to its frequency, from equations (C.3) and (C.4)

$$\Delta t_2 \approx \Delta t_1 \left(1 + \frac{\Delta \phi}{c^2} \right), \tag{C.5}$$

or

$$\Delta t_1 \approx \Delta t_2 \left(1 - \frac{\Delta \phi}{c^2} \right), \tag{C.6}$$

where $\Delta\phi$ is the difference in gravitational potential between Clocks 1 and 2. According to equation (C.5), Clock 2, which is at rest at the higher altitude, goes at a faster rate than Clock 1, which is at rest at the lower altitude. If the caesium clocks 1 and 2 were synchronized, according to equation (C.5), they would not remain synchronized, since the clock at the higher altitude would go at the faster rate. This shows that one cannot use spatially separated clocks to measure the times of events in a gravitational field, and, strictly, special relativity should not be applied. Fortunately, the effects predicted by equation (C.5) are generally so small in terrestrial experiments that they can be ignored, and special relativity can be applied.

C.4 BENDING OF THE PATH OF LIGHT IN A GRAVITATIONAL FIELD

Consider a spaceship moving with uniform acceleration $\boldsymbol{a}$ relative to an inertial reference frame Σ. If light enters a small hole in the spaceship, as shown in Fig. C.3(a), it travels in a straight line relative to Σ. However, relative to the accelerating spaceship, the light appears to travel in a curved path, as shown in Fig. C.3(b). According to the principle of equivalence, observations in the accelerating spaceship should be the same as observations carried out by a stationary observer in a uniform gravitational field, of such a strength that the acceleration due to gravity in the gravitational field is equal and opposite to the acceleration of the spaceship. Hence light rays should be curved in a gravitational field, the rays being bent in the direction of the acceleration due to gravity, as shown in Fig. C.3(c). As an example, Einstein considered the deflection of light from a distant star when the light passes close to the periphery of the Sun. Measurements carried out during solar eclipses have confirmed Einstein's predictions, made using the full theory of general relativity. Experiments on the deflection of radio waves from quasars by the Sun are also in agreement with Einstein's predictions. Near the Earth, the gravitational field is so weak that the deflection of light rays by the Earth's gravitational field can generally be ignored.

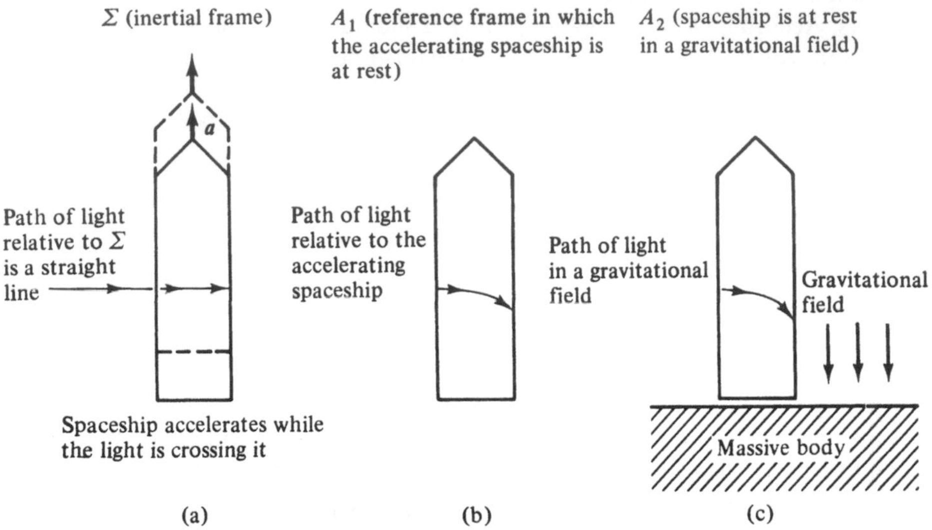

Fig. C.3 (a) Light enters an accelerating spaceship. The light path is a straight line relative to the inertial reference frame Σ. (b) Relative to the accelerating spaceship, the path of the light is curved. (c) Hence, according to the principle of equivalence, light is deflected in a gravitational field.

It has been shown that in the presence of gravitational fields light does not travel in straight lines, so that light rays cannot be used in these conditions to mark out a rectangular Cartesian coordinate system in which Euclidean geometry can be applied. It has been shown that the rates of clocks depend on the gravitational potential. So do the measured lengths of rods. One cannot use one single Cartesian reference frame with rulers and synchronized clocks to describe the *whole* universe. In the general theory of relativity non-Euclidean geometry is used, and the geometrical properties of space and time vary, depending on the distribution of matter.

C.5 LOCAL INERTIAL REFERENCE FRAMES

In the presence of a gravitational field one can approximate to a local inertial reference frame by considering a reference frame falling freely under gravity. As an example, consider the spaceship shown in Fig. C.4, which is at a distance r from the centre of the Earth and which is falling freely under gravity towards the centre of the Earth. If the variation of the acceleration due to gravity over the dimensions of the spaceship is ignored, particles 1 and 2 will fall with the same acceleration g_0 relative to the Earth, where

$$g_0 = \frac{GM}{r^2}, \tag{C.7}$$

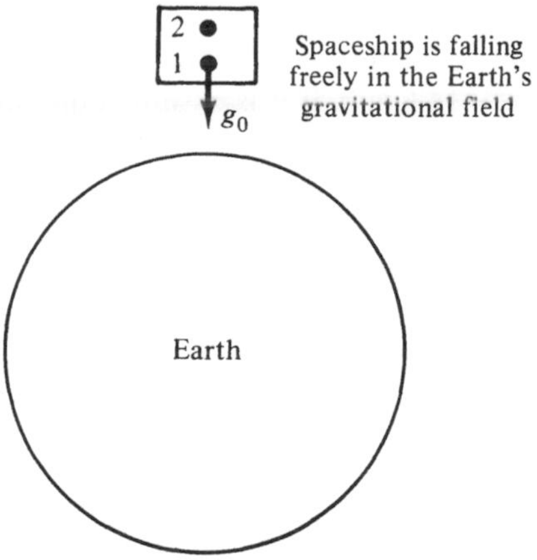

Fig. C.4 A spaceship is falling freely in the Earth's gravitational field. Particle 1 is nearer the Earth than particle 2 and has the greater acceleration. Relative to the spaceship, particles 1 and 2 move apart.

with M being the mass of the Earth. If an astronaut inside the spaceship let go of an apple then the astronaut, spaceship and apple would have the same acceleration g_0, so that the apple would not fall relative to the inside of the spaceship. These conditions are generally described by saying that there is "zero g" inside the spaceship. The spaceship is, to a good approximation, an inertial reference frame. For example, if the astronaut threw the apple, it would carry on moving with uniform speed in a straight line relative to the spaceship. The whole spaceship is not, however, an exact inertial reference frame. For example, since in Fig. C.4 particle 1 is nearer the centre of the Earth than particle 2, the acceleration due to gravity of particle 1 is greater than the acceleration of particle 2. Hence, relative to the spaceship, particles 1 and 2 will move apart. If g_0 is the acceleration due to gravity at the position of particle 1, which is at a distance r from the centre of the Earth, and if particle 2 is a distance h further away from the centre of the Earth than particle 1 then, according to Newton's law of universal gravitation, the value of g at the position of particle 2 is

$$g = \frac{GM}{(r+h)^2} = \frac{GM}{r^2}\left(1 + \frac{h}{r}\right)^{-2}$$
$$\approx g_0\left(1 - \frac{2h}{r}\right). \qquad \text{(C.8)}$$

Particle 1 moves faster towards the Earth than particle 2. The relative acceleration of particles 1 and 2 in Fig. C.4 is equal to $2hg_0/r$.

Consider a spaceship that starts from rest from an altitude of 600 km where $r = 600\text{ km} + 6400\text{ km} = 7000\text{ km}$ and where, according to equation (C.7), $g_0 = 8.2\text{ m s}^{-2}$. Assume that initially $h = 5$ m; that is, particle 2 is 5 m further from the centre of the Earth than particle 1 is. According to equation (C.8), the relative acceleration of particles 1 and 2 is $2g_0h/r = 1.17 \times 10^{-5}$ m s^{-2}. Fifty seconds after the spaceship starts falling under gravity, the separation of particles 1 and 2 increases by $\frac{1}{2}(2g_0h/r)\,t^2 = 14.6$ mm, if it is assumed that the relative acceleration is constant. In this time interval the spaceship travels a distance of about 10 km towards the centre of the Earth. The change in the separation of the particles would be an observable effect. If an astronaut in the falling spaceship held out two apples with outstretched arms and let them go, the apples would accelerate along slightly different paths towards the centre of the Earth and would get closer together relative to the spaceship. This would again be an observable effect. These results show that even a freely falling reference frame, of only the dimensions of a spaceship, cannot be treated as one inertial reference frame.

In this appendix, we have only given a brief insight into the principle of equivalence to illustrate some of the limitations of the theory of special relativity and inertial reference frames in the presence of gravitational fields. These phenomena are best interpreted using the full theory of general relativity. Those interested in a popular account of general relativity are referred to Berry (1989). A rigorous, comprehensive account of general relativity is given by Misner, Thorne and Wheeler (1973).

REFERENCES

Berry, M. (1989) *Principles of Cosmology and Gravitation*. Adam Hilger, Bristol.
Misner, C.W, Thorne, K.S. and Wheeler, J.A. (1973) *Gravitation*. Freeman, San Francisco.
Pound, R.V. and Rebka, G.A. (1960) *Phys. Rev. Lett.* **4**, 274

PROBLEMS

C.1 Calculate the fractional difference between the frequency of an atomic process on the top of the spire of Salisbury Cathedral, altitude 124 m, and the frequency of the same process at ground level. Take g to be 9.81 m s^{-2}. [*Hint*: Use equation (C.2).]

C.2 Calculate the difference in the readings of clocks at sea level and at the top of Mount Everest respectively after one year. Take $g = 9.81$ m s^{-2} and the altitude of Mount Everest as 8842 m. [*Hint*: Use equation (C.2).]

Answers to Problems

CHAPTER 1

1.2 (a) 25.5 m s^{-1} in a direction arctan 5 East of North.
(b) 27.8 m s^{-1} in a direction arctan 6.3 East of North.

1.3 (a) $x' = 0, y' = -19.6$ m, (b) $x = 30$ m, $y = -19.6$ m.

CHAPTER 2

2.1 (a) $1 + 3.9 \times 10^{-14}$; (b) $1 + 3.9 \times 10^{-12}$; (c) 1.005;
(d) 1.155; (e) 2.294; (f) 7.089; (g) 22.4; (h) 71.

2.2 (a) $0.196c$; (b) $0.549c$; (c) $0.94c$.

2.3 u is (a) $0.428c$; (b) $0.996c$; (c) $0.999\,999\,6c$.
p is (a) 445 MeV/c; (b) 10 897 MeV/c; (c) 1 000 938 MeV/c.
E is (a) 1038 MeV; (b) 10 938 MeV; (c) 1 000 938 MeV.

2.4 (a) 0.079 MeV; (b) 0.2616 MeV.

2.5 256 kV; $0.75c$.

2.6 (a) 144 MeV/c^2; (b) 938.1 MeV/c^2.

2.7 930 MeV/c^2; $0.76c$.

2.8 2.4×10^{-10} s.

CHAPTER 4

4.1 (a) (i) $x' = (5 - 1.35 \times 10^9)$ m, $t' = (7.5 - 10^{-8})$ s;
(ii) $x' = 4.25 \times 10^8$ m, $t' = 0.75$ s;
(iii) $x' = 7.43 \times 10^{10}$ m, $t' = -146$ s.
(b) (i) $x = (12.5 + 9 \times 10^8)$ m, $t = (5 + 2.5 \times 10^{-8})$ s;
(ii) $x = 1.21 \times 10^{10}$ m, $t = 27.5$ s;
(iii) $x = 1.36 \times 10^{11}$ m, $t = 312.5$ s.

4.2 63.7 mm.

4.3 $0.87c$.

4.4 In Σ, $L = L_0(\gamma^{-2}\cos^2\theta' + \sin^2\theta')^{1/2}$; $\tan\theta = \gamma\tan\theta'$.

4.6 (a) 1.36 km; (b) 4.6 km; (c) 14.7 km.

4.7 (a) 8.7 m; (b) 15.7 m; (c) 54 m; (d) 170 m; (e) 244 m.

4.8 2.17×10^{-8} s.

4.9 $\Delta\lambda = 0.656$ nm.

4.10 (a) 453 nm; (b) 115 nm.

4.11 £ 185 000 000.

4.13 (a) 6.96×10^{-5} s; (b) 20.86 km; (c) 660 m.

4.14 (a) $0.5c$; (b) 5.2 s.

4.15 (a) $0.8c$; (b) $0.89c$; (c) $0.65c$.

4.16 (a) $x' = 0.75c$, $t' = 2.75$ s; $v = 0.75c$.
(b) $x' = 2.75c$, $t' = 0.75$ s; $v = 0.75c$.

CHAPTER 5

5.1 (a) $u_x = 30(1 - 2.22 \times 10^{-15})$ m s^{-1}, $u_y = 0$;
(b) $u_x = 20$ m s^{-1}, $u_y = 10(1 - 2.22 \times 10^{-15})$ m s^{-1}.

5.2 (a) $u_x = 0.917c$, $u_y = 0$;
(b) $u_x = 0.1c$, $u_y = 0.896c$.

5.3 $u_x'' = 0.87c$, $u_y'' = 0.33c$.

5.4 $0.933c$.

5.6 (a) $1.6c$; (b) $0.8c$; (c) $0.976c$.

5.11 10.6 m.

5.12 75° 30′.

5.13 4.24×10^{18}, 3.76×10^{18}, 3.39×10^{18}, 3.08×10^{18} and 2.83×10^{18} Hz.

CHAPTER 6

6.1 83.2°.

6.4 (a) $\dfrac{2(h\nu_0)^2 + 2h\nu_0 mc^2 + m^2c^4}{2h\nu_0 + mc^2}$;

(b) $\dfrac{mc^2 h\nu_0}{2h\nu_0 + mc^2}$.

6.5 (a) $E_3 = 1.546$ MeV; (b) 3.74×10^{20} Hz; (c) $E_4 = 1.565$ MeV; (d) $\phi = 31.5°$.

6.6 27.2 MeV.

6.7 − 1.19 MeV.

6.8 4.2×10^9 kg.

6.10 110.5 MeV; $u = 0.83\,c$.

6.11 $m_1 = 142$ MeV/c^2; $m_2 = 945$ MeV/c^2. (a) 1208 MeV/c^2; (b) approx. 500 MeV/c; (c) approx. 99 MeV.

6.12 52.3 MeV.

6.16 (a) 0.6813 MeV/c; 0.8517 MeV.
(b) 0.2129 MeV/c; 0.5537 MeV.

6.17 0.4831 MeV/c; 0.7033 MeV.

6.18 1.024 MeV; $\theta' = 77.6°$.

6.19 $0.91c$.

6.21 $0.348c$.

6.23 18.1 GeV.

CHAPTER 7

7.5 (a) 145 MeV; (b) 172 MeV.

CHAPTER 9

9.1 (a) $0.8c$; (b) 10 years; (c) $t_{\text{lab}} = 20$ years; $t_{\text{spaceship}} = 12$ years.

9.2 $0.995c$.

9.3 (a) 670; (b) 18.

APPENDIX C

c.1 $\Delta\nu/\nu = 1.35 \times 10^{-14}$.

c.2 3×10^{-5} s.

Index

Milton Keynes UK
Ingram Content Group UK Ltd.
UKHW031129141024
449569UK00006B/322

9 780850 668384